Vacuum Technology

Nagamitsu Yoshimura

Vacuum Technology

Practice for Scientific Instruments

Springer

Dr. Nagamitsu Yoshimura
3-22-75 Fujimoto
Kokubunji, Tokyo
185-0031 Japan

ISBN 978-3-540-74432-0 ISBN 978-3-540-74433-7 (eBook)

Library of Congress Control Number: 2007933832

Cover design: WMX Design, Heidelberg

Printed on acid-free paper

9 8 7 6 5 4 3 2 1

springer.com

Preface

Many scientific instruments for analyzing specimen surface such as the electron microscope and the Auger electron spectrometer require clean, ultrahigh vacuum. The electron microscope and Auger electron spectrometer need fine electron probe, requiring a field emission emitter which can well work under ultrahigh vacuum. In the electron microscope and the ion microscope, microdischarges due to applying high voltage to electrodes sometime occur, resulting in deterioration of image quality. Microdischarges are related with the gas molecules on the surfaces of insulators and electrodes.

For many scientific instruments such as the electron microscope, a very clean, ultrahigh vacuum is necessary in the vicinity of the specimen and the electron emitter. So, ultrahigh vacuum technology is essential for microscope engineers and microscope users.

This book consists of the following chapters:
Chapter 1 Designing of Evacuation Systems
Chapter 2 Vacuum Pumps
Chapter 3 Simulation of Pressures in High-Vacuum Systems
Chapter 4 Outgassing
Chapter 5 Phenomena Induced by Electron Irradiation
Chapter 6 Vacuum Gauges
Chapter 7 Microdischarges in High Vacuum
Chapter 8 Emitters for Fine Electron Probes

Some important articles on the subject of every chapter are reviewed, and their discussions and conclusions are presented in rather high detail. The author believes that Chaps. 1, 2, 3, 4, 5, 6 would help the engineers who engage in designing the vacuum systems of ultrahigh vacuum scientific instruments such as electron microscopes. Chapters 7 and 8 will help the users of the scientific instruments using electron probes to understand the key technology intrinsic to electron beam systems, like high-voltage discharge and narrow electron beam emitter.

Acknowledgement is made to Dr. Frank Gualtieri, Professor of University of Maryland University Collage Asia, who kindly read the manuscript through and made useful correction. He gave me some useful suggestion throughout this review.

I would like to express my appreciation to Dr. Tatsuo Okano, Professor of Tokyo University, for encouraging me to publish this book.

Due thanks are given to the coworker of JEOL Ltd., especially to Mr. Haruo Hirano. It is a nice memory that we all did our best in working on the electron beam system technology related to vacuum.

I would like to express my appreciation to JEOL Ltd. for permission to present its know-how technologies based on our experiments.

09 January 2007
Nagamitsu Yoshimura

Profile of author
Nagamitsu Yoshimura

1965; Graduated from Osaka Prefecture University, Engineering Division. Entered JEOL Ltd.
Engaging in research and development of vacuum related technology in electron microscopes for 35 years at JEOL Ltd.

2000; Retired at the age limit from JEOL Ltd.

1985; Received Doctor of Engineering degree from Osaka Prefecture University. Thesis for degree: Research and development of the high-vacuum system of electron microscopes (in Japanese).

Contents

Chapter 1
Designing of Evacuation Systems

Many kinds of scientific instruments require a clean high vacuum. So, ultrahigh vacuum technology is essential for microscope engineers and microscope users.

Thin-film coaters have a large bell jar containing evaporating materials, presenting a large gas load. Such instruments must be evacuated from atmospheric pressure to a high vacuum in a short-time cycle. On the other hand, Auger electron spectrometers utilizing micro·nano electron probes must be evacuated to a very clean, ultrahigh vacuum.

Electron microscopes (EMs) have features of both thin-film coaters and Auger electron spectrometers. The camera chamber of an EM contains many sheets of micrograph film evolving a large amount of water-vapor molecules. At the same time, a very clean, ultrahigh vacuum is required in the vicinity of the specimen. A field emission electron gun (FEG) must be operated in an ultrahigh vacuum (UHV) of 10^{-8} Pa order. And, keeping the microscope column vibration-free is essential for photographing EM images at high resolution. Therefore, we must select vibration-free high vacuum pumps to evacuate the microscope column.

Consider the pressure level and vacuum quality required for the vacuum instrument.

Diffusion Pump System

When the vacuum chamber uses lubricant grease of hydrocarbons, you can use an oil diffusion pump (DP) system as a high vacuum pumping system. For this case, pneumatically driven high vacuum valves using lubricant grease, such as butterfly valves, can be adopted. For clean vacuum polyphenylether fluid is recommended as working fluid of the DP.

DP systems are still widely used in scientific instruments such as scanning electron microscopes (SEM) because DP systems are reliable in performance and comparatively silent during operation.

Note: Polyphenylether fluid has high boiling temperature and high viscosity, comparing with other commonly used working fluid of DP. Therefore, in order to use polyphenylether fluid one must ascertain that the DP can work well with polyphenylether fluid beforehand.

Turbo-molecular Pump System

For the greaseless vacuum chamber, a turbo-molecular pump (TMP) system is recommended. For this case, you must use greaseless high vacuum valves such as gate-type valves. When using elastomer seals in high-vacuum systems, water-vapor molecules in the atmosphere permeate elastomer seals, resulting in high H_2O pressure [1-1].

Sputter Ion Pump and Getter Pump System

When the vacuum chamber is sealed with metal gaskets such as Conflat® copper gaskets and metal O-ring gaskets, sputter ion pumps (SIPs) and getter pumps such as titanium sublimation pumps (TSPs) are suitable for final evacuating pumps. SIPs or getter pumps need a DP system or a TMP system as an auxiliary pumping system.

Selection of Pumping Speed

Pumping-down Characteristics

Ascertain that the desired high-vacuum pressure in the chamber is achieved within a specified evacuation time, using the following volume-evacuation equation.

When a chamber of volume V (in litter L) is evacuated by a pumping speed S (in L/s):

$$P = -\frac{V}{S_e}\left(\frac{dP}{dt}\right) \tag{1.1}$$

$$\log\left(\frac{P_2}{P_1}\right) = -\frac{1}{2.3}\frac{S_e}{V}(t_2 - t_1) \tag{1.2}$$

where outgassing of the chamber walls is neglected.

Effective pumping speed S_e:

$$\frac{1}{S_e} = \frac{1}{S_P} + \frac{1}{C}$$

where S_P is the pumping speed of the pump and C is the conductance of the evacuation pipe.

Actually, however, pumping-down time is affected much by the outgassing of the chamber walls, resulting in much longer comparing with the time calculated by Eq. (1.2).

Dayton (1959) [1-2] described from his experience on vacuum coaters that the net pumping speed in liters per second (L/s) at the exhaust port on the chamber should be two or three times the volume of the chamber in liters. He presented a rule that the speed of the diffusion pump in L/s should be five times the volume of the chamber in liters

In vacuum coaters, a high pumping speed is desirable for repeated pumping cycles from atmospheric pressure to high vacuum. On the other hand, the chambers of ultrahigh vacuum (UHV) scientific instruments such as Auger spectrometers and UHV scanning tunneling microscopes keep UHV for many days. Therefore, for UHV scientific instruments, the factors like the ultimate pressure of the pump and the pumping speed under UHV are more important than the rated pumping speed under high vacuum.

Steady-State Evacuation

The pressure P at an equilibrium state is expressed by the equation

$$P = \frac{Q}{S_e} \tag{1.3}$$

where Q is the net outgassing rate of the chamber under high vacuum.

When a specified gas of the species i is introduced into the chamber,

$$P_i = \frac{Q_i}{S_{ei}}. \tag{1.4}$$

Notes: The net pumping speed of the pump varies with the pressure. The net outgassing rate of the chamber Q also varies with the pressure.

Roughing System

In a DP system the time to reach the switching pressure P_S from atmospheric pressure becomes 2 to 3 times of the calculated time based on Eq. (1.2) due to the outgassing of the chamber walls. The pumping speed of the roughing pump is selected so that the switching pressure P_S is obtained within the specified roughing time.

Backstreaming of RP Oil Vapor

As Holland [1-3] presented, the backstreaming rate of RP oil vapor much increases when the pressure inside the roughing pipe of 1 in. diameter reduces below 0.1 Torr, as shown in **Fig. 1.1**.

As seen in **Fig. 1.1**, when the pump exhausts gas at its normal flow rate above 10^{-1} Torr the backstreaming rate is as small as a negligible amount.

Yoshimura et al. (1984) [1-4] evaluated the RP roughing systems with respect to backstreaming of RP oil vapor, as follows.

> Various systems have been proposed to minimize the oil vapor backstreaming from a running RP [1, 2, 3, 4]. Typical systems are presented in **Fig. 1.2**. The system **(a)** with a by-pass valve line with a sorption trap T [4] had the disadvantage of practical complexity in operation and is expensive. The system **(b)** with deliberate leakage L [1, 4] has a disadvantage because the leakage unnecessarily prolongs the roughing time. The system **(c)** with an additional isolation valve V_i, located just above a running RP, has been evaluated to be the most reasonable system for electron microscopes (EMs). In this system, V_i is opened in the roughing mode only, and closed when the pressure in the chamber reaches a switching pressure P_s which is higher than 13 Pa. Thus, most of the entire roughing pipe line is not evacuated below P_s with the continuously running RP, resulting in a maintenance-free clean system [1-4].

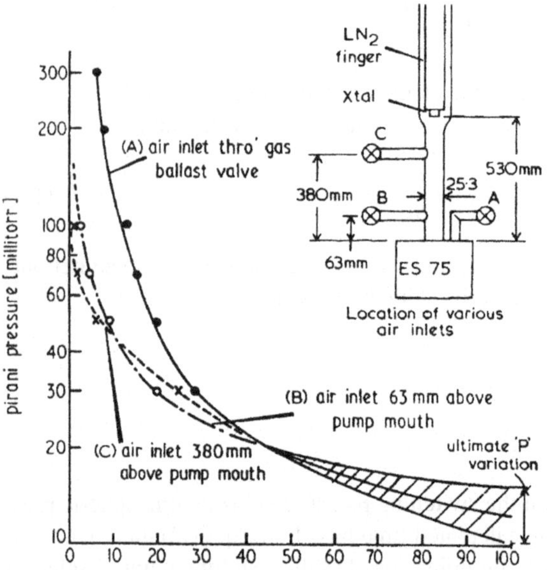

Fig. 1.1 Backstreaming rates as a function of pressure during gas flow into a 75 L/min rotary pump charged with No 16 Oil. The air inlets are in the pump line. Rates measured with a stationary gas in the same pressure range are plotted for comparison (Holland, 1971) [1-3]

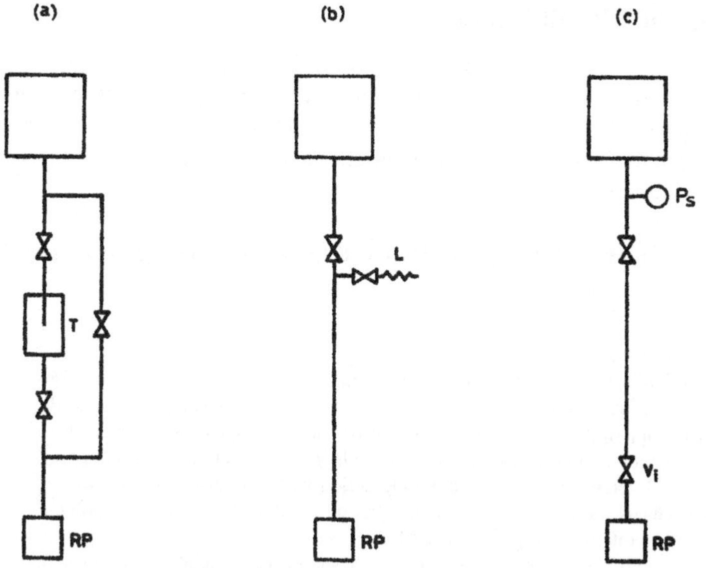

Fig. 1.2 Typical rough evacuation systems (Yoshimura et al., 1984) [1-4]

References

1. D. J. Santeler, *J. Vac. Sci. Technol.* **8**, 299 (1971).
2. M. A. Baker and G. H. Staniforth, *Vacuum* **18**, 17 (1968).
3. L. Holland, *Vacuum* **21**, 45 (1971).
4. N. T. M. Dennis, *Electron Components* **28**, 67 (1972).

Key points for preventing backstreaming of RP oil vapor:

1) For the case of a roughing pipe of 1 m length and 20 mm inner diameter, switch roughing to fine pumping at about 13 Pa in order to minimize the backstreaming rate of RP oil vapor. Select the pumping speed of a rotary pump to reach 13 Pa in the specified roughing time, taking into account Eq. (1.2).
2) Take into consideration that roughing time becomes 2 to 3 times of the calculated time due to the outgassing of the chamber walls.
3) Install an isolation valve just above the rotary pump besides the roughing valve positioned at the port of chamber, in order to prevent the oil vapor backstreaming into the roughing pipe line.
4) For strictly preventing the backstreaming of oil vapor, install a sorption trap containing porous alumina balls (Fulker, 1968) [1-5]. Porous alumina balls sorb much water vapor during pumping-down cycles from atmospheric pressure. And, such water vapor is evolved when the balls are exposed to vacuum of 10 Pa order, resulting in a prolonged roughing time. Replace the alumina balls absorbed much water vapor with fresh, degassed alumina balls for keeping a speedy roughing.

Backstreaming of DP Oil Vapor

In a steady state of fine pumping, a small pumping speed of a backing pump can evacuate the foreline of DP to sufficiently low pressure. However, we must take into account the transient condition just after switching over the evacuation modes from roughing to fine pumping.

Hablanian and Steinherz (1961) [1-6] described the backstreaming characteristics of DP systems in detail.

Forepressure Tolerance [1-6]

Forepressure tolerance is usually specified as a specific value of discharge pressure above which the pumping action stops (inlet and discharge pressures become essentially equal). Some diffusion pumps have an abrupt point at which this happens and in such cases it is easy to determine a maximum discharge pressure which should not be exceeded. Other pumps have a gradual deterioration of the pumping action when the discharge pressure is increased and in such cases a 10% increase in inlet pressure may be selected as the point at which the forepressure tolerance is specified [1-6].

Figure 1.3 represents some typical cases. Curve A shows a common performance in high vacuum pressure range and in this case the forepressure tolerance may be said to be 2×10^{-1} Torr. The performance of the same pump in the ultrahigh vacuum region is represented by the curve B.

The curves C and D indicate that some pumps behave normally in the ultrahigh vacuum region and have more or less distinct forepressure tolerance. The curve D is taken from [1] and it was obtained in a baked system by trapping the ionization gauge by two

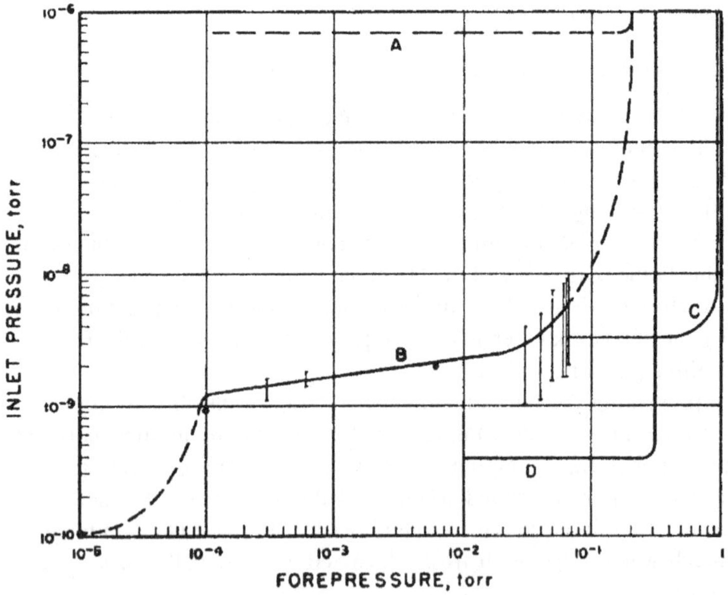

Fig. 1.3 Inlet pressure vs. forepressure for typical diffusion pumps. The vertical lines at curve B indicate pressure fluctuations (Hablanian and Steinherz, 1961) [1-6]

liquid nitrogen traps in series. The curve C was obtained with the new 6 in. pump (HS6-1500) without baking or trapping using polyphenylether oil. The pump was operated with an increased heat input (2 kW instead of standard 1.5 kW) which explains the unusually high forepressure tolerance.

It may be concluded that it is important to determine the inlet-discharge pressure relationship of a diffusion pump particularly if it is used for ultrahigh vacuum work. In some cases forepressures obtained by only a mechanical pump may not be sufficient as it may be necessary to reduce the forepressure below 10^{-4} Torr. This is conveniently accomplished by a second diffusion pump in series using the same pumping fluid to minimize the chances of contaminating the main pump with mechanical pump or booster pump oil.

Backstreaming [1-6]

All diffusion pumps have a certain amount of backstreaming. Without auxiliary devices such as baffles or cold traps surrounding the top nozzle the best pumps have a backstreaming rate of about 0.015 mg/min per cm^2 of inlet area.

In many cases the backstreaming rate can be sufficiently reduced by the use of cold caps surrounding the top nozzle [2]. This device intercepts a large amount of backstreaming molecules without significantly reducing the pumping speed. Various designs of such cold caps were tried and a simple water cooled guard extending beyond the nozzle edge found to be most effective. Depending on the particular pump, the reductions obtained are typically between 50 and 100 times less than the value obtained without the cold cap. **Figure 1.4** shows some of the results obtained with various designs.

Most measurements of backstreaming rate are generally conducted at the blank-off operation of the pump. The results are plotted in **Fig. 1.5** and it can be seen that the rate does not change significantly as long as the inlet pressure is below about 10^{-3} Torr. This is the

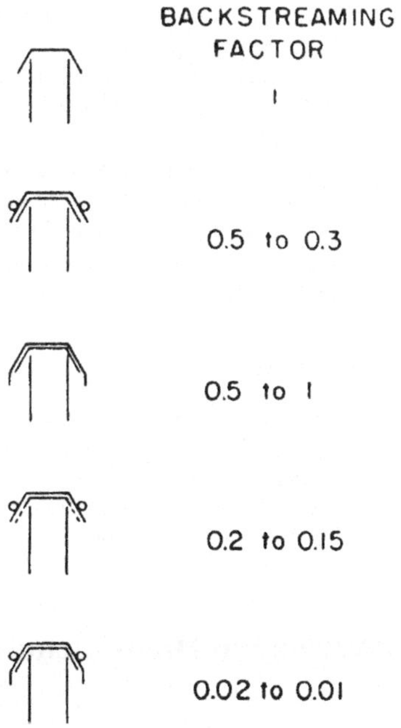

Fig. 1.4 Effect of cold cap configuration on backstreaming (Hablanian and Steinherz, 1961) [1-6]

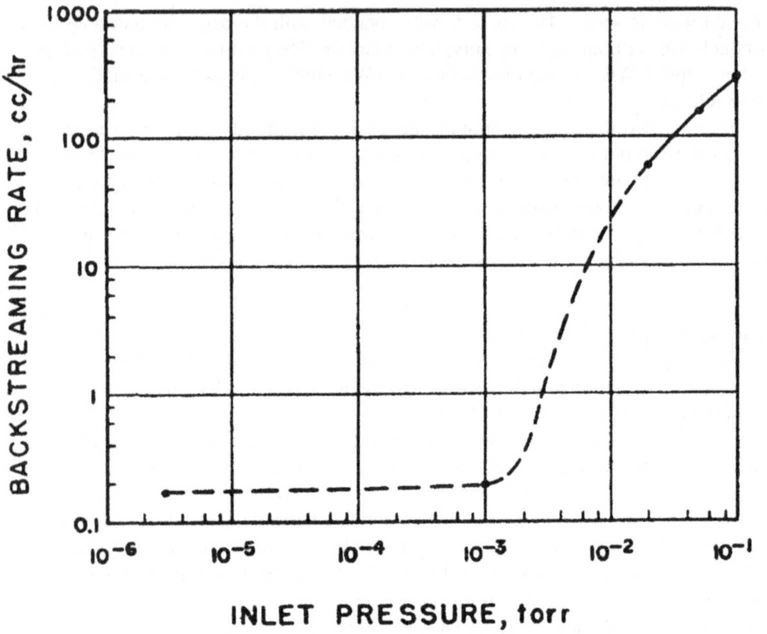

Fig. 1.5 Backstreaming rate vs. inlet pressure for 32 in. pump (fluid: Narcoil 40) (Hablanian and Steinherz, 1961) [1-6]

point where most diffusion pumps have an abrupt reduction in speed indicating that the top jet essentially stops pumping.

The sudden increase of backstreaming at inlet pressures above 10^{-3} Torr points out that a diffusion pump should not be operated in this range unless the condition lasts a very short time.

It is common practice in vacuum system operation to open the high vacuum valve after the chamber has been rough pumped to about 10^{-1} Torr exposing the diffusion pump to this pressure. In well proportioned systems this condition lasts only a few seconds. Otherwise extended operation at high pressure can direct unacceptably high amounts of oil into the vacuum system [1-6].

References

1. W. G. Hendfrson, J. T. Mark, and C. S. Geiger, "Evaluation of large diffusion pumps and traps for the ultrahigh vacuum system for the Model-C Stellarator", *1960 Vacuum Symposium Transactions*.
2. B. D. Power and D. J. Crawley, "Problems arising in the attainment of low pressure", U. S. Patent No. 2919061.

Overload in High-Vacuum Evacuation Systems

Hablanian (1992) [1-7] discussed how to prevent the **overload in high-vacuum systems**.

Abstract [1-7]: To prevent overloading high-vacuum pumps during and immediately after switching from pre-evacuation to high-vacuum pumping, a simple rule must be observed: the crossover must be performed when the gas mass flow from the vacuum chamber is less than the maximum throughput capacity of the high-vacuum pump. Typically, at the end of the pre-evacuation period, there are two somewhat distinct gas quantities associated with the vacuum chamber, the gas in the space of the chamber and the quasisteady outgassing rate. The overloading of the high-vacuum pump due to the space gas can be prevented by opening the high-vacuum valve slowly or by using a parallel low-conductance bypass. However, the overload due to the outgassing rate can only be prevented by following the golden rule of mass flow limitation. An immediate corollary of matching mass flows is that the larger the roughing pump, the lower the crossover pressure must be.

Quoted from Hablanian, 1992 [1-7]: It may be expected that the outgassing rate will increase when the pressure drops after crossover, switching over from roughing to high vacuum pumping.

The use of some safety factors is advisable. One "rule of thumb" may be noted: if, after crossover, the pressure does not drop quickly, then the high–vacuum pump is overloaded. The pressure reduction should be almost immediate and it should be according to the ratio of pumping speeds of the high–vacuum pump and the rough pump.

The mass flow emanating from the chamber can sometimes be easily determined by measuring the rate of the rise of pressure in an isolated chamber. This is illustrated in **Fig. 1.6**. The chamber should be fully loaded with process fixtures and materials. It should be evacuated to the temporarily chosen crossover pressure and the roughing valve closed. The measured mass flow, $V\frac{\Delta P}{\Delta t}$ must be less than the $Q_{\max}$ of the high–vacuum pump. If it is higher, then a lower crossover pressure must be chosen and the experiment repeated.

This is particularly important for diffusion pumps because each vapor jet has its own mass flow limit. It is possible for the inlet stage to be completely overloaded and, therefore, producing excessive amount of backstreaming, while the rest of the pump is still functioning. The situation is better in turbomolecular pumps because all stages rotate together on a common shaft.

Hablanian states in Abstract [1-7] that using a parallel low conductance bypass is not so effective against overloading on cross over. However, Yoshimura recognizes that slow high vacuum pumping using a low conductance by pass valve is effective against cross-over overload conditions. This is due to the following two reasons: one is keeping the mass flow rate into the high vacuum pump small on cross-over, the other is slowly reducing the pressure in the champer, resulting in preventing the sudden increase of the outgassing ratio of the champer walls from occurring on cross-over.

Iwami (2006) [1-8] presented an article, "Outgassing rate for One-minute exhaust."

Abstract [1-8]: With increasing demand for short-TAT (Turn Around Time) vacuum process, high-speed exhaust is becoming critical. Exhaust time depends on vacuum pump, plus volume and gas discharge amount of vacuum chamber. Precise estimation of gas discharge amount, in particular, is a must. Discussed in

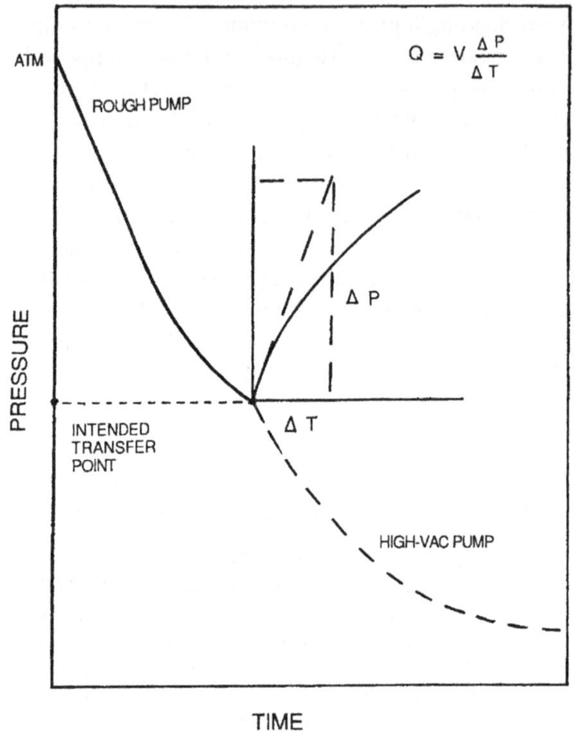

Fig. 1.6 Finding the proper crossover mass flow (and pressure) by rate-of-pressure-rise measurement (Hablanian, 1992) [1-7]

this article is the comparison of estimation methods for gas discharge amount, exhaust characteristics method vs. build-up method. Gas discharge rate after 1 minute exhaust was estimated, using "alumite" which generally has large gas discharge rate, and is easy to evaluate. Gas discharge rate estimated by the build-up method is 1/3–1/6 smaller than that estimated by exhaust characteristic method, presumably because of recombination of discharged gas to chamber wall. Also, gas discharge rate of "alumina" was found to be 5 times of published data exterpolated, $0.5\,\mathrm{Pa} \cdot \mathrm{m}^3 \cdot \mathrm{s}^{-1} \cdot \mathrm{m}^{-2}$ (approx.) after 1 minute vacuum exhaust. Alumite's structure is considered to be among the effects. When to design "One-Minute Exhaust System," estimating gas discharge rate by exhaust characteristics measurement is appropriate.

The experimental device outline is presented in **Fig. 1.7**. The outgassing rates of alumite measured by the build-up method and pump-down curve method are presented in **Fig. 1.8**.

Yoshimura et al. (1984) [1-4] discussed typical DP systems tolerant of excess gas load in transitional evacuation.

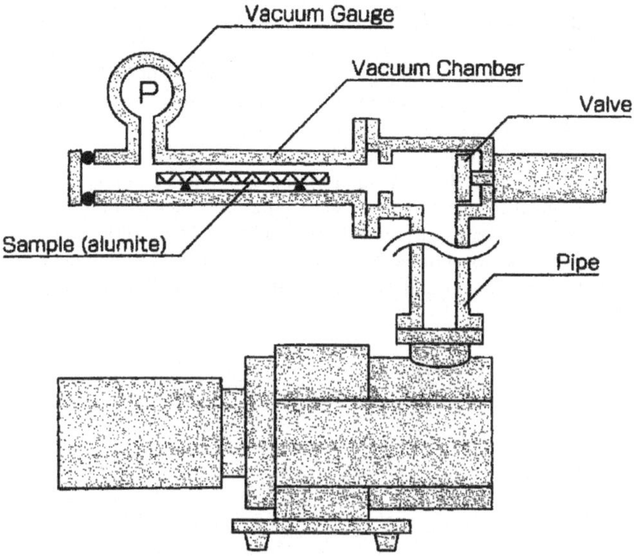

Fig. 1.7 Experimental device outline (Iwami, 2006) [1-8]

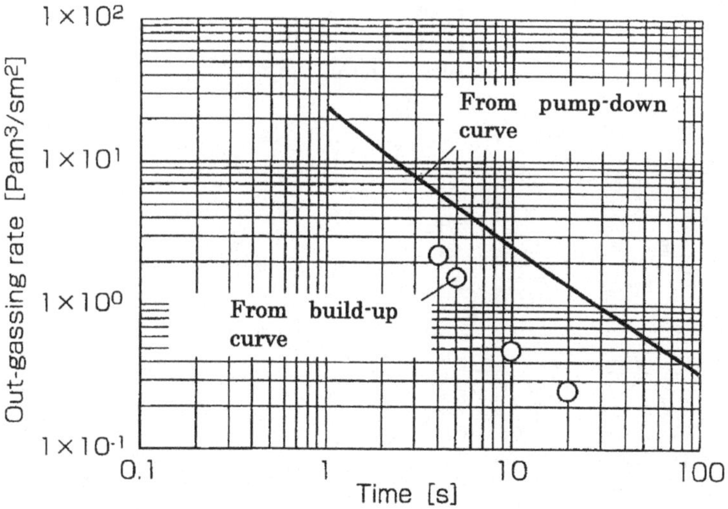

Fig. 1.8 Rate of gas discharge of alumite material (Iwami, 2006) [1-8]

Typical DP Systems Tolerant of Excess Gas Load (Yoshimura et al., 1984) [1-4]

Transitional evacuations from rough to fine pumping are often repeated in EM operation. The transitional evacuation for the camera chamber can present a severe load to the high vacuum system, because the camera chamber has a large volume and accommodates many photographic films with a high outgassing rate. Photographic films evolve a maximum amount of gases soon after switching over the evacuation from rough to fine pumping.

DP systems are usually designed for the maximum steady state gas load, and they are in many cases not provided against a temporary excess gas load for a short period. In a dynamic evacuation, a very large transitional throughput flows, for instance, several times a day, into the DP system just after switching over the evacuation from rough to fine pumping.

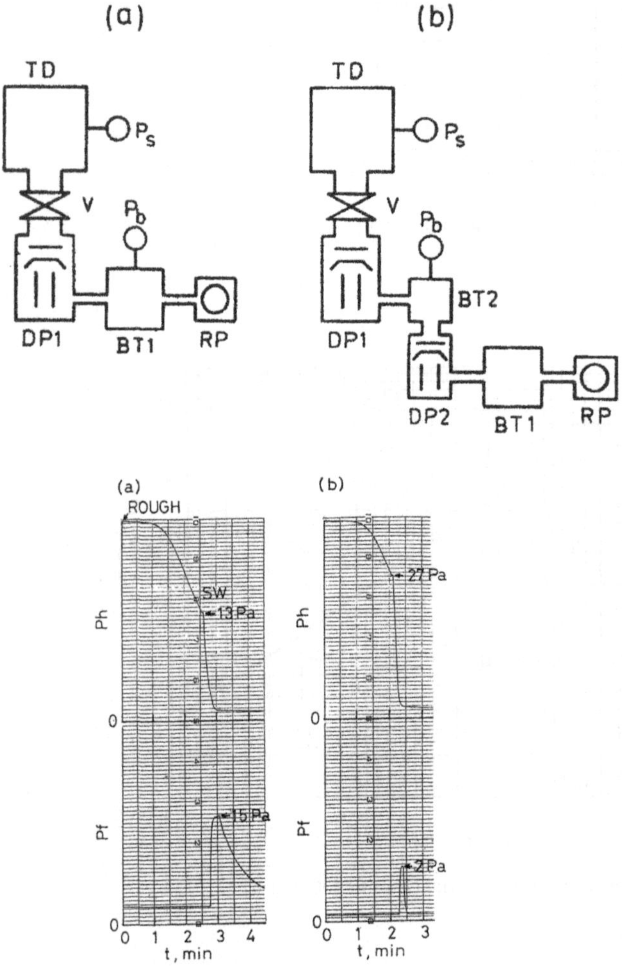

Fig. 1.9 Pressure rises of P_b in a DP1-BT-RP system **(a)** and DP1-DP2-BT-RP system **(b)**. TD: test dome of approx. 20 L, BT1: buffer tank of 10 L, BT2: buffer tank of 1 L, DP1: 4 in. (conventional one, Santovac-5), DP2: 2.5 in. (conventional one, Santovac-5), RP: 100 L/min, V: butterfly valve, 4 in (Yoshimura et al., 1984) [1-4]

Against such an extremely large gas load for a short period (typically a few seconds), the use of a large mechanical forepump (RP) would be of almost no use because it has a pumping speed much lower than the DP.

In a single DP system with a small-volume foreline, the backing pressure P_b of DP is apt to rise above its critical backing pressure P_c for a short period after switching over the evacuation mode. When P_b rises above P_c, the compressed gases at the foreline flow backwards together with oil vapor into high vacuum side. The abnormal backstreaming in this situation has been clearly described by Hablanian [1].

The rise of P_b of the first DP in a DP1-DP2-BT RP system [**Fig. 1.9 (b)**] was compared with that of P_b of DP1 in a DP1-BT-RP system [**Fig. 1.9 (a)**]. The rises of P_b after switching over the evacuation mode in both systems are presented in **Fig. 1.9**, where the switching pressure P_S in system **(b)** is approx. 27 Pa, and P_S in system **(a)** is approx. 13 Pa. The critical backing pressures P_c of both DPs (DP1 and DP2) are approx. 40 Pa. As seen in **Fig. 1.9**, the rise of P_b of DP1 in system **(b)** is much smaller than that of P_b in system **(a)**, despite the fact that the P_S in system **(b)** is double that in system **(a)**. It should be noted that the second DP in system **(b)** is a conventional small diffusion pump, but not an ejector type with a high maximum throughput.

Figure 1.10 presents typical DP systems, which meet the requirements for transitional evacuation, and their transitional characteristics of the pressure P_h in the high vacuum side and P_b in the forevacuum side.

In system **(a)** of **Fig. 1.10**, only a by-pass valve V2 with small conductance is opened at first during transition. Adequate sequential control of V1 and V2 suppresses the rise of P_b below P_c, as shown in the lower figure. A system with a slowly opening valve [1, 2, 3]

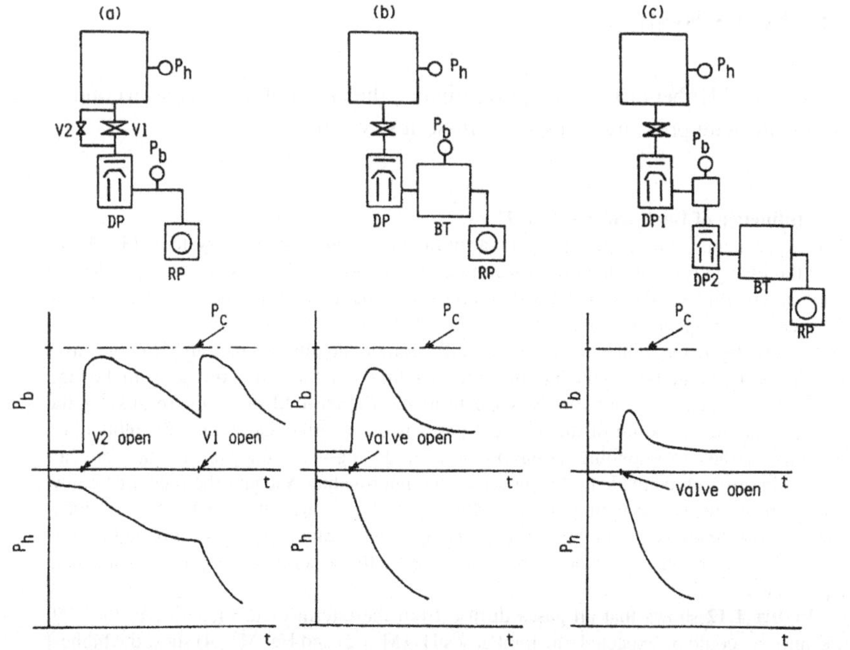

Fig. 1.10 Transitional characteristics of typical DP systems **(a)** System with by-pass valve V2, **(b)** system with a large buffer tank (BT), and **(c)** system backed by DP2 with a large BT (Yoshimura et al., 1984) [1-4]

would be almost equivalent to system **(a)**. In system **(b)** of **Fig. 1.10**, a large buffer tank (BT) suppresses the rise of P_b. The necessary volume of BT depends on the chamber volume, outgassing rate of the chamber including parts and materials, the functional characteristics of DP such as S and P_c, and the switching pressure P_S from rough to fine pumping. In general, the necessary volume of BT is almost half of the chamber volume. This system has additional advantages, the rapid pumping down of P_h and the slow rise of P_b, as shown in the lower figure. The latter advantage is really important for establishing a reliable safety system driven by P_b signals at an overload emergency. In system **(c)** of **Fig. 1.10**, DP2 with BT evacuates the foreline of DP1 under the minimum influence of the oil vapor from the backing pump. The rise of P_b is considerably reduced by the use of DP2, as shown in **Fig. 1.9**. The system has an additional advantage that the foreline of DP1 is held in a sufficiently clean high vacuum in almost the entire operation period, resulting in very clean fine pumping [4]. The control of the system is simple because both DPs can simultaneously be switched on/off.

References

1. M. H. Hablanian, *Industrial Research/Development*, Aug. **84** (1979).
2. M. H. Hablanian, *J. Appl. Phys. Suppl. 2, Pt. 1*, **25** (1974).
3. N. T. M. Dennis, L. Laurenson, A. Devaney, and B. H. Colwell, *J. Vac. Sci. Technol.* **20**, 996 (1982).
4. J. Hengevoss and W. K. Huber, *Vacuum* **13**, 1 (1963).

DP In-Series System

Hengevoss and Huber (1963) [1-9] investigated the effect of fore-vacuum conditions upon the ultimate pressure using a DP in-series system.

The Influence of Back Diffusion [1-9]

The experiments were made with a metal ultrahigh vacuum pumping system (**Fig. 1.11**). Two three-stage fractionating oil diffusion pumps connected in series are used. As driving fluid, silicon oil DC 704 is used. Each pump has a nominal pumping speed of 60 L/s. In the following experiments the cold trap was not cooled.

To investigate the influence of the back diffusion on the ultimate pressure, various gases were passed separately into the intermediate vacuum while the partial pressures in the high vacuum were measured with the mass spectrometer. **Figure 1.12** shows the results. On the abscissa the total pressure of the different gases in the intermediate vacuum is plotted, on the ordinate the corresponding partial pressures in the high vacuum. The pressures are calibrated in nitrogen equivalents (N_2-equ.). In the intermediate vacuum the pressures below 10^{-2} Torr are measured with a Bayard-Alpert gauge, pressures above 10^{-3} Torr are measured with a Pirani gauge. In the overlapping region the Pirani gauge was calibrated against the ion gauge, so that the whole pressure scale of the abscissa corresponds to nitrogen equivalents.

Figure 1.12 shows that all gases diffuse from the intermediate vacuum into the high vacuum. As could be expected, the light gases H_2 (M = 2) and He (M = 4) show the highest diffusion rate. For the inert gases Ne (M = 20), Ar (M = 40) and Kr (M = 84) the diffusion rate apparently decreases in inverse proportion to the molecular weight. The curves for the other gases do not seem to show this dependence with the same accuracy, e.g. it might be expected that the CH_4 curve (M = 16) would lie higher than the Ne curve (M = 20) but

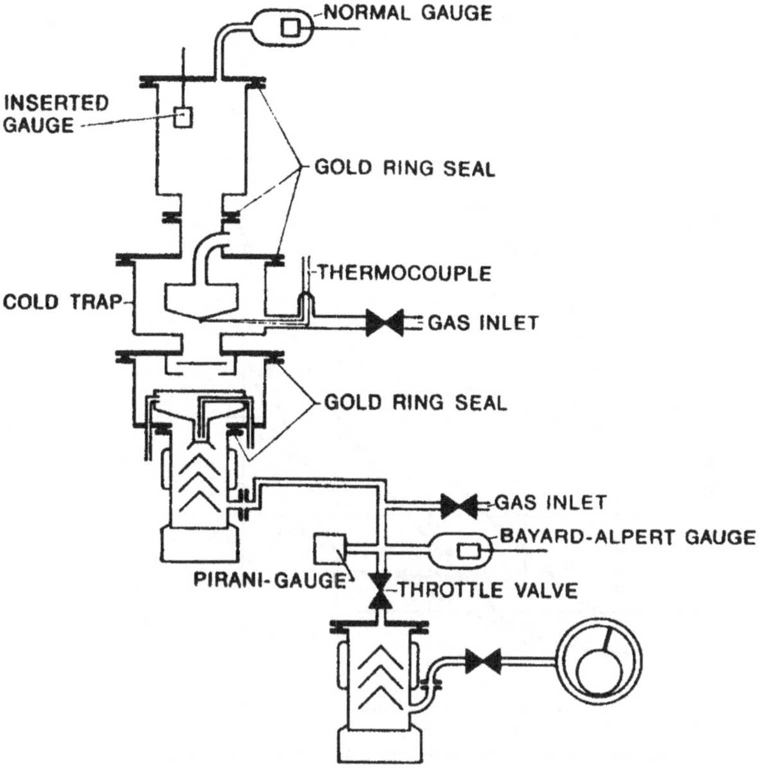

Fig. 1.11 Experimental apparatus. Metal pump system with creep baffle and cold trap. Ultimate vacuum without refrigeration about 1×10^{-9} Torr. Ultimate vacuum with refrigeration about 5×10^{-10} Torr (Hengevoss and Huber, 1963) [1-9]

as seen in **Fig. 1.12** the opposite is the case. The reason for this is not quite clear. Possibly solution effects of the gas in the diffusion pump oil are responsible for this deviation from the predicted behavior.

The following rules for production of ultrahigh vacua can be derived from the examination of **Fig. 1.12**.

In order to produce an ultimate pressure of 1×10^{-9} Torr, the partial pressure of the light gases whose molecular weight lies between 1 and 20, must not be higher at the fore-vacuum side of the main diffusion pump than about 2×10^{-4} Torr N_2 equivalent. In order to produce an ultimate pressure of 10^{-10} Torr, the partial pressure of these gases must not be higher than 1×10^{-5} Torr N_2 equ. At this point it must be noted that water vapor is one of the lighter gases which limit the ultimate pressure by back diffusion. This is an important fact because water vapor is one of the most abundant components of the fore-vacuum atmosphere, if the fore-vacuum part of the pumping system is not baked, which is the normal case.

Turning Off the Booster Diffusion Pump [1-9]

To estimate the influence of the booster diffusion pump on the ultimate pressure, this pump was turned off. As a result, the total pressure increased from 5×10^{-9} Torr

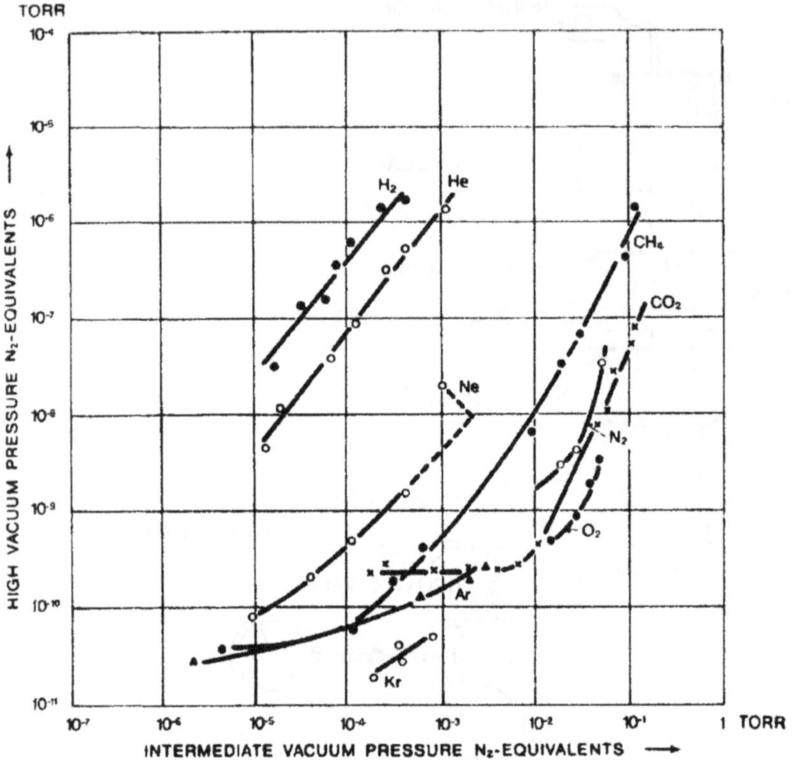

Fig. 1.12 Back diffusion from the intermediate vacuum into the high vacuum. The various gases are passed separately into the intermediate vacuum and the partial pressure of each measured in the high vacuum (Hengevoss and Huber, 1963) [1-9]

to 6×10^{-8} Torr. **Figure 1.13** shows the spectrum of the remaining gases before and after the pump operation had been stopped. Only the mass lines up to 78 are shown. It was observed that after the booster diffusion pump was turned off, particularly the mass peaks 2 (H_2), 15 (CH_4) and 78 increased, and also 28 (CO) and 44 (CO_2) both increased a little. It is especially noteworthy that after the booster diffusion pump had been turned off, many new peaks appeared in the mass spectrum.

In **Table 1.1** the increase in the partial pressure of the most important residual gases are summarized.

Experimental results show that the two-DP in-series system creates a very clean, ultra-high vacuum.

Norioka and Yoshimura (1991) [1-10] reported the performance of a DP in-series system.

Static Performance of In-Series DP System [1-10]

DP system can encounter accidental events, such as the breakdown of the DP heater and the cooling water supply trouble.

Considering the actual conditions of accidents, some experiments were conducted on the DP in-series system shown in **Fig. 1.14**. The DP1 (4 in.) and DP2 (2.5 in.) were both fractionating types with four nozzles including an ejector one, charged with Santovac-5

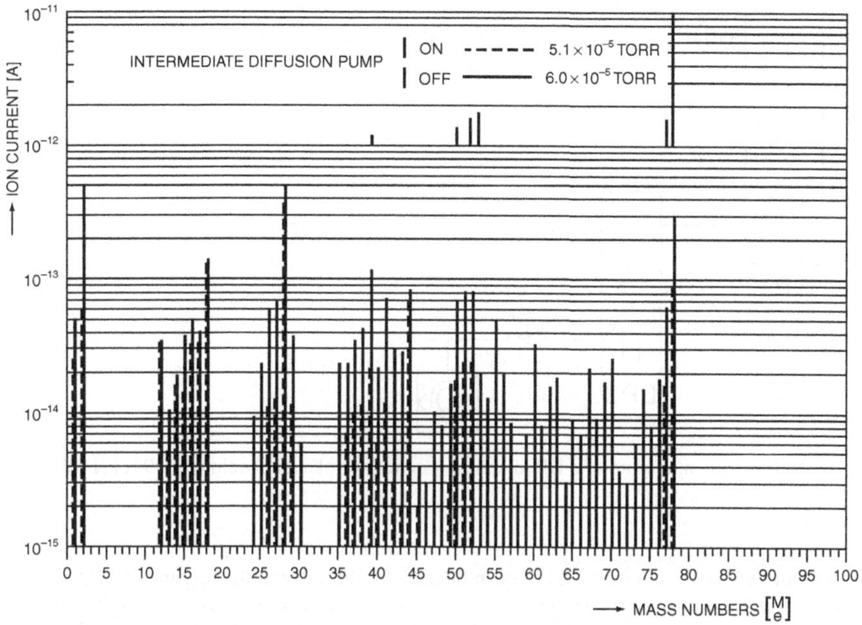

Fig. 1.13 Turning off the booster diffusion pump. The six typical peaks of benzene are noted (Hengevoss and Huber, 1963) [1-9]

(polyphenylether). The tolerable forepressure of both pumps was 38 Pa. The Bayard-Alpert gauge (BAG) was a glass-tube type with W filament, operated at 1 mA of emission current. The quadrupole mass-filter (MS, W filament) was operated at 0.5 mA. In the experiment of a single DP system the DP2 was kept off.

The chamber (type304 stainless steel) was evacuated for a long period, following an in situ bakeout (100 °C, 3 days). A residual gas spectrum obtained for the chamber evacuated continuously for one month is presented in **Fig. 1.15 (1)**, showing a very clean vacuum. The spectrum is characterized by the high peaks of water vapor (peaks 17 and 18), which is due to the water vapor permeation through Viton seals [1]. Then, the power to the heaters

Table 1.1 Increase in the partial pressure after turning off the booster diffusion pump (Hengevoss and Huber, 1963) [1-9]

Gas	Partial pressure in Torr N_2-equ.	
	Before turning off	After turning off
H_2	6.0×10^{-10}	50.0×10^{-10}
CH_4	2.3×10^{-10}	6.0×10^{-10}
H_2O	12.0×10^{-10}	12.0×10^{-10}
CO	23.0×10^{-10}	33.0×10^{-10}
CO_2	7.0×10^{-10}	9.0×10^{-10}
M78	41.0×10^{-10}	140.0×10^{-10}
Total	8.1×10^{-9}	2.5×10^{-8}
Reading of the immersed gauge	5.0×10^{-9}	6.0×10^{-8}

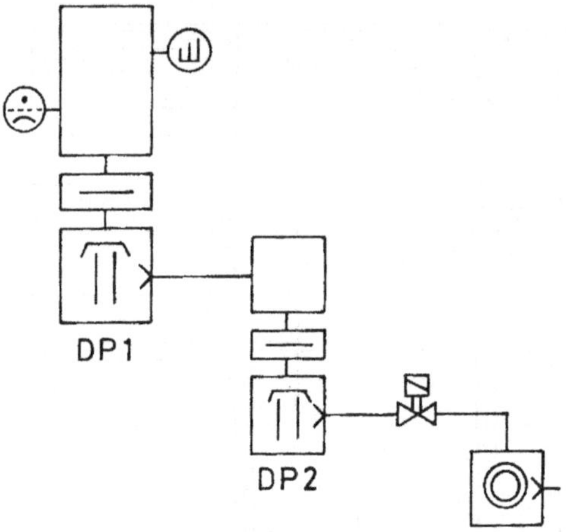

Fig. 1.14 Experimental setup of a DP in-series system (Norioka an Yoshimura, 1991) [1-10]

of both DPs was shut down, and a few minutes later the valve at the foreline of DP2 was closed. After a 1-h cool down, both DPs were simultaneously switched on, just after the valve was opened. The MS (0.5 mA) and BAG (1 mA) were switched on 1 h after supplying DP heater power. Residual gas spectra analyzed 4 and 26 h after switching on the DP heaters are presented in **Fig. 1.15 (2)** and **(3)**, respectively. A number of hydrocarbon peaks were

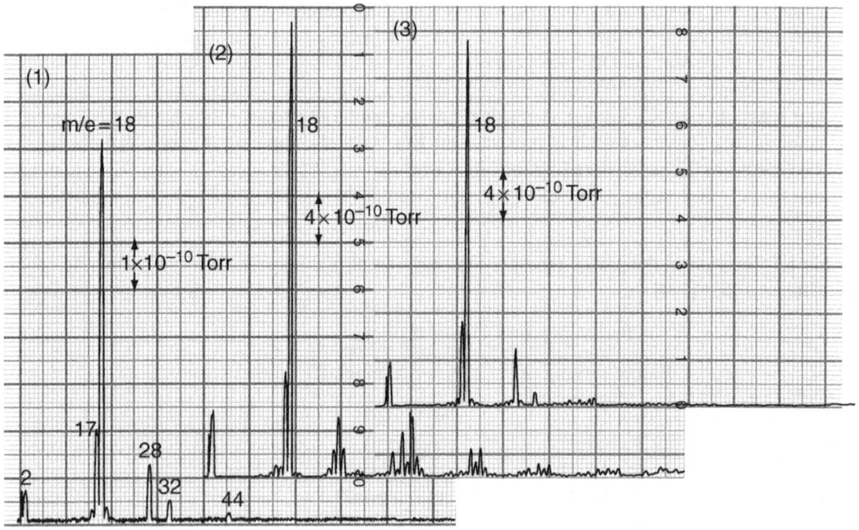

Fig. 1.15 Residual gases depending on the evacuation time t_e after DPs were again switched on in the DP in-series system. **(1)** Before switching off both DPs, **(2)** 4 h of t_e, and **(3)** 26 h (Norioka and Yoshimura, 1991) [1-10]

detected at a short elapsed time after supplying heater power. It takes more than two days to recover the clean vacuum without bakeout treatment. Continuous pumping for many days is essential for DP systems to produce a clean vacuum.

Next, only the power to the DP1 heater was switched off on the DP in-series system (DP1-DP2-RP system). The pressure in the chamber rose gradually but smoothly from 4.0×10^{-9} Torr (5.3×10^{-7} Pa) up to an equilibrium pressure of 5.5×10^{-8} Torr (7.3×10^{-6} Pa) produced by the DP2. On the other hand, the pressure on the single DP system (DP1-RP system) rose rapidly to a low vacuum range soon after the power to DP1 was switched off. The pressure rise characteristics of the respective systems are presented in **Fig. 1.16 (a)** and **(b)**. In such a way, the worst situation is avoided with the DP in-series system when the DP1 heater breaks down.

The input voltage to the DP heater often varies.

Residual gases were analyzed by changing the input voltage on the single DP system (DP2 kept off) and on the DP in-series system (DP1-DP2-RP system). The regular input voltage for both DPs is 200 V. The input voltages to both DPs were simultaneously changed from 200 V to 180 V and 220 V, and kept constant for 30 min, respectively. Residual gases were analyzed just before changing the input voltage on the respective system, as shown in **Fig. 1.17 (a)** and **(b)** [(1) 200 V, (2) 180 V, (3) 220 V].

In the single-DP system **(a)**, the peak of H_2 ($m/e = 2$) increased at 180 V, compared with the peak height of H_2 at 200 V. And at 220 V, hydrocarbon peaks of m/e 39, 41, and 43 increased due to the thermal cracking of working fluid (Santovac-5, polyphenylether).

In the DP in-series system on the other hand, the peak height of every peak did not vary after the input voltage was changed. At 180 V, the limiting compression ratio of the DP1 decreases due to the power reduction. Nevertheless, thanks to the DP2, the H_2 pressure at the foreline of the DP1 was low enough not to increase the H_2 pressure on the high vacuum side. At 220 V, thermal cracking of the working fluid of the DPs must be accelerated due to somewhat higher boiler surface temperature. However, the hydrocarbon cracking products were quickly evacuated by the DP2, with the result that there was no increase of hydrocarbon peaks.

The DP system sometimes encounters accidents in cooling water lines. On both systems, the high-vacuum pressure indicated by the BAG was recorded after the cooling water line to both DPs was shut off. Experimental results are presented in **Fig. 1.18**, where the pressure range of the recorder chart for both systems was switched to the $\times 10^{-8}$ Torr range at about 9 minutes of elapsed time.

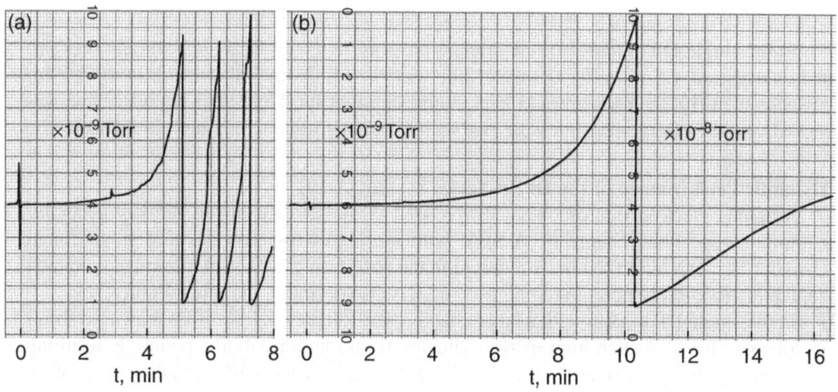

Fig. 1.16 Pressure-rise characteristics after the DP1 was switched off. **(a)** In the DP1-RP system and **(b)** in the DP1-DP2-RP system (Norioka and Yoshimura, 1991) [1-10]

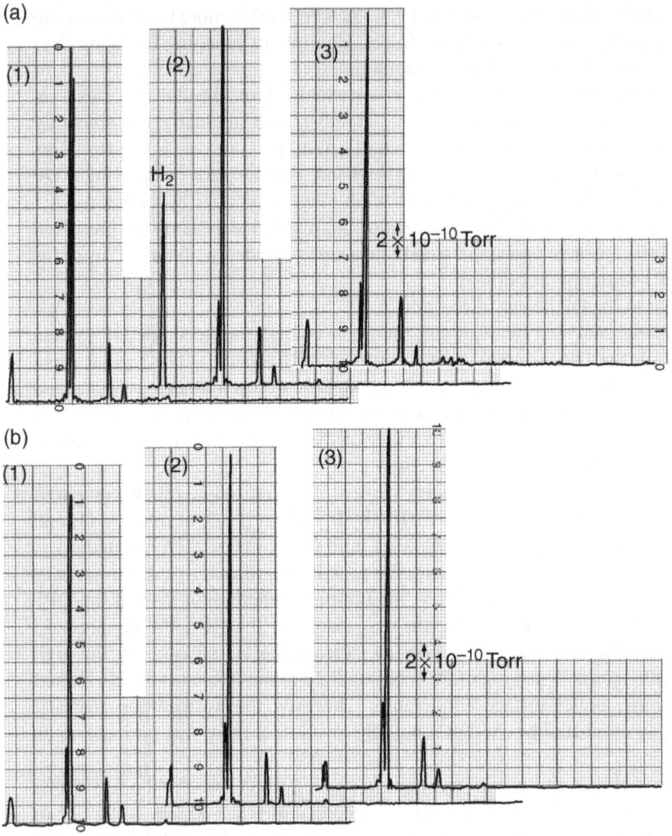

Fig. 1.17 Residual gases depending on the voltage inputs to the DP heaters. **(a)** On the DP1-RP system with (1) 200 V, (2) 180 V, (3) 220 V, and **(b)** on the DP1-DP2-RP system with (1) 200 V, (2) 180 V, (3) 220 V (Norioka and Yoshimura, 1991) [1-10]

In the single DP system **(a)**, the gradual pressure rise with large, intermittent pressure pulses occurred 1.5 min after the water line was closed. On the other hand, in the DP in-series system **(b)**, the pressure was kept almost constant for 4 min after the water line was closed, and then the gradual pressure rise with large, intermittent pressure pulses occurred.

Dynamic Performance of the DP In-Series System [1-10]

Scanning electron microscopes (SEMs) used in the semiconductor industry generally equip large specimen chambers. A SEM with x-ray spectrometers has a large specimen chamber. A clean high vacuum is required in the specimen chamber to minimize specimen contamination. Rapid pumping to high vacuum is also desirable for general efficiency.

In the single DP system with a large chamber the forepressure P_f of DP may rise above its tolerable limit P_t when improperly switching from roughing to fine pumping, resulting in increasing the backstreaming of DP oil vapor [2, 3]. Therefore, the DP system of an instrument of clean high vacuum must be basically controlled by a P_f signal detected at the foreline of the DP, as has been discussed by Yoshimura et al. [2]. Nevertheless, the system tolerance to overload is necessary.

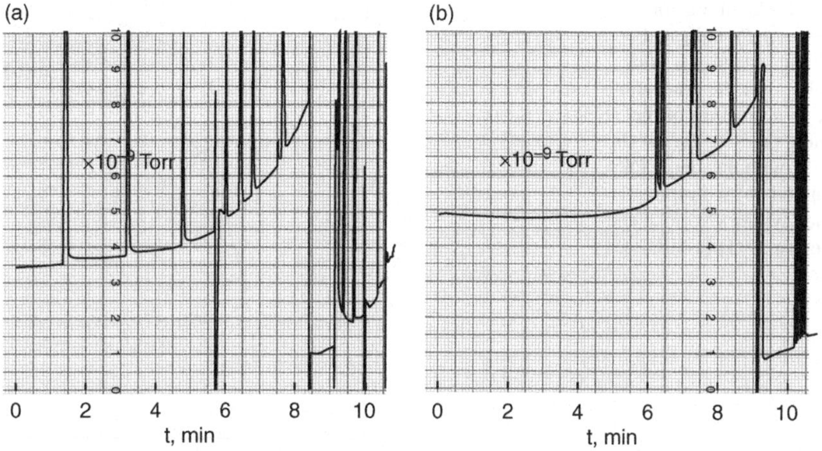

Fig. 1.18 Pressure changes after cooling water was stopped. **(a)** In the DP1-RP system and **(b)** in the DP1-DP2-RP system (Norioka and Yoshimura, 1991) [1-10]

A DP in-series system with a buffer tank (BT) meets the requirements for the SEM vacuum system mentioned above. The vacuum system diagram of a SEM with a DP in-series system is presented in **Fig. 1.19**. The specimen chamber equips three x-ray spectrometers, having an effective volume of 42 L. The chamber also equips the standard movable mechanisms, such as a large specimen stage and a specimen exchanging mechanism. Such movable mechanisms were lubricated with Apiezon grease, showing considerable outgassing when evacuated after being exposed to the atmosphere. The DP1 (4 in.) and DP2 (2.5 in.) were charged with Santovac-5. The volumes of BT1 and BT2 are 1 L and 10 L, respectively. The trap above the DP1 was not cooled throughout the following experiments. For the experiments on the SEM with the single DP system, the DP2 and BT1 were removed, and the foreline of the DP1 was directly connected to the BT2. Note that the pumping speed of the mechanical rotary pump (RP) was rather small (100 L/min) for the large specimen chamber.

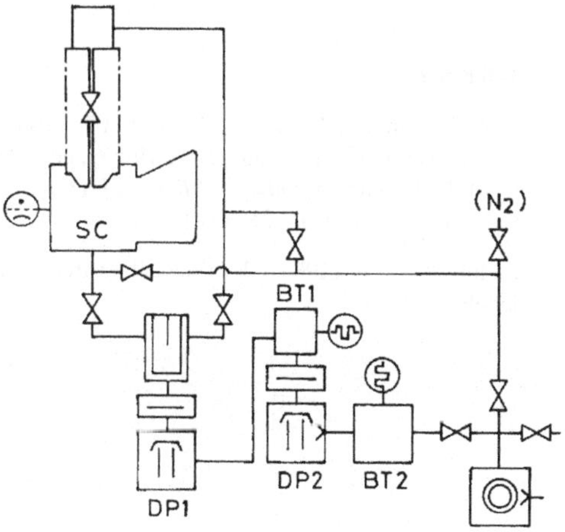

Fig. 1.19 Vacuum system diagram of a SEM equipped with a DP in-series system (Norioka and Yoshimura 1991) [1-10]

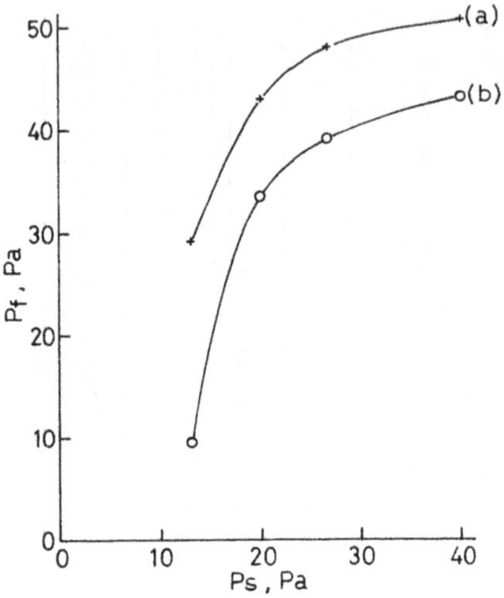

Fig. 1.20 Maximum fore pressures of P_f indicated by a Pirani gauge just after switching over the evacuation mode, as a function of the switching pressure P_s. (**a**) On the SEM with the single DP system and (**b**) with the DP in-series system (Norioka and Yoshimura 1991) [1-10]

On the respective systems the fore-pressures P_f of the DP1, indicated by a Pirani gauge, were recorded when the evacuation mode was switched. The P_f on the single DP system **(a)** was measured at the BT2, and the P_f on the DP in-series system **(b)** was measured at the BT1. Experimental results showing the maximum fore pressures of P_f, as a function of switching pressure P_s, are presented in **Fig. 1.20**.

On the single DP system the maximum rise of P_f was much higher than that on the DP in-series system, although the volume of the BT2 (10 L) was much larger than that of the BT1 (1 L).

The critical backing pressure P_c of DP1 is 38 Pa. According to the data of **Fig. 1.20**, in which the slow response of the Pirani gauge was ignored, the chamber on the single DP system must be evacuated in roughing down to 17 Pa (P_s) in order not to raise P_f above P_c just after switching over to fine pumping. On the other hand, on the DP in-series system roughing to 24 Pa (P_s) is sufficient enough not to raise P_f above P_c. Practically speaking, for the SEM chamber of 42 L the roughing time down to 24 Pa was 20% shorter than the time to 17 Pa. Also, switching the evacuation modes at a higher pressure is desirable in view of the backstreaming of RP oil vapor [4].

References

1. N. Yoshimura, *J. Vac. Sci. Technol. A* **7**, 110 (1989).
2. N. Yoshimura, H. Hirano, S. Norioka, and T. Etoh, *J. Vac. Sci. Technol. A* **2**, 61 (1984).
3. M. H. Hablanian, *Industrial Research/Development*, August 1979, p. 84.
4. L. Holland, *Vacuum* **21**, 45 (1971).

Yoshimura et al. (1984) [1-4] described the safety system of the cascaded DP system in details.

Differential Evacuation Systems [1-4]
A very clean vacuum is required in the column of an analytical electron microscope to minimize specimen contamination. The whole column can be considered to be divided into two parts by a small conductance aperture positioned just below the projector lens pole-piece. Thus, a differential evacuation system is necessary between the column and the camera chamber.

Typical differential evacuation systems, **(a)** independent system, **(b)** parallel system with a common backing line, and **(c)** cascade system are presented in **Fig. 1.21**, where the chamber A corresponds to the transmission electron microscope (TEM) column, and B to the camera chamber. All systems have a DP-DP in-series system for the chamber A to obtain a clean vacuum and a buffer tank BT for proper transitional evacuation. These systems are evaluated with respect to correlation (mutual relationship) in gas flow between the chambers A and B [1-4].

Consider a large leakage occurring in the chamber A in the system **(b)**. Leaked gases are compressed immediately into the common foreline, and flow backwards with dense oil vapor into the chamber B through DP2 when P_b rises above the critical backing pressure P_c. A leakage in the chamber B also causes similar backstreaming into the chamber A through DP1. On the other hand, this kind of backstreaming does not occur in the independent system **(a)**. Correlation in gas flow is therefore related to system resistance to emergency due to leakage. In the cascade DP system **(c)**, a leakage in the chamber A does not cause such backstreaming of oil vapor into the chamber B, whereas leakage at B may cause backstreaming into A through DP1. In the operation of a TEM, leakage may occur in the column, and almost never at the camera chamber, because most movable mechanisms, such as the specimen holder and apertures, are located in the column, associated with the chamber A. Considering such actual circumstances, correlation in gas flow between A and B is evaluated as tolerable for the system **(c)**. For system cost and occupied space, the cascade DP system of **(c)** is best, which can be easily recognized by comparing the number of system parts with each other.

The cascade DP system with a BT is clearly evaluated to be the most practical and reasonable for a TEM, where the column is evaluated by DP1 backed by DP2, and the camera chamber by DP2 with a large volume foreline backed by RP.

Safety Systems in Cascade DP System [1-4]
DP systems must be basically controlled by P_b (the backing pressure) signals detected by a gauge at the foreline of DP [1]. When P_b rises to the reference pressure P_r which is set to be lower than the critical backing pressure P_c in any evacuation mode, the vacuum system is rapidly driven to a safe state. This system is called the **"basic" safety system** here. It

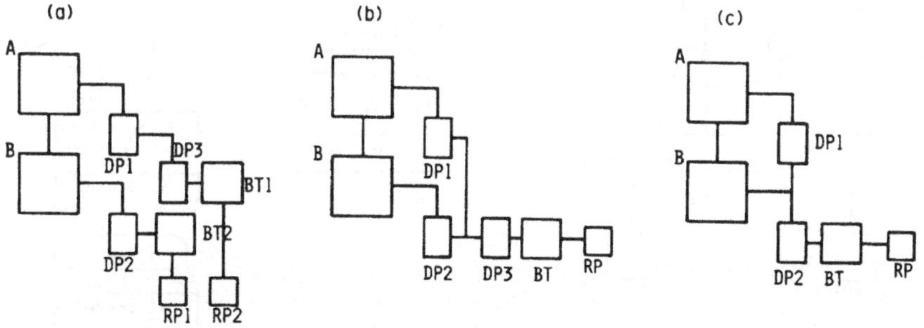

Fig. 1.21 Differential DP evacuation systems for TEM. **(a)** Independent system, **(b)** parallel system with a common backing line, and **(c)** cascade DP system (Yoshimura et al., 1984) [1-4]

serves well against the following emergencies: (1) leakage at any part of the whole column, (2) leakage at any part of the DP system, (3) failures of the backing RP, and (4) failures of gauges for switching the evacuation mode.

In addition, the vacuum system is provided with the **"normal" safety system** which drives the system to a safe state using signals which have detected failures on system parts such as the power supply including fuses, cooling water system, pneumatic air system for valves, and DP heaters. The **"normal" safety system** is activated in all evacuation modes including warming up.

Moreover, the cascade DP system is provided with a safety valve which is closed temporarily, regardless of the existing failures, to protect the vacuum system just before switching the evacuation mode. This system is called the **"standby" safety system**. The system is presented in **Fig. 1.22**, where V_s is the safety valve. Consider that air is admitted into the chamber B while the chamber A is continuously evacuated by DP1. First, consider a system without the safety valve V_s on the foreline of DP1. If sealing of V2 is insufficient to isolate the chamber B, the introduced air rapidly flows backwards through DP1 into the chamber A, resulting in serious backstreaming. Similar trouble may occur if P_h in the chamber B is incorrectly detected in switching the evacuation mode. In the cascade DP system with the **"standby" safety system**, V_s is closed temporarily just before venting the chamber B and switching the evacuation mode. With the association of the **"basic" safety system**, the evacuation system is rapidly put into a stop state when P_b rises above P_r, so the **"standby" safety system** protects the vacuum system against the above-mentioned faults. The flow sequences of this system are given in **Fig. 1.22 (b)**, one is for venting of the chamber B and the other for evacuation of the chamber B.

Figure 1.23 presents the vacuum system diagram of an EM (JEM-1200EX, JEOL Ltd.), provided with a refined cascade DP system. The column and the gun chamber are evacuated by DP1 backed by DP2, while the camera chamber is evacuated by DP2 with the tank RT2 backed by RP. The bypass valve V1 with a low conductance is used for the camera chamber. The volume of the tank RT2 is rather small (approx. 10 L) due to the help of the low-conductance bypass valve V1. The tank serves as a buffer in transitional evacuation as

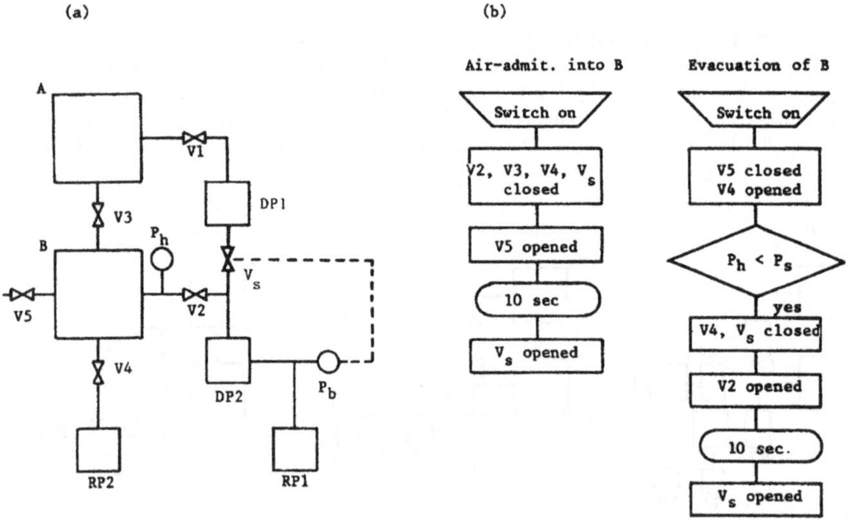

Fig. 1.22 "Standby" safety system. **(a)** System diagram and **(b)** sequence flows: one for venting of the chamber B and the other for evacuation of the chamber B (Yoshimura et al., 1984) [1-4]

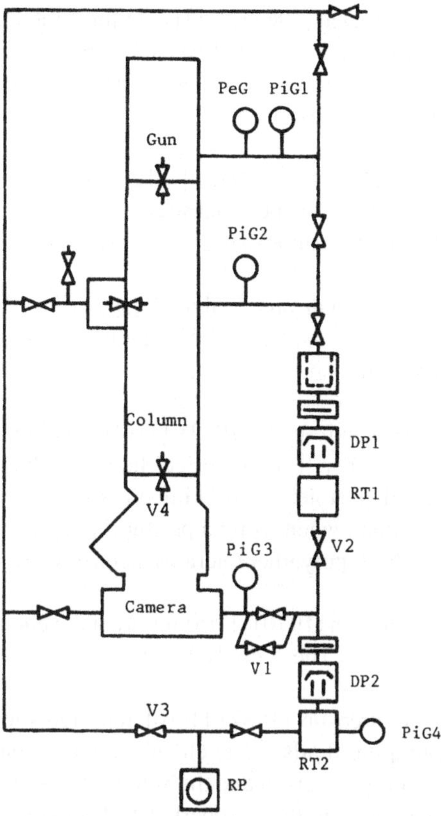

Fig. 1.23 Vacuum system diagram of EM with a cascade DP system. The liquid-nitrogen cold trap [1] is optional (Yoshimura et al., 1984) [1-4]

well as a reservoir in roughing. Signals from the Pirani gauge PiG4 at the foreline control the **"basic" safety system**. The valve V2 is a safety valve in the **"standby" safety system** which is closed temporarily just before the vacuum system is put into the air-admittance state and the switching the evacuation mode to fine pumping for the camera chamber. The valve V3 on the roughing line is opened in the roughing mode only. A cold trap can be installed above DP1 in the main pumping line.

Key Points to Minimize the Backstreaming of DP Oil Vapor

1) The diffusion pump meets an overload condition several seconds after switching the evacuation mode from roughing to fine pumping. During this overload period, the backstreaming rate of oil vapor is increased abnormally.
2) The forepressure of the diffusion pump rises above its critical forepressure several seconds after switching the evacuation mode from roughing to fine pumping, thus increasing the backstreaming abnormally.
3) The diffusion pump system must be designed so that the overload period passes rapidly by keeping the forepressure even sufficiently low after switching the evacuation mode.

Kumagai (1961) [1-11] examined the DP system having two fractionating pumps in series and a specially designed backing pipe of the DP. "In the system, impurities of the oil in the DP oil boiler are quickly removed and oil-vapor backstreaming is much suppressed."

Dennis et al. (1982) [1-12] investigated the effect of DP-valve opening speed upon the oil-vapor backstreaming rate. "Opening the DP-valve slowly is very important for suppressing the pressure-rise in the fore-vacuum side."

Hablanian and Maliakal (1973) [1-13] reviewed the advance of DP technology. They emphasized that the use of polyphenylether makes a DP achieve a clean vacuum.

Holland et al. (1972) [1-14] obtained a clean high vacuum using a perfluoropolyether. They reported the basic characteristics of a DP, such as the pumping speed, the tolerable backing pressure and the residual gas spectra in the high-vacuum chamber, when using a perfluoropolyether. They described that the system using a perfluoropolyether tolerates a considerably high backing-pressure.

Lawson (1970) [1-15] reported the chemical features of perfluoropolyether fluids in detail. "These fluids do not polymerize when irradiated with electrons."

Laurenson (1982) [1-16] reported the technology and applications of typical pumping fluids. "The chemical and physical properties of the fluids employed in vacuum are given and the suitability of these fluids for various vacuum applications in relation to their properties is discussed."

Rettinghaus and Huber (1974) [1-17] presented an article, "Backstreaming in diffusion pump systems." "With a molecular beam detector based on a mass spectrometer, backstreaming phenomena are measured which are not due to mere evaporation but to molecular scattering within the baffle. The influence of a number of parameters is studied (among others the type of the pump and the pressure). The contribution of backstreaming to system contamination appears negligible, except at higher pressures in the 10^{-4} Torr range where special precautions are recommended to meet exceptional high requirements for cleanliness."

Harris (1977) [1-18] presented an article, "Diffusion pump backstreaming." "This review deals exclusively with contamination which can be attributed to backstreaming of diffusion pump working fluid. It discusses the origins of this type of contamination, surveys methods used to prevent and control it and shows that it is possible to produce a clean vacuum using diffusion pumps."

Baker and Staniforth (1968) [1-19] presented an article, "The development of an adsorption foreline trap suitable for quick cycling vacuum systems." "By using 5–10 mm diameter activated alumina balls in an inline column geometry trap the adsorption of water vapor during pumping was reduced to a minimum. With this material and trap design unlimited repeat pumpdown cycles to less than 100 millitorr were possible on a typical 116 liter volume chamber. The pump-down time in this case was extended from 3.5 minutes without the trap to 6.5 minutes with the trap."

Ultrahigh Vacuum Electron Microscopes

Recently, transmission electron microscopes with several sputter ion pumps (SIPs) have become widely used. A field-emission electron gun (FEG) is built inside the small SIP for JEOL transmission electron microscopes.

SIPs are suitable for ultrahigh-vacuum (UHV) instruments because they can be baked up to about 200 °C for degassing. An SIP with a sufficient pumping speed in UHV is required for the FEG chamber.

Harada and Yoshimura (1987) [1-20] reported the development of JEOL UHV electron microscopes. The evacuation system of a typical UHV-TEM, the JEOL JEM-2000FXV, is also introduced. Three types of the evacuation systems of standard transmission electron microscopes (JEOL TEM) are schematically presented in **Fig. 1.24**. The UHV transmission electron microscope, JEM-2000FXV (JEOL) is also presented in **Fig. 1.25**.

Tomita (1990) [1-21] reported the evacuation system of a UHV-scanning transmission electron microscope (UHV-STEM) and its vacuum characteristics. A W <100> thermal-field-emission (TFE) emitter is used in the UHV-STEM, whose vacuum system diagram is presented in **Fig. 1.26**.

"Key technologies for the UHV electron-microscope column are metal-gasket sealing for the entire microscope column and baking the microscope column at up to 150 °C under high vacuum. Metal O-rings are used for sealing the electron-optics (EO) lens column and chambers in order to achieve assembling accuracy. On the other hand, vacuum parts such as valves and gauges are sealed using ICF ConFlat® flanges. The EO lens column, the specimen chamber and the gun chamber of the UHV-STEM can be baked up to 160 °C." [1-21]

Kondo et al. (1991) [1-22] developed the 200 kV UHV-high-resolution transmission electron microscope (UHV-HRTEM) for in situ surface observation:

> With use of metal sealings, the pressure at the specimen position was 2.6×10^{-7} Pa without any cooling traps, and less than 1.5×10^{-7} Pa with the built-in traps cooled by liquid nitrogen. A cantilever-type goniometer, a specimen holder and a high-resolution objective lens were newly developed for high-resolution transmission (HR-TEM) and reflection (HR-REM) to observe specimen surfaces with a point-to-point resolution of 0.21 nm. The specimen can be tilted up to ±20 °. Crucible-type and direct-heating specimen holders were constructed for in situ experiments on thin films and small particles, and on bulk crystalline surfaces, respectively. The usefulness of this microscope for surface studies is shown by images of clean surfaces of Au and Si prepared in the microscope [1-22].

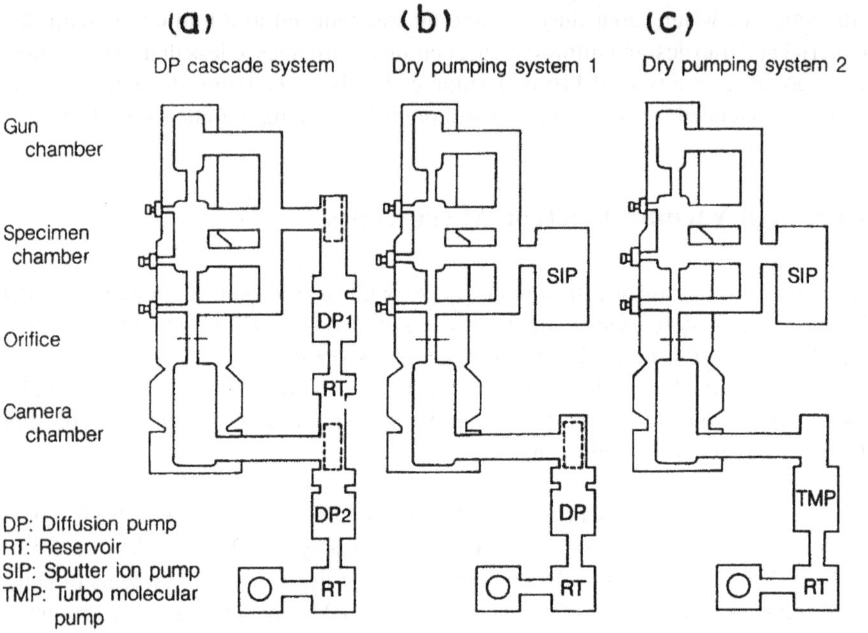

Fig. 1.24 Three types of standard transmission electron microscopes (JEOL TEM) (Harada and Yoshimura, 1987) [1-20]

The block diagram of the vacuum system of the UHV-TEM/REM is presented in **Fig. 1.27**.

Okamoto and Naruse (1998) [1-23] reported the key technology on UHV-electron microscopes, such as the evacuation system, the specimen-changing mechanism, and the bakeout system.

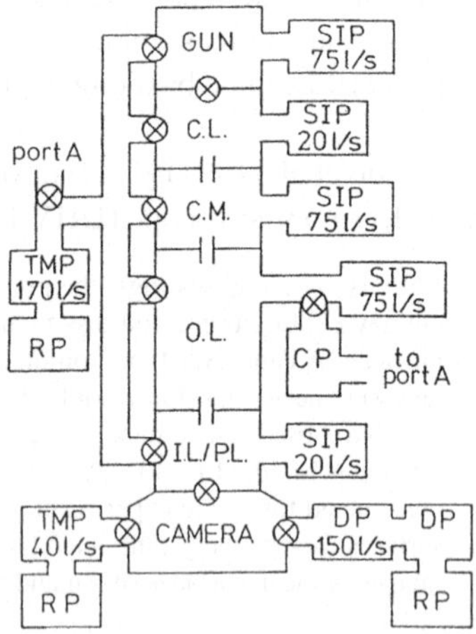

Fig. 1.25 Vacuum system of UHV transmission electron microscope, JEM-2000FXV (Harada and Yoshimura, 1987) [1-20]

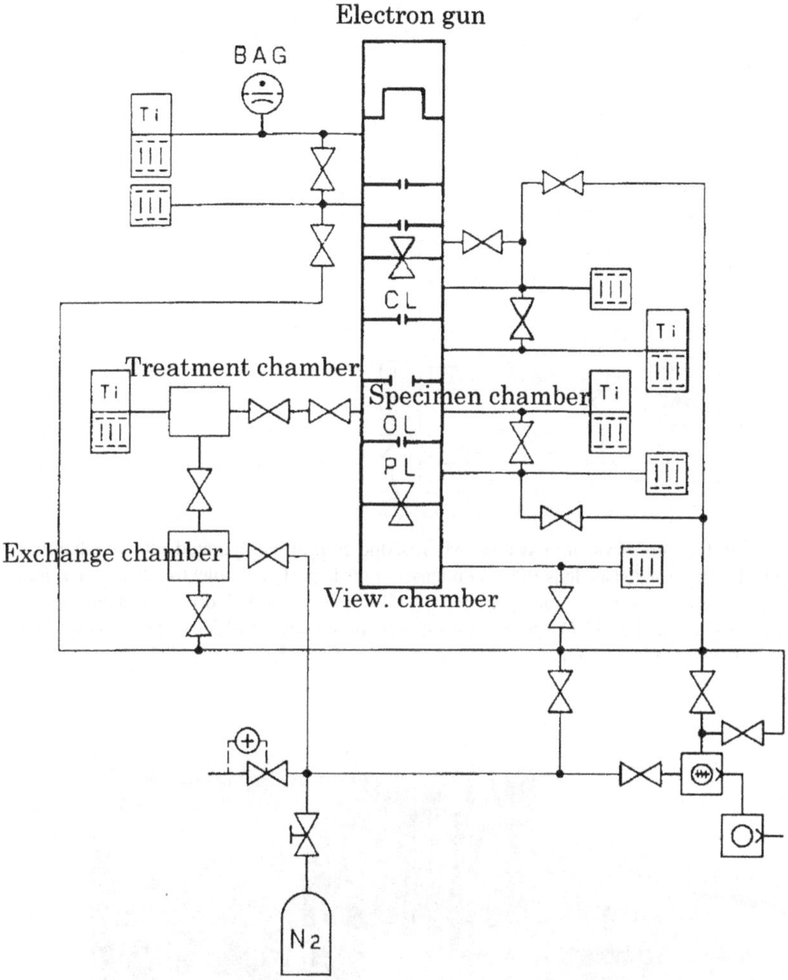

Fig. 1.26 Vacuum system diagram of a JEOL UHV-STEM (Tomita, 1990)[1-21]

Poppa (2004) [1-24] presented a review, "High resolution, high speed ultrahigh vacuum microscopy," in which UHV scientific instruments such as a UHV transmission electron microscope (UHV-TEM) systems and a UHV-HRTEM/STM system are introduced in photographs. **Figure 1.28** shows the fully integrated surface analysis and HRTEM/STM system at the National Institute for Materials Research in Tsukuba, Japan.

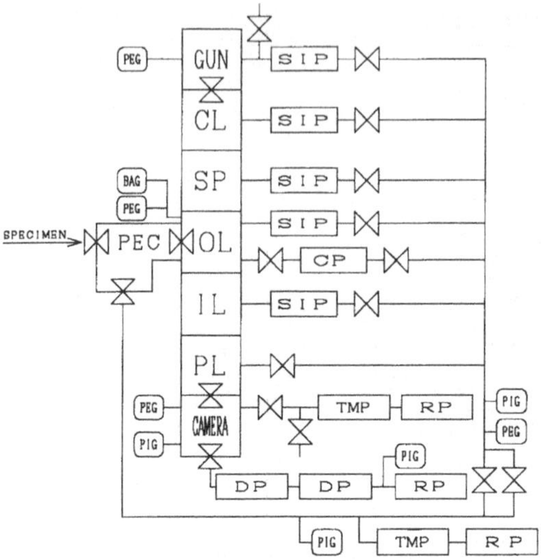

Fig. 1.27 Block diagram of vacuum system. Microscope column is separated into six chambers, electron gun (GUN), condenser lens (CL), condenser mini-lens (CM), objective lens (OL), intermediate and projector lens (IL/PL) and camera (CAMERA) chambers. Individual vacuum pumps are attached to each chamber. PEC: specimen pre-evacuation chamber, PEG: Penning gauge, PIG: Pirani gauge, BAG: B-A gauge, CP: Cryogenic pump (Kondo et al., 1991) [1-22]

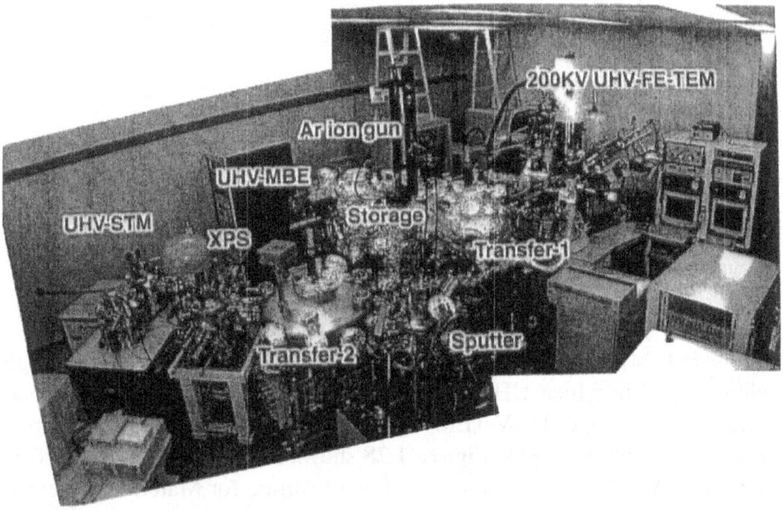

Fig. 1.28 Fully integrated surface analysis and HRTEM/STM system (UTSICS UHV-TEM-STM-integrated characterization system) at the National Institute for Materials Research in Tsukuba, Japan (Poppa, 2004) [1-24]

Know-how Technology in Designing UHV Evacuation Systems

(1) Penning gauge head with a radioisotope source

Electron microscope evacuation systems with an SIP have a Penning gauge for switching the SIP on and off at about 10^{-2} Pa. An FEG system is controlled by the SIP ion current showing a pressure measure. When a Penning gauge head is contaminated with hydrocarbon gases, Penning discharge becomes difficult to initiate. To prevent this trouble, a radioisotope source is built into the Penning gauge head.

(2) Leakage flux from SIP magnet

When an SIP is attached to the gun chamber, magnetic-field leakage from it might cause deflection of the electron beam. As seen in **Fig. 1.29**, a field emission (FE) emitter is built inside the small SIP, which is composed of three constituent SIPs arranged symmetrically on the axis of the electron beam. In this arrangement, magnetic-field leakage flux from the three constituent magnet pairs is cancelled on the axis of the electron beam.

(3) Ion leakage from SIP

Many electrons and ions exist inside an SIP vessel. From a diode-type SIP with a cathode at ground potential, some ions easily escape from the SIP vessel. Such ions, entering the accelerating tube, might induce microdischarge. (refer to Chap. 7). Opaque plates at ground potential must be inserted inside the pumping pipe to prevent the ions from a diode-type SIP from entering the microscope column.

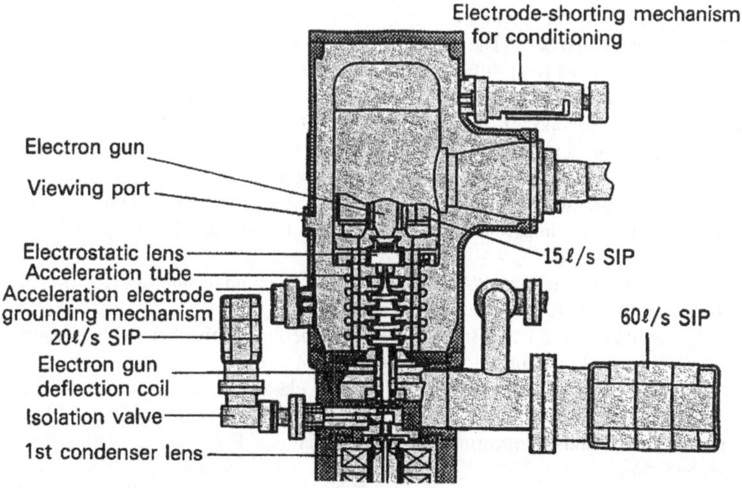

Fig. 1.29 Arrangement of SIPs for the acceleration tube of the JEM-2010F (Courtesy of JEOL Ltd.)

(4) Particle contamination

In order to achieve high productivity, it is important in wafer-inspection SEMs and semiconductor-manufacturing instruments not to contaminate wafer surfaces or semiconductor surfaces with dust particles. For this purpose, pipes and chambers with mirror-polished inner surfaces are used so that dust particles do not stick to their inner surfaces. Also slow gas introduction and slow evacuation in the first stage of roughing are important in order not to peel off dust particles that have adhered to their inner surfaces.

References

1-1. N. Yoshimura, "Water vapor permeation through Viton O ring seals", *J. Vac. Sci. Technol. A* **7** (1), pp. 110–112 (1989).

1-2. B. B. Dayton, "Relations between size of vacuum chamber, outgassing rate, and required pumping speed", *1959 6th National Symposium on Vacuum Technology Transactions* (Pergamon Press 1960), pp. 101–119.

1-3. L. Holland, "Vacua: How they may be improved or impaired by vacuum pumps and traps", *Vacuum* **21** (1/2), pp. 45–53 (1971).

1-4. N. Yoshimura, H. Hirano, S. Norioka, and T. Etoh, "A cascade diffusion pump system for an electron microscope", *J. Vac. Sci. Technol. A* **2** (1), pp. 61–67 (1984).

1-5. M. J. Fulker, "Backstreaming from rotary pumps", *Vacuum* **18** (8), pp. 445–449 (1968).

1-6. M. H. Hablanian and H. A. Steinherz, "Testing performance of diffusion pumps", *Transactions of the 8th National Vacuum Symposium* 1961 (Pergamon Press, New York, 1962), pp. 333–341.

1-7. M. H. Hablanian, "Prevention of overload in high-vacuum systems", *J. Vac. Sci. Technol. A* **10** (4), pp. 2629–2632 (1992).

1-8. M. Iwami, "Outgassing rate for one-minute exhaust", *SHIBAURA Mechatronics Technical Journal* 6-2006, pp. 19–24 (in Japanese).

1-9. J. Hengevoss and W. K. Huber, "The influence of fore-vacuum conditions on ultrahigh-vacuum pumping systems with oil diffusion pumps", *Vacuum* **13** (1), pp. 1–9 (1963).

1-10. S. Norioka and N. Yoshimura, "Practical advantages of a cascade diffusion pump system of a scanning electron microscope", *J. Vac. Sci. Technol. A* **9** (4), pp. 2384–2388 (1991).

1-11. H. Kumagai, "On an oil self-purifying diffusion pump system", *Transactions of the 8th National Vacuum Symposium 1961* (Pergamon Press, New York, 1962), pp. 327–332.

1-12. N. T. M. Dennis, L. Laurenson, A. Devaney, and B. H. Colwell, "The effect of the inlet valve on the ultimate vacua above integrated pumping groups", *Vacuum* **32** (10/11), pp. 631–633 (1982).

1-13. M. H. Hablanian and J. C. Maliakal, "Advances in diffusion pump technology", *J. Vac. Sci. Technol.* **10** (1), pp. 58–64 (1973).

1-14. L. Holland, L. Laurenson, and P. N. Baker, "The use of perfluoropolyether fluids in vapour stream pumps", *Vacuum* **22** (8), pp. 315–319 (1972).

1-15. N. D. Lawson, "Perfluoroalkylpolyethers: A unique new class of synthetic lubricants", Presented at The American Institute of Chemical Engineers, 67th National Meeting and Biennial Materials Sciences and Engineering Division Conference, Feb. 15–18, 1970, Atlanta, Georgia, pp. 535–543.

1-16. L. Laurenson, "Technology and applications of pumping fluids", *J. Vac. Sci. Technol.* **20** (4), pp. 989–995 (1982).

1-17. G. Rettinghaus and W. K. Huber, "Backstreaming in diffusion pump systems", *Vacuum* **24** (6), pp. 249–255 (1974).

1-18. N. S. Harris, "Diffusion pump back-streaming", *Vacuum* **27** (9), pp. 519–530 (1977).
1-19. M. A. Baker and G. H. Staniforth, "The development of an adsorption foreline trap suitable for quick cycling vacuum systems", *Vacuum* **18** (1), pp. 17–20 (1968).
1-20. Y. Harada and N. Yoshimura, "Electron microscope", *Shinku (J. Vac. Soc. Japan)* **30** (12), pp. 985–988 (1987) (in Japanese).
1-21. T. Tomita, "Ultrahigh-Vacuum scanning electron microscope", *Electron Microscope* **25** (2), pp. 114–117 (1990) (in Japanese).
1-22. Y. Kondo, K. Ohi, Y. Ishibashi, H. Hirano, Y. Harada, K. Takayanagi, Y. Tanishiro, K. Kobayashi, and K. Yagi, "Design and development of an ultrahigh vacuum high-resolution transmission electron microscope", *Ultramicroscopy* **35**, pp. 111–118 (1991).
1-23. K. Okamoto and M. Naruse, "Fundamental techniques of the ultrahigh vacuum electron microscope", *Shinku (J. Vac. Soc. Japan)* **41** (11), pp. 921–926 (1998) (in Japanese).
1-24. H. Poppa, "High resolution, high speed ultrahigh vacuum microscopy", *J. Vac. Sci. Technol. A* **22** (5), pp. 1931–1947 (2004).

Other Articles

*1-1. T. Kojima, Y. Nagahama, N. Yoshimura, and H. Oikawa, "Improved vacuum system for the electron microscope", *Shinku (J. Vac. Soc. Japan)* **16** (6), pp. 217–224 (1973) (in Japanese).
*1-2. N. Yoshimura, S. Ohmori, Y. Nagahama, and H. Oikawa, "A new vacuum system for an electron microscope", *Proc. 6th Internl. Vacuum Congr. 1974, Japan. J. Appl. Phys. Suppl. 2, Pt, 1*, 1974, pp. 249–252 (in English).
*1-3. T. Someya, T. Goto, Y. Harada, K. Yamada, H. Koike, Y. Kokubo, and M. Watanabe, "On development of a 100 kV field emission electron microscope", *Optik* **41** (3), pp. 225–244 (1974) (in English).
*1-4. A. Tonomura, T. Matsuda, J. Endo, H. Todokoro, and T. Komoda, "Development of a field emission electron microscope", *J. Electron Microsc.* **28** (1), pp. 1–11 (1979).
*1-5. Y. Tsutsumi, S. Ueda, M. Ikegawa, and J. Kobayashi, "Prevention of oil vapor backstreaming in vacuum systems by gas purge method", *J. Vac. Sci. Technol. A* **8** (3), pp. 2764–2767 (1990).
*1-6. A. D. Feinerman, D. A. Crewe, D. C. Perng, S. E. Shoaf, and A. V. Crewe, "Sub-centimeter micromachined electron microscope", *J. Vac. Sci. Technol. A* **10** (4), pp. 611–616 (1992).
*1-7. K. Y. Lee, S. A. Rishton, and T. H. P. Chang, "High aspect ratio aligned multilayer microstructure fabrication", *J. Vac. Sci. Technol. B* **12** (6), pp. 3425–3430 (1994).
*1-8. E. Kratschmer, H. S. Kim, M. G. R. Thomson, K. Y. Lee, S. A. Rishton, M. L. Yu, and T. H. P. Chang, "Sub-40 nm resolution 1 keV scanning tunneling microscope field-emission microcolumn", *J. Vac. Sci. Technol. B* **12** (6), pp. 3503–3507 (1994).
*1-9. P. D. Hoh, "Quantitative particle contamination studies utilizing reduced turbulence pumping and venting", *J. Vac. Sci. Technol. A* **2** (2), pp. 198–199 (1984).
*1-10. G. Strasser, H. P. Bader, and M. E. Bader, "Reduction of particle contamination by controlled venting and pumping of vacuum loadlocks", *J. Vac. Sci. Technol. A* **8** (6), pp. 4092–4097 (1990).
*1-11. D. Chen and S. Hackwood, "Vacuum particle generation and the nucleation phenomena during pumpdown", *J. Vac. Sci. Technol. A* **8** (2), pp. 933–940 (1990).
*1-12. J. L. Dobson and B. R. F. Kendall, "Controllable leaks using electrically pulsed valves", *J. Vac. Sci. Technol. A* **8** (3), pp. 2790–2794 (1990).
*1-13. J. F. O'Hanlon and H. G. Parks, "Impact of vacuum equipment contamination on semiconductor yield", *J. Vac. Sci. Technol. A* **10** (4), pp. 1863–1868 (1992).

*1-14. J. Sullivan, S. Schaffer, S. King, D. Manos, and H. F. Dylla, "Contamination control in the design and manufacture of gas flow components", *J. Vac. Sci. Technol. A* **10** (4), pp. 1869–1874 (1992).
*1-15. J. S. Logan and J. J. McGill, "Study of particle emission in vacuum from film deposits", *J. Vac. Sci. Technol. A* **10** (4), pp. 1875–1878 (1992).
*1-16. G. Y. Hsiung, J. R. Huang, D. J. Wang, J. G. Shyy, J. C. Lee, R. J. Sheu, H. S. Tzeng, S. N. Hsu, S. Y. Perng, K. M. Hsiao, W. D. Wey, J. R. Chen, and Y. C. Liu, "Low contamination ultrahigh vacuum system of the Synchrotron Radiation Research Center 1.3 GeV electron storage ring". *J. Vac. Sci. Technol. A* **13** (5), pp. 2569–2573 (1995).
*1-17. H. Miyashita, T. Kikuchi, Y. Kawasaki, Y. Katakura, and N. Ohsako, "Particle measurements in vacuum tools by in situ particle monitor", *J. Vac. Sci. Technol. A* **17** (3), pp. 1066–1070 (1999).
*1-18. H. F. Dylla, G. Biallas, L. A. Dillon-Townes, E. Feldl, G. R. Myneni, J. Prakinson, J. Preble, T. Siggins, S. Williams, and M. Wiseman, "Design and installation of a low particle, ultrahigh vacuum system for a high power free-electron laser", *J. Vac. Sci. Technol. A* **17** (4), pp. 2113–2118 (1999).
*1-19. L. Rosai and M. Borghi, "Purification of rare gases and the measurement of low residual impurity levels", *J. Vac. Sci. Technol.* **11** (1), pp. 347–349 (1974).
*1-20. H. Y. Kumagai, "Hazardous gas handling in semiconductor processing", *J. Vac. Sci. Technol. A* **8** (3), pp. 2865–2873 (1990).
*1-21. H. Hirano and N. Yoshimura, "Thermal loss of a cold trap", *Shinku (J. Vac. Soc. Japan)* **24** (4), pp. 167–169 (1981) (in Japanese).

Chapter 2
Vacuum Pumps

The mechanical rotary pump and the oil diffusion pump are widely used in many kinds of high vacuum systems. The DP system with a mechanical rotary pump is reliable in performance, silent in noise and small in vibration, and cheap in price.

The sputter ion pump (SIP) is suited to ultrahigh vacuum because it can be baked to degas up to about 200 °C. Getter pumps such as the titanium sublimation pump (TSP) and the non-evaporable getter (NEG) pump can also be used for ultrahigh vacuum (UHV) due to the same reason. A well designed SIP indicates accurate pressures in ultrahigh vacuum region.

The SIP and the getter pump need an auxiliary pump. The turbomolecular pump and the dry vacuum pump (DVP) are used as an auxiliary pump in the UHV evacuation system.

Many kinds of dry vacuum pumps (DVPs) are widely used in the film-deposition equipment such as chemical vapor deposition (CVD) systems. Dry vacuum pumps (DVPs), without using any lubricating oil inside the pumps, are resistant to reactive gases used in CVD systems.

Mechanical Pumps

The mechanical rotary pump and the mechanical booster pump belong to this category. The rotary vane pump is exclusively used as a backing pump and/or a roughing pump in the oil diffusion pump (DP) system. A mechanical booster pumps is often used as a backing pump for the high-pumping-speed DP system in deposition systems.

Harris and Budgen (1976) [2-1] presented an article, "Design and manufacture of modern mechanical vacuum pumps," which covers the design and manufacturing aspects of rotary vane, rotary piston and mechanical booster pumps.

Quoted from Harris and Budgen, 1976) [2-1]:
Figure 2.1 shows the constructional principles of the three different pumps. **Figure 2.1(a)** shows a rotary vane pump where vanes housed in slots in a cylindrical rotor

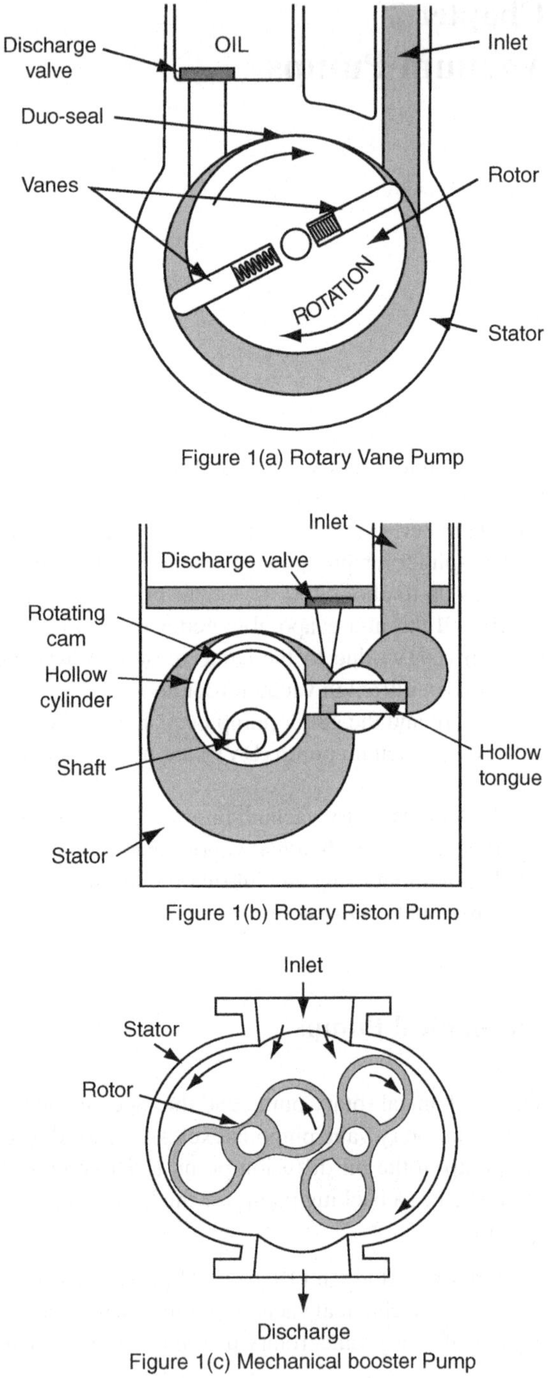

Fig. 2.1 (**a**) Rotary vane pump. (**b**) Rotary piston pump. (**c**) Mechanical booster pump (Harris and Budgen, 1976) [2-1]

are urged by springs to press against the cylindrical stator wall while being carried round the rotor. **Figure 2.1(b)** shows a rotary piston type of pump where the piston (a hollow cylinder with a hollow tongue attached) is made to process around inside a circular stator by means of the rotating cam within it. Both of these pumps are oil sealed. **Figure 2.1(c)** shows a mechanical booster pump where the "figure-of-eight" rotors are synchronized by external gears and rotate in the oval stator so as to maintain small clearances between each other and between the stator walls. The rotors run dry of lubricant.

Rotary Pumps [2-1]

Although today the rotary vane type is mainly used for pumps of low and medium pumping speed whilst the rotary piston is preferred for higher pumping speeds, clear cut advantages or disadvantages cannot be formulated for either type. One advantage of the rotary vane design is the comparatively small out-of-balance forces consequent upon rotation. The only eccentrically rotating masses are the blades, the center of mass of a blade pair describing an oval path twice per rotor revolution. With the use of various plastic blades, which are somewhat lighter than steel, this imbalance becomes insignificant.

In the two stage operation of a rotary vane pump the gas load is compressed in two stages in series from the inlet vacuum to interstage pressure by the high vacuum or first stage, and then from interstage pressure to discharge (i.e. atmospheric) pressure by the low vacuum or second stage. Oil for lubricating and sealing the first stage is outgassed by the second stage before being passed to the first stage. This together with the sharing of compression between two stages gives very much lower pressure than does a single stage pump.

Direct Drive Pumps: The use of direct drive i.e. at 1500 rpm for 50 Hz or 1800 rpm for 60 Hz is made practical by the use of reinforced plastic blades and by further specialized techniques for quieting the operation of the pump.

Figure 2.2 shows an internal view of such a pump, this example is two-stage, and is fitted with an advanced design of hydraulically operated inlet valve. This device consists of a piston valve which opens against spring pressure on pump start-up and close on pump

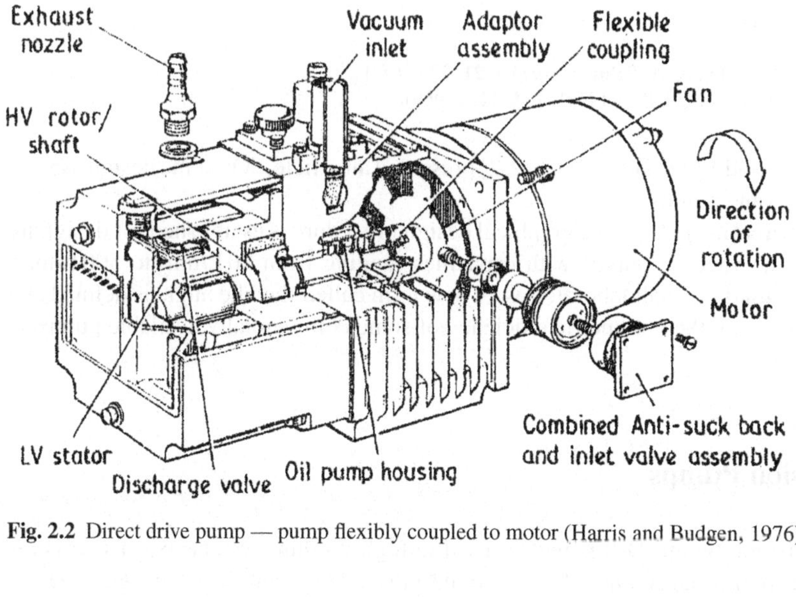

Fig. 2.2 Direct drive pump — pump flexibly coupled to motor (Harris and Budgen, 1976) [2-1]

shut-down, isolating the pump from the system at the same time admitting air into the pump itself. Oil "suck-back" is thus prevented.

Mechanical Booster Pump [2-1]
General Operation: Mechanical booster pumps are usually operated at fairly high rotational speeds i.e. 1400–4000 rpm depending on size. This is permissible as there is no mechanical contact of the pumping components. Problems associated with mechanical boosters are:

(1) Severe heating (particularly rotors) can occur with large gas loads at high compression ratios. This heating can lead to closure of working clearances and consequent mechanical failure. This problem may be overcome by cooling the rotors with oil passing through their shafts or more conveniently by providing a water-cooled heat exchanger in the gas discharge area close up to the rotors.
(2) The torque input to the mechanical booster is dependent upon the pressure difference across it. At full rotational speed this becomes unacceptably high above 10–20 Torr inlet pressure, depending on displacement ratio of mechanical booster and backing pump. There are several ways of overcoming this difficulty:
    (a) by pressure switching such that the pump is only switched on below 10–20 Torr.
    (b) by providing a pressure relief by-pass such that the maximum pressure differential that the motor can drive is not exceeded.
    (c) by providing hydrokinetic drive [1, 2] between motor and pump so arranged that motor full load torque is not exceeded. The pump slows down at high pressures maintaining full pressure differential without overheating and without overloading the motor.

Mechanical boosters fitted with water-cooling and with hydrokinetic drive may be continuously rated at all working pressures and are normally switched on with the backing pump, thus providing a valuable contribution to pumping speed at all pressures.

References

1. H. Wycliffe and A. Salmon, *Vacuum* **21**, 223 (1971).
2. B. D. Power, UK Patent No1, **154**, 341 (1969).

Ramprasad and Radha (1973) [2-2] presented an article, "On some design aspects of rotary vane pumps."

Abstract [2-2]: A simple graphical design procedure is established for designing rotary vane vacuum pumps with capacities ranging from 20 L/min to 500 L/min, normally used in small laboratories. A new term called the idle angle is defined and its influence on the volumetric efficiency of some commercially available pumps is discussed.

Diffusion Pumps

The diffusion pump (DP) is widely used in high vacuum systems because it is reliable in performance and silent in sound noise. The main disadvantage of DP is backstreaming of oil vapor.

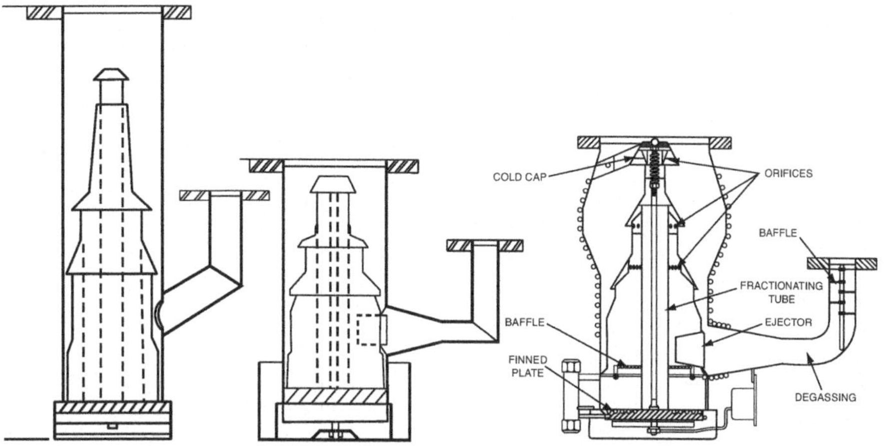

Fig. 2.3 An illustration of the progress in pump design: **left–prior** to 1958; **center**–1961; **right**–1965 (Hablanian and Maliakal, 1973) [2-3]

Hablanian and Maliakal (1973) [2-3] presented an article, "Advances in diffusion pump technology." They presented an illustration of the progress in DP design (**Fig. 2.3**). The latest designed pump shows the high throughput characteristics in pumping.

Forepressure Tolerance and Pressure Ratio [2-3]

Modern diffusion pumps have a boiler pressure of about 1 Torr. Approximately half of this initial pressure is recovered in the form of forepressure tolerance. When the forepressure tolerance is measured, a conventional non-ultrahigh vacuum system usually does not reveal the dependence between the discharge and the inlet pressures. This is because the maximum pressure ratio for air is usually not exceeded unless the experiment is conducted under ultrahigh vacuum inlet pressures. The dependence of inlet pressure on discharge pressure is shown in **Fig. 2.4**. The dashed curves show older pumps having poorer pressure ratio capability. The various horizontal levels represent different ultimate pressure at which the tests were conducted. This behavior is obviously tied to the maximum pressure ratio which can be sustained by the pump.

The pressure ratio can be sufficiently small for the light gases to reveal the dependence of inlet pressure on the discharge pressure. Measurements of pressure ratio for various gases have been reported as follows: hydrogen 3×10^2–2×10^6, helium 10^3–2×10^6, neon 1 or 2×10^8, CO and argon 10^7, oxygen and krypton 3–5×10^7, and hydrocarbons $(nC_2\,H_3)7 \times 10^8$ [1-5]. In modern pumps the helium pressure ratio is closer to 10^7 and it can be increased even as high as 10^{10} by doubling the heat input [3]. In practice even an ion gauge operated in the foreline can produce sufficient hydrogen to cause an increase of the input pressure [6]. Exactly the same occurrence has been observed in a turbomolecular pump system. The supply of hydrogen apparently comes from the mechanical backing pump oil.

References

1. A. Fujinaga, T. Hanasaka, and H. Tottori, *Trans. Amer. Vac. Soc.* (AIP for AVS, New York, 1962), p. 390.
2. J. Hengevoss and W. K. Huber, *Vacuum* **13**, 1 (1963).

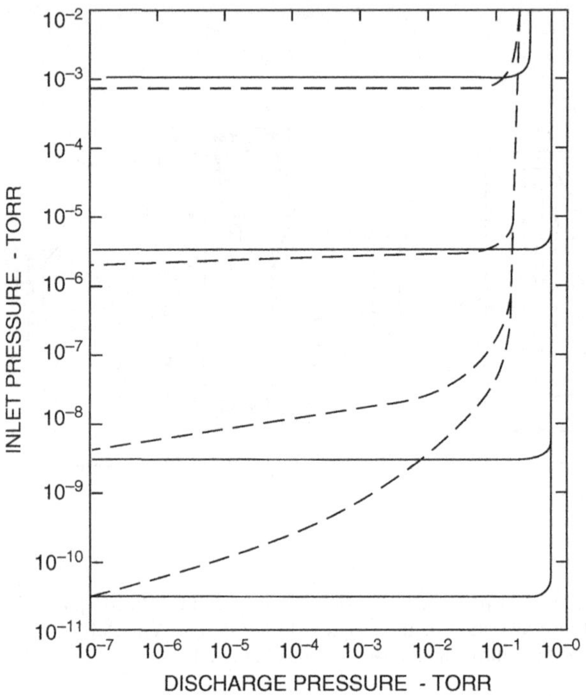

Fig. 2.4 Relationship between inlet and discharge pressures. The *horizontal lines* represent ultimate pressures in the measurement system. The *dashed curves* show older pumps (Hablanian and Maliakal, 1973) [2-3]

3. M. H. Hablanian and P. L. Vitkus, *Trans. Amer. Vac. Soc.* (AIP for AVS, New York, 1963), p. 140.
4. W. M. Langdon and E. G. Fochtman, *Abs. Trans. Amer. Vac. Soc.* (AIP for AVS, New York, 1965).
5. H. Okamoto and Y. Murakami, *Abs. Amer. Vac. Soc. Symp.* (AIP for AVS, New York, 1967), p. 67.

Comments (by Yoshimura):

The modern DP shown in **Fig. 2.3** has features of high heater power, high boiler pressure, long taper back-side pipe, and thin walls of pump body. This type of pump has a relatively large vibration, which is due to the bombardment of oil vapor jet onto the thin pump wall and the oil boiler. Yoshimura tried installing this type of DP on a high-resolution electron microscope, resulting in causing vibration problem due to the newly installed DP.

Turbomolecular Pumps

For UHV-analytical instruments such as UHV electron microscopes (UHV EM) and Auger electron spectrometers (AES), oil-free evacuation systems using turbomolecular pumps (TMP) and/or dry vacuum pumps (DVP) are increasingly used to eliminate specimen contamination. Obstacles for using such oil-free pumps (TMP and DVP) are their vibration and sound noise.

TMPs of the magnetic-bearing type are available as options for evacuating the camera chamber of a UHV-EM. On the other hand, the vibration level of DVPs is too high to use as a continuously running pump. When a DVP is used for roughing, it must be stopped during specimen-image observation.

The pumping performance of the TMP is similar to that of the diffusion pump (DP), so an evacuation system similar to the DP evacuation system can be designed.

Since a TMP with a backing pump (RP or DVP) works from atmospheric pressure, the simplest evacuation system with no roughing line and no isolation valve can be realized, as shown in **Fig. 2.5**.

Notes on the system of **Fig. 2.5** are listed below.

(1) Switching on the power makes the vent valve close and both pumps start operating.
(2) Switching off the power makes both pumps stop operating, and a few seconds later the vent valve having a low conductance opens. Slow leakage through the vent valve is important to minimize the damage to the rotating blades of the TMP.
(3) Selection of the pumping speed of RP: Select the pumping speed so that the TMP rotation speed reaches at least 80% of its rated speed before the pressure in the chamber decreases to about 13 Pa, in order to keep oil vapor backstreaming from the RP very low.

Two types of turbomolecular pumps are commercially available, a mechanical bearing type and a magnetic bearing type. The magnetic bearing type does not use any

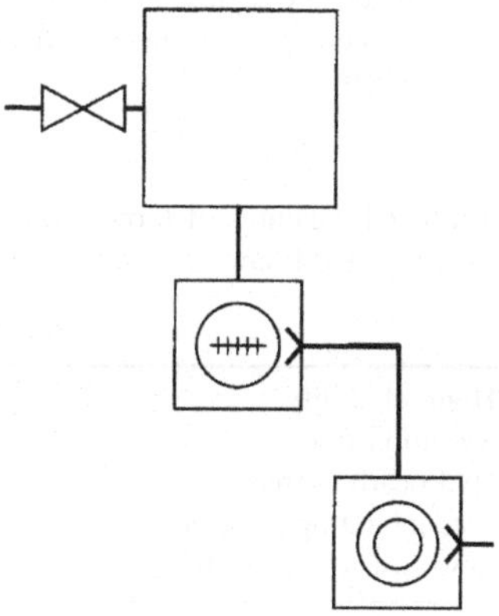

Fig. 2.5 Simple TMP system with no roughing line and no isolation valve

lubricant oil inside the pump, and its vibration is much smaller than that of the mechanical bearing type. For applying the magnetic-bearing-type to the high-resolution transmission electron microscopes a vibration absorption damper is still necessary between the pump and the evacuation pipe.

Kebelitz and Fremerey (1988) [2-4] presented an article, "Turbomolecular vacuum pumps with a new magnetic bearing concept."

As described in Abstract [2-4], a turbomolecular pump with a new concept in magnetic bearings has the technical operating advantages, and the new bearing allows lower manufacturing costs.

The Use of Turbomolecular Pumps with Magnetic Bearings [2-4]

If turbomolecular pumps with mechanical bearings are compared with those having magnetic bearings or turbopumps with mixed bearing systems, one finds that turbopumps with magnetic bearings always fulfill best the above-mentioned technical requirements (**Table 2.1**).

The main reason is that magnetic bearings are free from wear. This means reliability. No lubricant is required because of the lack of mechanical contact, this leads directly to an elimination of maintenance and of contamination by hydrocarbons.

The low vibration is due to the fact that the rigidity of magnetic bearings is much smaller than that of mechanical bearings. The absence of a lubricant also means that it is not necessary to pay attention to the influence of gravity on the lubricant deposition and transportation. Any mounting position can be easily realized.

Levi (1992) [2-5] tested combination of turbomolecular stages and molecular drag stages.

Abstract [2-5]: Molecular drag stages of a new design have been incorporated in a conventional turbomolecular pump without adding to its dimensions. From these tests it appears possible to have high compression ratios with the foreline pressures up to 10 mbar: the forevacuum pump could be a conventional, double-stage rotary pump or a less conventional diaphragm oil-free pump. The behavior for different gases has been analyzed and advantages of this arrangement are examined.

The Experimental Pump and Experimental Tests [2-5]

A schematic representation of the entire pump is shown in **Fig. 2.6**. Geometrical parameters of turbomolecular stages and molecular drag stages are presented in **Table 2.2** and **2.3**, respectively.

Table 2.1 Features of turbomolecular pumps and their fulfillment by different bearing systems (Kebelitz and Fremerey, 1988) [2-4]

	Ball bearing	Combined system	Magnetic bearings
High reliability		+	+ +
Vibration-free			+
Hydrocarbon-free		+	+ +
Free mounting position			+
Low costs of production	+ +	+	

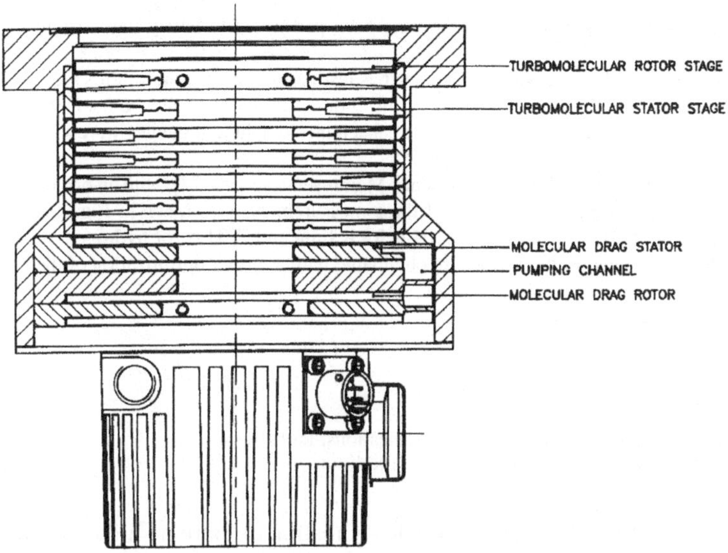

Fig. 2.6 General view of the pump (Levi, 1992) [2-5]

Table 2.2 Geometrical parameters of turbomolecular stages (Levi, 1992) [2-5]

Stage	Type	Blades angle	Disk height (mm)	Number of blades
1	Rotor	42°	3.5	36
2	Stator	36°	3.5	36
3	Rotor	28°	3.5	40
4	Stator	26°	3.5	28
5	Rotor	24°	2.5	46
6, 8	Stator	20°	2.5	36
7, 9,	Rotor	20°	2.5	46
10, 12, 14	Stator	10°	2.5	18
11, 13, 15	Rotor	10°	2.5	20

Table 2.3 Geometrical parameters of molecular drag stages (Levi, 1992) [2-5]

Stage	Section of the pumping channel (mm^2)	Length of the pumping channel (mm)	Disk height (mm)	Clearances (mm)
1	100	233	2.5	0.2
2	70	233	2.5	0.2
3	40	233	2.5	0.2

Tests on an experimental combined pump have been performed to verify the vacuum characteristics.

Nitrogen, helium, and hydrogen have been used as test gases. Compression ratios were measured following PNEUROP standards [1]; ionization gauges were used both on the high vacuum side and on the foreline side, where capacitance diaphragm gauges were also installed.

Pumping speed was measured using the Fischer-Mommsen test dome [2], with an orifice 10 mm in diameter and 1 mm thick. Results of these measurements are shown in **Fig. 2.7**. Obtained values were 230 L/s for nitrogen, 210 L/s for helium and 170 L/s for hydrogen, in agreement with calculations. Better results are obtainable with a further optimization of the pump geometry.

References

1. PNEUROP, Vacuum Pumps Acceptance Specifications, Part III.
2. E. Fisher and H. Mommsen, *Vacuum* **17**, 309 (1967).

Cho et al. (1995) [2-6] presented an article, "Creation of extreme high vacuum with a turbomolecular pumping system: A baking approach."

Abstract [2-6]: A small molecular drag pump in conjunction with the membrane pump turns out to be an effective means of achieving low backing pressures. The stainless-steel test chamber of 6800 cm^2 surface area has been prebaked to 450 °C. An ultimate pressure of 1×10^{-12} Torr is obtained on the extractor gauge and to date this is the lowest pressure attained with turbomolecular pumps only. This system

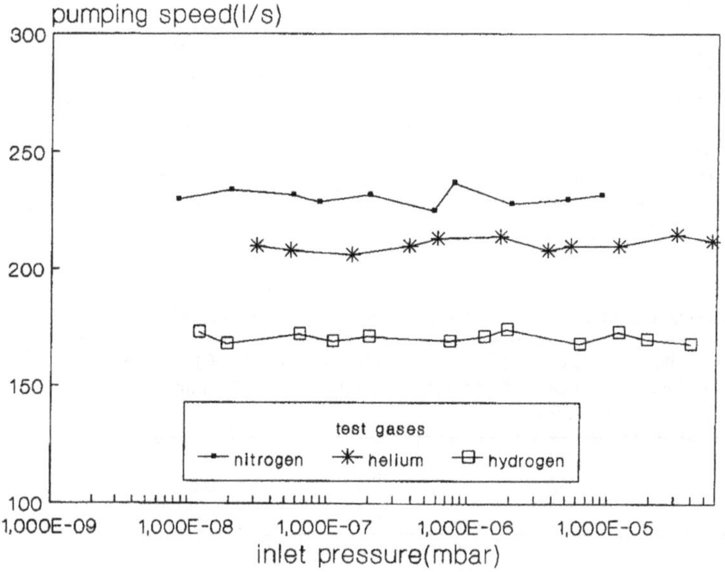

Fig. 2.7 Pumping speed vs. inlet pressure (Levi, 1992) [2-5]

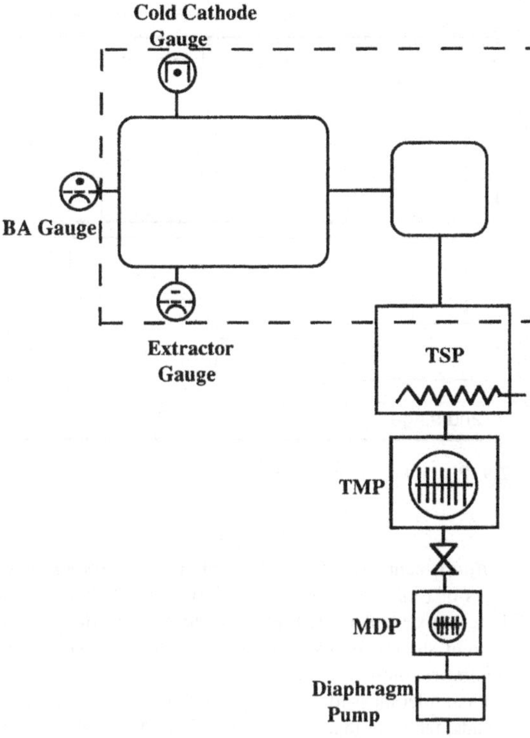

Fig. 2.8 The schematic diagram of the experimental setup (Cho et al., 1995) [2-6]

offers simplicity of arrangement, ease of operation, rather low cost, and fast pump-down to ultrahigh vacuum.

The experimental setup is presented in **Fig. 2.8**.

Dry Vacuum Pumps

Many types of dry vacuum pumps (DVPs) are used mainly for evacuating semiconductor-manufacturing equipment using corrosive gases. The types are the piston [2-7], Roots [2-7], scroll [2-7], screw [2-7, 2-8], spiral-grooved turbobooster [2-9], and diaphragm [2-11].

Lessard (2000) [2-7] presented an article, "Dry vacuum pumps for semiconductor processes: Guidelines for primary pump selection."

Pumping Technologies Overview [2-7]

The following is a brief overview of those that have been used in semiconductor applications:

Piston technology resembles a car engine with opposed pistons; typically both the cylinder walls and piston surface feature a low friction coating to reduce power loss. The most popular style has four pistons (two opposed sets) with clever internal porting to minimize flow losses (**Fig. 2.9**). The piston technology is heavy but simplified by being air cooled.

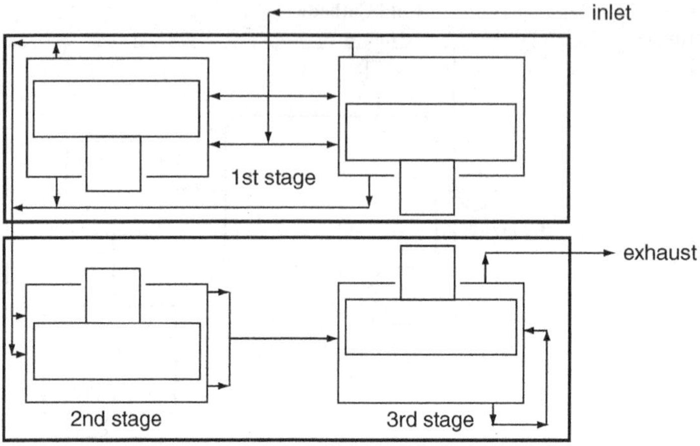

Fig. 2.9 Piston type pump (Lessard, 2000) [2-7]

Roots technology uses intermeshing lobed rotors rotating at high speed. Several sets or stages of roots lobes are mounted on parallel shafts to provide a high compression ratio. There are several related advances to roots technology where the interacting rotors have different shapes. In addition to having colorful names, such as *hook and claw*, or *tongue and groove*, these rotor variations can be combined with ordinary roots stages and tight mechanical tolerances to optimize the pumping action as desired. These pumps are dependable and adaptable for many processes, but tend to need difficult engineering adaptations for the very harshest applications (**Fig. 2.10**).

Scroll designs adapt an old compressor technology for vacuum use. A pair of helical scrolls, one fixed and one orbiting, trap a volume of gas and progressively squeeze it to the exhaust, producing a vacuum. Scrolls tend to be limited to clean applications because of tight tolerances, but they have significant advantages in low capital and maintenance costs (**Fig. 2.11**).

Screw technology is an adaptation of arguably the oldest type of pumping — Archimedes screws. In the case of vacuum pumps, the most popular design has two meshing screws rotating in opposite directions in a case, or stator. Screw pumps have advantages in the very harshest processes because of their rugged and simple design, and also because their displacement design allows water flushing in place for cleaning, but the technology is not yet in widespread use (**Fig. 2.12**).

Semiconductor Process Primary Vacuum Requirements [2-7]

The processes needed in the production of semiconductor devices generically divided into four operations: (1) modification or doping of material, (2) removal of material,

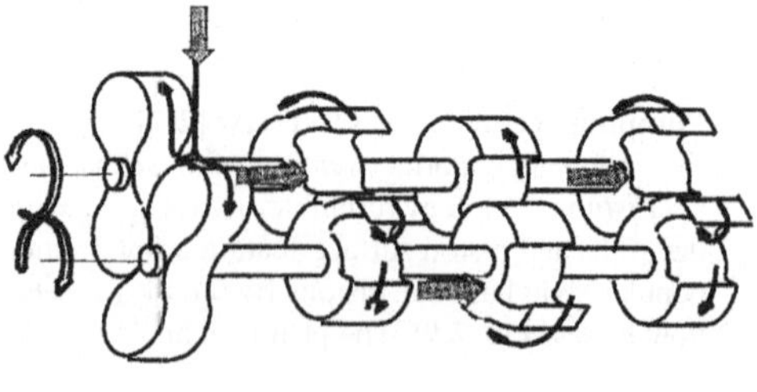

Fig. 2.10 Roots type pump (Lessard, 2000) [2-7]

Fig. 2.11 Scroll type pump (Lessard, 2000) [2-7]

(3) deposition of material, and (4) selection of where to do these operations. The last, lithography, generally does not use vacuum, except for electron beam technology used to produce the masks and reticles used in the production of patterns. The others, with certain exceptions in the deposition category, use at least a rough or primary vacuum. Each application imposes different requirements on the vacuum system.

Ion Implantation [2-7]

Nearly every surface on a wafer must be doped with low levels of suitable material, usually boron, arsenic, or phosphorus, to produce desirable electrical characteristics. This doping is almost universally done using ion implanters, which produce a charged beam of the dopant and drive the beam into the substrate material to control dopant amount, depth, and profile.

There are three separate sections of an implanter — source, beamline, and end station — each with different environments. The source generates a charged beam of the dopant with vacuum on the order of 10^{-3} Torr. Since the feed gases used to make these beams are usually boron trifluoride, arsine, and phosphine, which are hazardous, the roughing pump must be of hermetic design and able to handle the gases. This application is moderate, save for some processes using phosphorus, which can form a solid and plug the system, where a harsh service pump would be desirable.

The beam line is typically maintained at high vacuum by a turbopump, which needs a primary pump to back it. This application could be either clean or moderate, depending on the degree of vacuum isolation of the beam line from the source. The primary gas load should be hydrogen from the end station.

The end station is maintained at high vacuum almost exclusively by cryopumps. The major gas load is from the breakdown products of the photoresist on the wafers being fractured by the ion beam — mostly hydrogen with some nitrogen and hydrocarbons [1]. This is a clean application. The end station may have wafers transferred to it from a load lock, which is usually just rough pumped. As the gas load is primarily air and water vapor, this is also a clean application.

Etching [2-7]

Etching involves physical or chemical removal of material from the surface of a substrate in order to clean it or to transfer patterns from a mask via photoresist onto the silicon surface. In the latter case, the wafer is coated with a photo-sensitive resist, then exposed via

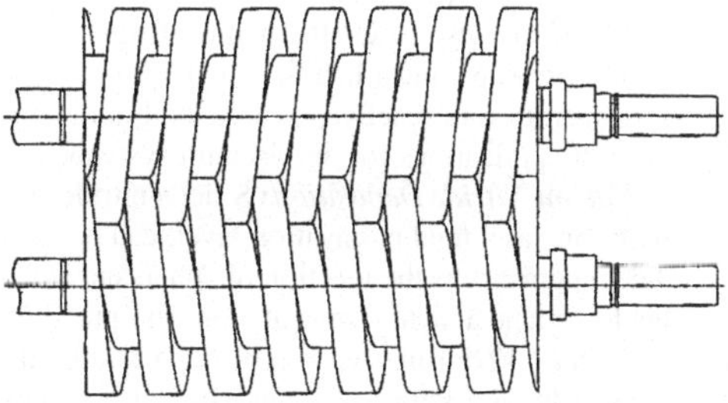

Fig. 2.12 Screw type pump (Lessard, 2000) [2-7]

photolithography through a mask, processed and washed to form a pattern on the wafer. Nearly all etch processes presently are dry and done in a vacuum. Wet chemical etch is too isotropic for use in ultralarge scale integration (ULSI); essentially equal vertical and horizontal etch rates result in undercutting of the masking layer underneath the photoresist mask.

***Physical Etching*:** Physical etching is much like physical vapor deposition (PVD, see below), except that the wafer is in the position of the source and the ionized gas, usually argon, is directed at the wafer to sputter etch the surface by physical displacement of the solid surface. This process is often used to remove the native oxide of a metal layer prior to additional deposition on the metal layer. Typical operating pressure is 10–100 mTorr. This application requires a clean service dry pump, unless hydrogen fluoride (HF) is used to accelerate the etch chemically, in which case a moderate service pump is needed to confine the HF.

***Reactive Etching*:** Reactive etching techniques (e.g., plasma etch, reactive ion etch, reactive ion beam etching) use chemical reactions, often combined with the physical etching of a directed plasma, to remove material from the selected portions of the substrate surface. Halogen gases are typically used as feed stocks for etching silicon and metal because the chloride or fluorides of the parent material are volatile (and thus can be pumped away) at high temperatures. Pressures are typically less than 100 mTorr for plasma etching and can be considerably lower for some ion beam processes. These applications may require only a moderate duty pump when etching silicon and its oxide.

Chemical Vapor Deposition (CVD) [2-7]

In CVD, material is deposited on a substrate (the wafer or display element) by thermal decomposition of feed gases or by chemical reaction of the feed gases from the gas phase. CVD is the process of choice for the deposition of silicon, silicon dioxide, and silicon nitride and is competitive for the deposition of some metals, especially tungsten.

***Silicon Deposition*:** Silicon can be deposited by the low pressure CVD (LPCVD) thermal decomposition of silane.

The silane may be 100%, or diluted with nitrogen to 20%–30%. Silane is quite dangerous and represents an explosion hazard. The safety issues have resulted in an alternative CVD technique involving the decomposition of tetraethylorthosilicate (TEOS), but in either case a pump designed for harsh duty is needed; silane is dangerous and TEOS or its by-products can plug the pump physically.

***Silicon Diode Deposition*:** SiO_2 can be deposited by a number of reactions, primarily depending on the required temperature during deposition. If aluminum is under the silicon diode, the temperature should be kept below the silicon/aluminum eutectic temperature of 577 °C; hence the reaction of choice is the decomposition of silane in an oxygen atmosphere at a temperature between 300 and 500 °C.

However, better step coverage and uniformity follow from a higher deposition temperature. Therefore, the chemistry of choice for a film on a nonmetallized surface involves dichlorosilane reacting with nitrous oxide at about 900 °C.

Another popular choice, because of improved step coverage and uniformity, and especially because of safety considerations, is the thermal decomposition of TEOS [$Si(OC_2H_5)_4$] at a temperature of 650–750 °C.

In any of these cases, the pump (if used – some of these chemistries are done at atmospheric pressure) must handle silicon dioxide, which is essentially sand. The combination of dangerous and reactive species, high temperature and sand makes this one of the very harshest applications for dry vacuum systems.

***Silicon Nitride Deposition*:** Silicon nitride, a hard film used as an oxidation mask, a gate oxide, and as a final passivation layer, can be one of the very toughest processes on pumps. Older chemistry is the reaction of silane and ammonia, but more common is the reaction of dichlorosilane and ammonia at a process pressure of about 100 mTorr.

***Metal CVD*:** Tungsten can be deposited by the decomposition of tungsten hexafluoride or by reduction with hydrogen. Similar, titanium, molybdenum, and tantalum react with hydrogen to make metal layers. Note that the products of these reactions are extremely reactive – fluorine, hydrogen fluoride, and hydrogen chloride. This is a harsh process. Pressures may be as high as 100 Torr.

Physical Vapor Deposition (PVD) [2-7]
In PVD processes, a source of the film type to be deposited is vaporized when heated or bombarded with a high energy electron beam. The energetic molecules then flow to the substrate and form a film, either alone or together with a second feed gas to form oxides and nitrides. These evaporative type processes can be augmented or replaced by the addition of a plasma, which provides energetic ions to "sputter" the source, then direct it preferentially to the substrate. Generally, metal films of aluminum and the seed and barrier layers for copper films are produced with such sputtering tools operating in the 1–100 mTorr range.

Since there is little chemistry necessary, there is not an abundance of dangerous or reactive species. The normal gases present in these processes are argon, nitrogen, and oxygen. These applications require a clean duty primary pump.

Reference

1. "Photoresist Outgassing and UV Curing in High Energy Implantation," edited by M. Jones et al., XI International Conference on Ion Implantation Technology, 1996.

Akutsu and Ohmi (1999) [2-8] presented an article, "Innovation of the fore pump and roughing pump for high-gas-flow semiconductor processing."

Abstract [2-8]: A new screw pump has been developed as a useful backing pump for high-gas-flow semiconductor processing. The characteristic features of the screws are unequal lead and unequal slant angle, which realize higher pumping speed and smaller volume of the pumping system. This pump has a very wide dynamic range of 10^{-3}–760 Torr with a high pumping speed and low electric power consumption by gradational lead screw structure and check valve at outlet. Furthermore, this pump is free from byproduct deposition, corrosion of inner surface and oil degradation. The backdiffusion of oils can be suppressed by admitting the N_2 purge gas flow larger than 10 sccm to the inlet side of the pump.

Jou et al. (2000) [2-9] presented an article, "Designs, analyses, and tests of a spiral-grooved turbobooster pump."

Abstract [2-9]: This article shows a spiral-grooved turbobooster pump, which has a momentum transfer-type vacuum pump function, and is capable of operation in a discharge pressure range up to 1000 Pa. Transitional flow pumping speeds are improved by a well-designed midsection element. Pumping performance of this new pump is predicted and examined by both of the computational fluid dynamics method and the direct simulation Monte Carlo method. These two complementary approaches are successfully being used to investigate the whole pumping speed curve from the viscous high-pressure to the molecular low-pressure range. Also, the developed pump is tested on an evaluation system constructed according to JVIS-005 standard. Good agreement between analytical results and experimental data are shown. Furthermore, to achieve a clean, safe, and reliable pumping process, this pump is designed with and operated by an innovative digital magnetic-bearing system.

Table 2.4 Characteristic list of TMP, MDP, and CMP (Jou et al., 2000) [2-9]

	TMP	MDP	CMP
Pumping speed	High	Relatively small	High
Foreline pressure	Very low (< 1 Pa)	High (1000–2000 Pa)	High (1000–2000 Pa)
Transitional flow performance	Poor	Good	Good
Structure	Complex	Compact	Complex; with TMP element Compact; w/o TMP element
Manufacture	Difficult	Easy	Difficult; with TMP element Easy; without TMP element

A brief list of some basic characteristics of the TMP (turbomolecular pump), MDP (molecular drag pump), and CMP (compound molecular pump) is shown in **Table 2.4** [2-9].

Liepert and Lessard (2001) [2-10] presented an article, "Design and operation of scroll-type dry primary vacuum pumps."

Abstract [2-10]: Since being rediscovered three decades ago, scroll technology has been successfully adapted to vacuum use. Several innovations have allowed scroll-type pumps to produce vacuums lower than 10 mTorr; the pumps are used in many applications requiring an inexpensive, dry, long-lived pump. In this article, we detail the scroll design procedure, focusing on the trade offs-between the need for sufficient vacuum and low manufacturing cost. As an illustrative example, we consider the design requirements for a high helium compression scroll design for use in leak detectors. We also present guidelines for proper pump operation, including the need for gas ballasting, high water operation, and proper pump isolation schemes.

Cheng and Chiang (2003) [2-12] estimated the pumping performance of a turbo-booster vacuum pump (TBP) equipped with spiral-grooved rotor and inner housing by the computational fluids dynamics method."

Abstract [2-12]: The calculations show that the pumping performance of the TBP can be enhanced when the spiral-grooved inner housing is integrated. The spiral angle of the rotor, the groove number, and the spiral angle of the inner housing have to be designed carefully to increase the effective pumping length of the flow channel of the TBP to gain further energy from the rotational energy, overcome the subsequent increased friction force, and increase the pumping performance. The calculations also show that the compression ratio and the pumping speed in this case, which comprises the rotor with five grooves and 15° spiral angle and the inner housing with five grooves and 27° spiral angle, are the largest among all cases studied in this article.

Giors et al. (2006) [2-13] presented an article, "Computational fluid dynamic model of a tapered Holweck vacuum pump operating in the viscous and transition regimes. I. Vacuum performance."

Abstract [2-13]: We develop a Navier-Stokes model of a Holweck pump with tapered pumping channels, applying slip-flow boundary conditions, to predict vacuum performances with and without gas flow. The commercial computational fluid dynamic code FLUENT is used to solve the model equations and to predict the pressure profile along the grooves. A specifically designed experiment is presented, whose arrangement provides the boundary conditions as input to the model and whose results are used to validate it.

Davis et al. (2000) [2-14] presented an article, "Dry vacuum pumps: A method for the evaluation of the degree of dry."

Abstract [2-14]: A methodology developed for repeatable measurements of pump cleanliness will be discussed. It utilized residual gas analysis with carefully controlled pump conditions. This facilitates direct comparisons of the degree of cleanliness between pumps of the same and those of different design. Additionally, it allows for the assessment of methods (either in pump design or use) introduced to improve cleanliness.

Cryopumps

The cryopump has the advantages that it can evacuate a large gas-load in the high vacuum region and create a clean vacuum. These features are indeed a big advantage for thin film coaters such as sputtering plants and vacuum evaporators. However, the cryopump emits noise and vibration from the moving piston to compress the He gas. Also, the maintenance cost of cryopumps is relatively high. Due to the vibration from the cryopump, the cryopump is unsuitable for scientific instruments for analyzing the micro-area of specimen, such as the electron microscope and the scanning tunneling electron microscope.

Bentley (1980) [2-15] presented an article, "The modern cryopump."

Operation of Cryopumps [2-15]

Figure 2.13 shows a typical arrangement for cryopumping a vacuum chamber and it is seen that a rotary pump is still required in order to rough out the chamber initially. It is usually the duration of the pump-down cycle that determines the optimum size of rotary pump rather than the backing requirements. Unfortunately, it is, therefore, often the case that the introduction of a cryopump does not result in any significant savings in the cost of the rotary pump for a system.

Start-up [2-15]

Before starting the cooldown of a cryopump it is advantageous to seal it off under a good vacuum from the main chamber, since there is a tendency for the sorption panels to adsorb condensable gases during cooldown long before the main cryopanels are cold enough to initiate deposition. Clearly, the less gas is available to be adsorbed during the cool-down period the less will the charcoal be poisoned. In the arrangement shown in **Fig. 2.13** the correct start-up procedure is, therefore, to close the roughing valve V_1 and the main high vacuum isolation valve, and to open the regeneration valve V_2. The cryopump can then be roughed out in isolation period to commencing cooldown. The rotary pump should be run until the cryopump is able to hold its own internal vacuum steady at 10^{-3} mbar or below, after which V_2 may be closed. If contamination of the crypsorption panels is not considered to be important a higher initial pressure may be tolerated provided that it does not exceed the value given by:

$$P_C V_C \leq 0.1\,\text{mbar} \cdot \text{L} \tag{2.1}$$

where P_C is the initial pressure in the cryopump (mbar) and V_C its volume (L). If this condition is exceeded heat transfer by gas conduction will prevent the cryopanels from becoming

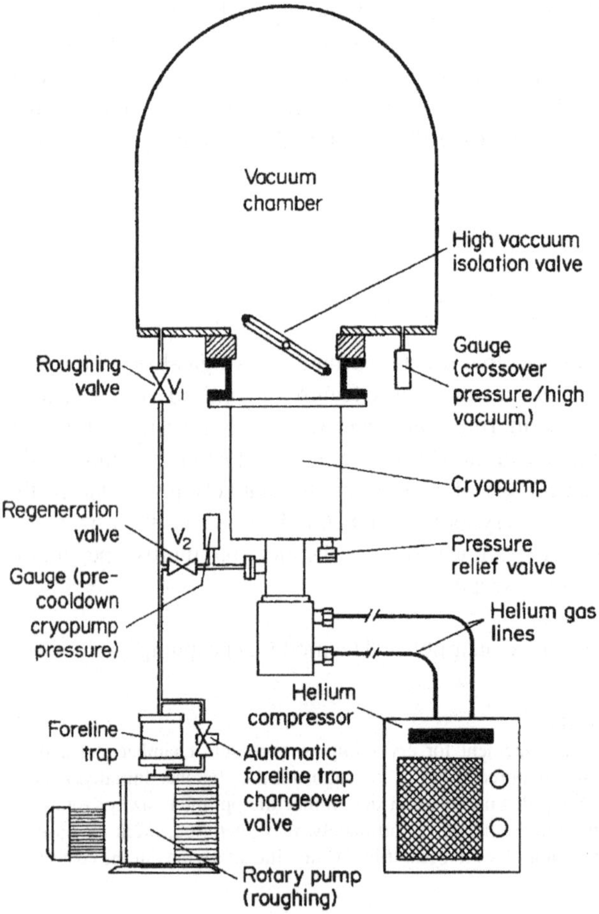

Fig. 2.13 Typical arrangement for cryopump in operation (Bentley, 1980) [2-15]

cold enough to initiate deposition. Cooldown normally takes between 45–120 min depending on make and size of cryopump.

Pumpdown [2-15]

Once the cryopump is operational and V_2 is closed, the roughing valve V_1 may be opened to evacuate the main chamber. The optimum crossover pressure is determined by several factors. A high crossover pressure will impose an unnecessarily high gas load on the cryopump during the initial stages of pumpdown and thus reduce the time between regeneration cycles. On the other hand, a low crossover pressure will result in a long roughing cycle, which is an important consideration in repetitive industrial processes. The optimum pressure must, therefore, be a compromise between these two conflicting criteria. If extreme cleanliness is important a third factor to consider is contamination from the rotary pump, which can only be avoided if roughing is carried out entirely in the viscous flow regime, and in consequence the crossover pressure must be kept above 0.2 mbar. The highest permissable crossover pressure is not determined by the continuous maximum throughput rating of the cryopump but rather by the thermal inertia of its cryopanels, which in general increases with size of cryopump. As a "rule of thumb" the maximum impulsive gas load that a cryopump can withstand, whether during pump-down or at any other time during a process cycle, is given by:

$$P_i\,(V_C + V_S)/W_2 \leq 40\,\text{mbar}\cdot\text{L/W} \tag{2.2}$$

where P_i (mbar) is the maximum impulsive pressure that can be sustained, V_C as before, is the internal volume of the cryopump and V_S is that of the system which it is pumping (L); W_2 is the second stage cooling capacity of the cryogenerator (W). Reference should always be made, however, to the manufacturer's recommendations in each particular case.

Since cryopumps are usually used in processes requiring extreme cleanliness it is prudent to fit a foreline trap to the rotary pump whenever or not the crossover pressure is held in the viscous flow regime. It is possible to omit the high vacuum isolation valve and adopt the cheaper alternative of roughing out the cryopump and vacuum chamber simultaneously. However, this is not generally advisable, since not only does the use of an isolation valve reduce contamination of the sorption panels as already explained, but it also enables the cryopump to be regenerated without releasing the captured gases back into the process chamber. If there is an inordinately high outgassing load from the chamber it may even be impossible to rough out to a low enough pressure to enable the cryopump to cool down at all [see Eq. (2.1)], unless an isolation valve is provided to seal off the cryopump. If it is essential to completely eliminate organic materials from the system all the valves, including the main isolation valve, should be metal sealed.

Regeneration [2-15]

Some examples of typical operational duties will demonstrate the ability of modern cryopumps to operate for quite long periods of time before requiring regeneration. Consider a cryopump of the popular 8 in. size for which representative pumping capacities could be as in **Table 2.5**. Let us take four hypothetical applications and in each case calculate the operational lifetimes between regeneration cycles.

(a) Steady state cryopumping of argon at a speed of 1000 L/s and a pressure of 10^{-5} mbar:
gas load: 0.01 mbar·L/s
operational lifetime: 580 days continuous running.

(b) Sputtering (without throttling the pump) at a speed of 1000 L/s and a pressure of 3×10^{-3} mbar:
gas load: 3 mbar. L/s
operational lifetime: 93 h continuous running.

(c) Evaporation process cycle in which a chamber is pumped down to 1×10^{-6} mbar from a crossover pressure of 1×10^{-2} mbar:

Table 2.5 Representative argon and hydrogen capacities at various pressures for a typical 8 in. commercial cryopump (Bentley, 1980) [2-15]

Gas	Capacity (mbar·L)	Pressure (mbar)
Argon	5×10^5	1×10^{-6}
Argon	1×10^6	1×10^{-3}
Hydrogen	3000	1×10^{-7}
Hydrogen	9000	5×10^{-6}

total gas load per cycle: 500 mbar·L (say)
number of operational cycles: 1000.

(d) Evaporative coating process cycle in which there is a high hydrogen gas load (e.g. electron beam evaporation of aluminum):
total H_2 gas load per cycle: 20 mbar·L (typically)
number of operational cycles: 150 at 1×10^{-6} mbar
450 at 5×10^{-5} mbar

It is evident from these examples that regeneration is unlikely to be a problem in most applications, and when necessary can conveniently be carried out at weekends. The need for regeneration will be apparent by the inability to attain the required pressures (and in the case of the non-condensable gases, by a significant reduction in speed). Another indication can often be provided by the temperature of the second stage of the cryogenerator as suggested in the previous section.

Regeneration is accomplished by sealing off the cryopump with the high vacuum isolation valve and opening the regeneration valve V_2 (**Fig. 2.13**). The rotary pump can then be used to exhaust the gases that are released as the cryopump warm up. An automatic foreline trap changeover valve is advisable in order to bypass the trap during the majority of the regeneration cycle, thus avoiding early saturation of the foreline trap sorbent with water vapor. A typical cryopump takes between 2–4 h to warm up if no external form of heating is supplied. An alternative technique is to circulate dry nitrogen through the cryopump during regeneration, which accelerates warm up and helps to remove residual water vapor. Once regeneration has commenced it should be allowed to proceed to completion, since partial warming of the cryopump will usually transfer a large amount of water vapor from the radiation shield to the cryopanels, where it will seriously impair subsequent pumping performance. One attractive feature of cryopumps is that as capture pumps they can be used to handle gases that would damage other types of pumps. Regeneration in such cases can be accomplished quite safely by closed-off venting and thorough flushing prior to re-evacuation and cooling down again. Particular care to avoid explosion should be taken when venting inflammable gases or gas mixtures.

Safety [2-15]

Since cryopumps are capture pumps a pressure relief valve should always be fitted if not already supplied as an integral part of the pump itself. Very dangerous pressures can build up if a cryopump is accidentally allowed to warm up in a sealed condition as, for example, during power failure. An 8 in. cryopump with an internal volume of about 10 L would pressurize to 60 atm if allowed to warm up without venting when loaded with 5×10^{-5} mbar · L of gas. Fortunately it is rare for a large amount of gas to be released instantaneously during regeneration; rather the solid gas sublimes at a controlled rate as the cryopanels warm up. The pressure relief valve need not, therefore, be of large diameter, though the advice of manufacturers should always be sought, and when in doubt it may be desirable to fit bursting discs in particular cases. Above all, the venting provisions for a cryopump must be fail-safe.

Klein et al. (1984) [2-16] presented an article, "Use of refrigerator-cooled cryopumps in sputtering plants."

Abstract [2-16]: The report especially deals with the topics, "usable for process pressure up to 2×10^{-2} mbar," and "rapid evacuation of the process chamber from the process pressure down to a pressure 5×10^{-7} mbar after process is finished." Here special modifications are necessary compared to common cryopumps. There are two possibilities to cover the demand of using the pump at high pressures: (1) Using a refrigerator with high cooling capacity, (2) Installing a throttling valve into the cryopump. Both are discussed.

Dennison and Gray (1979) [2-17] made performance comparisons using a quadrupole mass spectrometer between nominal 1000 L/s turbomolecular and cryogenic high vacuum pumps on a commercially available sputter deposition system.

Abstract [2-17]: The quadrupole analyzer was mounted in a differentially ion getter-pumped sampling chamber to allow gas analysis at sputtering pressures. The cryogenic pump without liquid nitrogen trapping showed pump-down performance comparable to the turbomolecular pump with liquid nitrogen trapping. We were unable to get repeatable meaningful results of residual gas behavior during sputtering operations when argon was let into the ion-getter-pumped sampling chamber.

Cheng and Shen [2-18] (2006) presented an article, "Effect of heat on the pumping performance of cryopump."

Abstract [2-18]: In the experiment, the heat was applied by two means, for the first-stage array by thermal radiation and for the second-stage array by applying directly. The experiment applied varied levels of electrical power to the heaters, introduced Ar to the test dome, and measured the corresponding dome pressure, gas throughput, and temperatures of the different arrays inside the cryopump. The results showed that the temperature of the introduced gas heated by the heater in the test dome would influence the measured pumping speed of cryopump. When the thermal load is applied primarily on the first-stage array, the temperature of the second-stage array varies significantly as the temperature of the second-stage array changes; when applied directly to the second-stage array, the temperature of the first-stage array varies little as the temperature of the second-stage array changes. The pumping speed of cryopump and temperature of the two-stage arrays increase gradually as gas throughput rises, and sharply when the gas in the test dome is nonmolecular flow.

Vapor Pressures for Gases

Vapor pressures for gases are basis for cryopumping.

Honig and Hook (1960) [2-19] presented an article, "Vapor pressure data for some common gases."

Figures 2.14 and **2.15** present the vapor pressure data in graphical form for the range 10^{-11}–10^{3} mmHg. Log p versus $\log_T$ plots, employed in the previous

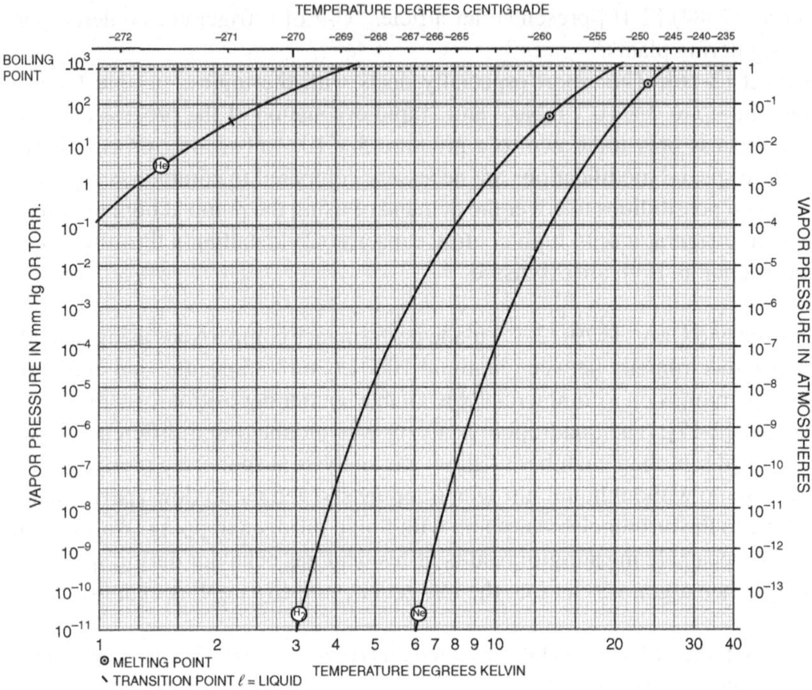

Fig. 2.14 Vapor pressure curves for common gases (1) (Honig and Hook, 1960) [2-19]

publication [1], were chosen again as the most suitable means of presenting these data over wide pressure and temperature ranges. **Figure 2.14** covers the range 1–40 K and includes He, H_2, and Ne. The remainder of the gases is plotted in **Fig. 2.15** for temperatures between 10 and 400 K. As in **Fig. 2.14** and **Fig. 2.15**, ⊙ and \ represent melting points and transition points, respectively. Where the melting point falls outside the plotted range, the letter "s" (solid) or "l" (liquid) have been appended to the chemical symbol [2-19].

Reference

1. K. K. Kelley, "Contributions to the Data on Theoretical Metallurgy," *U. S. Bureau of Mines Bulletin* 383, 1935.

Sputter Ion Pumps

Modern transmission electron microscopes have a sputter ion pump (SIP) for evacuating their microscope columns. Auger electron spectrometers equipped with an argon-ion gun for thinning the specimen, must be evacuated with a noble-type SIP with a large pumping capacity for noble gas.

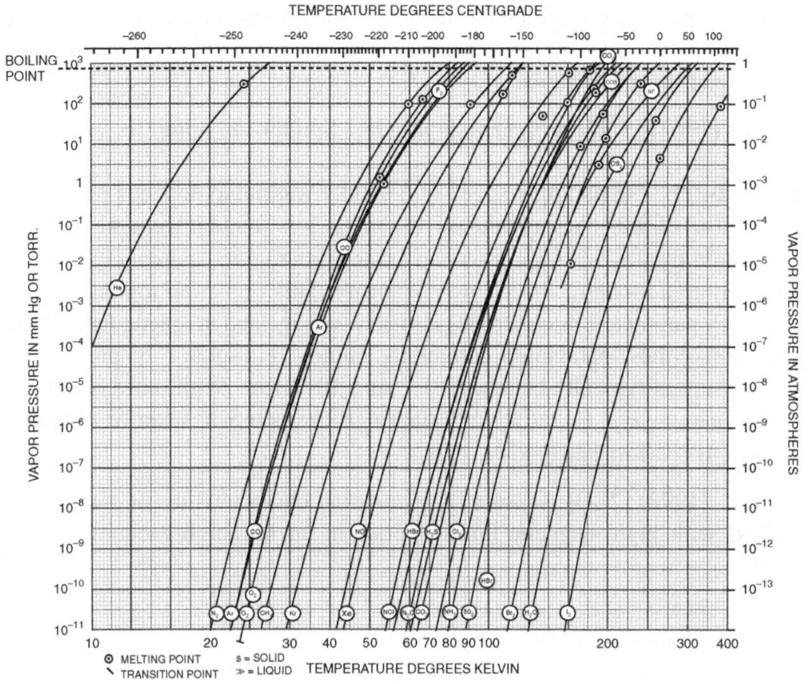

Fig. 2.15 Vapor pressure curves for common gases (2) (Honig and Hook, 1960) [2-19]

SIPs used in an ultrahigh vacuum must keep their pumping function in the 10^{-8} Pa range.

Andrew (1968) [2-20] reviewed the theory and practice of the Penning discharge.

Abstract [2-20]: Improvements in pump cell design, aimed at increasing the inert-gas pumping speed relative to the air speed, are described. The problem of unstable pumping of argon is discussed, and an explanation based on mode changes of the discharge is supported by experimental observations. Future refinements of the sputter-ion pump involving the use of new cathode materials and cell geometries may give additional benefits in performance.

Jepsen (1968) [2-21] presented an article, "The physics of sputter ion pumps."

Abstract [2-21]: The steady-state pumping behavior of a sputter ion pump depends on discharge intensity (I/P) and pumping efficiency (SP/I) (I, ion current; P, pressure; S, pumping speed). Within the approximations employed, it is shown that the discharge intensity is proportional to the anode voltage and anode length, but is independent of anode diameter for the product (magnetic field × anode diameter) = const.

It is proposed here that a significant fraction of the noble-gas ions incident upon the cathodes is scattered as energetic neutrals, and that these play a dominant role in the pumping of the noble gases.

Rutherford (1963) [2-22] presented an article, "Sputter-ion pump for low pressure operation."

Abstract [2-22]: The pumping speed of a sputter-ion pump depends on the intensity (I/P ratio) of the cold-cathode gas discharge. It is found in general that I/P is nearly constant from $\sim 10^{-4}$ Torr down to a transition region where the discharge changes rapidly to a mode of much lower intensity. This transition region is quite sensitive to such parameters as electrode geometry, voltage and magnetic field. With the combination of these parameters commonly employed in sputter-ion pumps, the transition pressure is typically between 10^{-8} and 10^{-9} Torr. With other combinations of these parameters, essentially constant I/P characteristics can be obtained to pressures of 10^{-11} Torr and below.

Ohara et al. (1992) [2-23] measured the pumping characteristics of sputter ion pumps (SIPs) with high-magnetic-flux densities in an ultrahigh-vacuum (UHV) range.

Pump Design Parameters [2-23]

A common pump body of type 304 stainless steel (SS304) with two discharge chambers and two high-voltage feedthroughs, was made by vacuum brazing and Ar-arc welding. Dimensions of both discharge chambers were $58 \times 250 \times 100$ mm. Three pump-element assemblies composed of two Ti plates (cathodes) and cylindrical anode cells (SS304) of different diameters, respectively, were designed and made, each of which was alternately assembled into the common pump body. Further, three magnet assemblies of different high-flux densities were made, whose specified flux densities at the center of their magnet pole gaps were 0.15, 0.2, and 0.3 T, respectively. The magnet of 0.3 T was made of a rare-earth magnetic material.

The magnetic-field distributions in the pole gaps of the respective magnet assemblies were measured, and are presented in **Fig. 2.16**. The maximum flux densities at the centers of their fields are just 0.15 T, slightly weaker than 0.2 T, and just 0.3 T, respectively. The broken curves represent the flux densities at places 30 mm from the x axis. For all magnet assemblies, the reduction of the flux densities at places 30 mm from the x axis is about 13%. The reduction of flux densities at places 60 mm from the y axis is about 20%.

Three pump elements of the following anode cells were designed and made for the common discharge chambers:

(1) 28 pieces with 17 mm-diameter cells (4 columns$\times$7 rows),
(2) 15 pieces with 24 mm-diameter cells (3 columns$\times$5 rows),
(3) 8 pieces with 29 mm-diameter cells (8 columns$\times$4 rows).

The lengths of the anode cells of all the pump elements were 28 mm. Cathode pairs of all the pump elements were composed of flat Ti plates of 2 mm thickness. The gap space between the anode cells and the cathode plates was 10 mm. The applied voltage between the anode and the cathode was 6.5 kV.

Pumping Speed and Discharge Intensity Characteristics [2-23]

Pump-down characteristics and discharge characteristics were measured after the baking treatment (250 °C for 2 days) by a throughput method using an orifice (3.3 mm diameter).

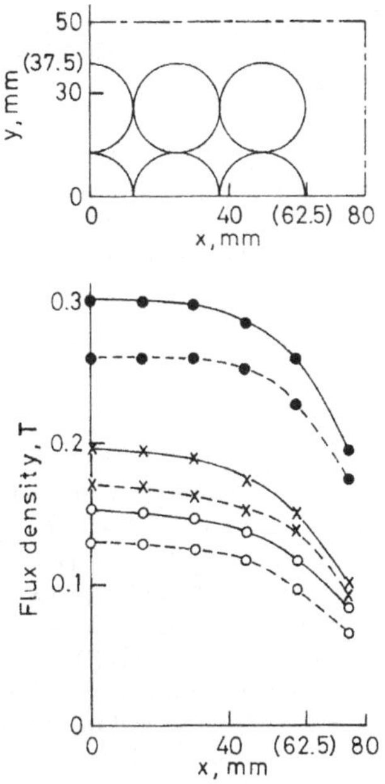

Fig. 2.16 Distributions of magnetic-flux densities in the pole gaps of the magnet assemblies: —,y = 0 mm; ---, y = 30 mm; -○-, magnet assembly of 0.15 T at the center of the pole gap; -×-, magnet assembly of 0.2 T; and -•-: 0.3 T (Ohara et al., 1992) [2-23]

In measuring the pumping speed the pressure was first increased step by step from a very low pressure of nearly 1×10^{-8} Pa to about 1×10^{-6} Pa by increasing the introduced N_2 gas load through the needle valve, and the pressure was then reduced step by step in the return course. After controlling the introduced gas load to set a one-point pressure, the needle valve was not handled for about 10 min. The pumping speed measured in the pressure-fall course was about 10% lower than the corresponding pumping speed measured in the pressure-rise course. The pressures in the figures were thus measured at the end of these 10-min periods in the pressure-rise course. Ion current characteristics were also measured at the same time in the pressure-rise course.

The pumping speed characteristics and the discharge intensity I/P characteristics of the pump with 17-mm diameter cells for N_2 are presented in **Fig. 2.17(a)**, and those of the pumps with 24- and 29-mm diameter cells are presented in **Fig. 2.17(b)** and **2.17(c)**, respectively.

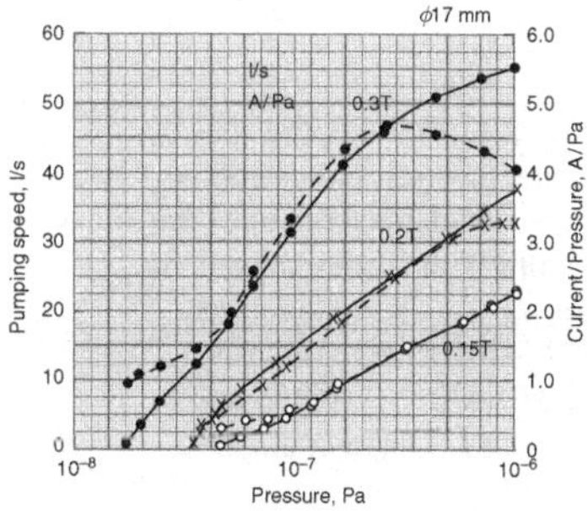

Fig. 2.17(a) Pumping speed characteristics and discharge intensity characteristics of the pumps with 17-mm diam cells for N_2. —, *pumping speed*; – – –, *discharge intensity*; -○-, 0.15 T; -×-, 0.2 T; and -•-, 0.3 T (Ohara et al., 1992) [2-23]

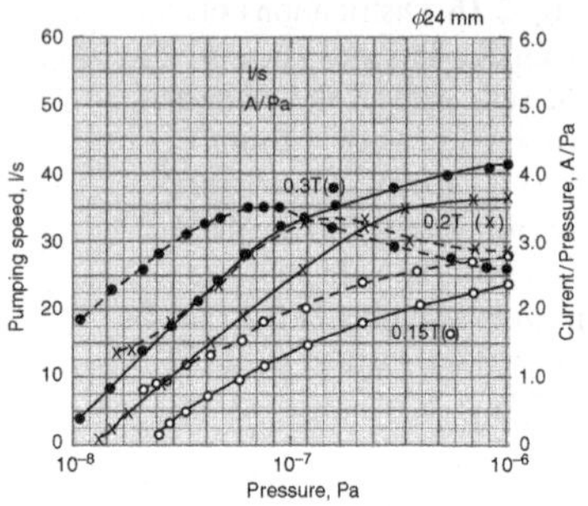

Fig. 2.17(b) Pumping speed characteristics and discharge intensity characteristics of the pumps with 24-mm diam cells for N_2. —, pumping speed; – – –, discharge intensity; -o-, 0.15 T; -×-, 0.2 T; and -•-, 0.3 T (Ohara et al., 1992) [2-23]

The characteristics of the pump with 24-mm diameter cells in **Fig. 2.17(b)** show the following features. Both the pumping speed and the discharge intensity of the pump of 0.15 T increase smoothly with pressure rise up to 1×10^{-6} Pa. 1.0 A/Pa of discharge intensity corresponds to pumping speeds smaller than 10 L/s for the pump of 0.15 T. The pumping speed curves of the pump of 0.2 T cross over the discharge intensity curve at 2.4×10^{-7} Pa. In the pressure range below 4×10^{-7} Pa the pumping speed of the pump of 0.2 T goes down with pressure decrease, and finally it becomes almost zero at 1.3×10^{-8} Pa. The pumping speed curve and the discharge intensity curve of the pump of 0.3 T cross each other at 1.2×10^{-7} Pa. The pump still shows a considerable pumping speed in the 10^{-8} range, though the pumping speed reduces with pressure decrease.

Comment: JEOL electron microscopes exclusively use sputter ion pumps having 24 mm-diameter cells and 0.15 T.

Noble Pumps for Inert Gases

Jepsen et al. (1960) [2-24] presented an article, "Stabilized air pumping with diode type getter-ion pumps."

Abstract [2-24]: The argon pumping speed has been increased by a factor of five through modifications of the geometry of the pump (slotted cathode pumps), although the diode structure has been retained. Pumps employing these modified geometries have been tested on air leaks resulting in various pressures up to

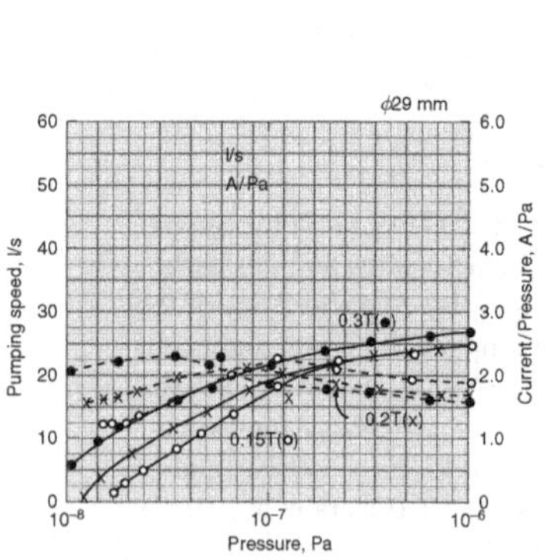

Fig. 2.17(c) Pumping speed characteristics and discharge intensity characteristics of the pumps with 29-mm diam cells for N_2. —, pumping speed; – – –, discharge intensity; -o-, 0.15 T; -×-, 0.2 T; and -•-, 0.3 T (Ohara et al., 1992) [2-23]

5×10^{-5} Torr, and have given no indication of argon instability for the equivalent of over 2000 hr at 1×10^{-5} Torr.

Jepsen et al. (1960) [2-24] described the "argon instability" in details.

Argon Instability (Jepsen et al., 1960) [2-24]

Under certain conditions of pumping against a continuous air leak, the conventional diode type of getter-ion pump may exhibit periodic pressure fluctuations. Since these fluctuations are associated with the argon which is normally present in air, they have been given the name "argon instability." Some of the characteristics of argon instability are the following:

(1) To get a new or freshly cleaned pump to go unstable, it is typically necessary to operate it at a pressure of 1×10^{-5} Torr for several hundred hours; at $P = 1 \times 10^{-6}$ Torr, several thousand hours would be required. There is, however, wide variation from pump to pump in the ease or difficulty with which argon instability occurs. In some cases it is necessary to feed the pumps argon in higher concentrations before they will go unstable.

(2) A representative pattern of argon instability is shown in **Fig. 2.18**. It should be remarked, however, that a considerable variety exists in the pressure vs. time curves that are observed.

(3) During an instability the pressure rises to a maximum value of about 2×10^{-4} Torr. For leak rates such that the pressure between instabilities is less than about 2 to 3×10^{-5} Torr, this maximum pressure is independent of leak rate.

(4) Time intervals between the pressure fluctuations are typically several minutes at a pressure of 1×10^{-5} Torr, and vary approximately inversely with leak rate for pressures in the range 5×10^{-7}–1×10^{-5} Torr.

(5) If the pressure exceeds about 3×10^{-5} Torr, instabilities no longer occur, even with pure argon (Note that this pressure is below the maximum attained during instability. This phenomenon is discussed in more detail in another paper [1]).

(6) The detailed nature of argon instability depends on such things as voltage-current characteristics of the pump power supply and size of system to which the pump is attached. On systems which are large relative to the size of the pump, the pressure fluctuations occur less readily because of "volume stabilization" of the pressure.

(7) Operation at sufficiently high pressures (as during the normal starting following rough pumping, for example) results in a substantial amount of pump cleaning and greatly reduces the tendency of a pump to exhibit argon instability. Thus pumps which are frequently cycled seldom go unstable.

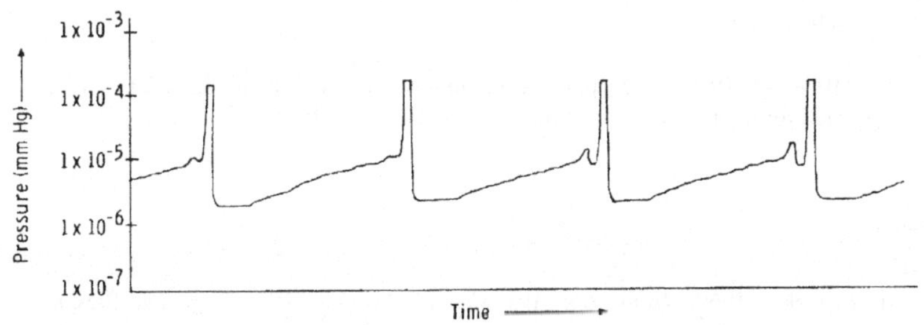

Fig. 2.18 Typical pattern of pressure vs. time for a getter-ion pump exhibiting argon instability (Jepsen et al., 1960) [2-24]

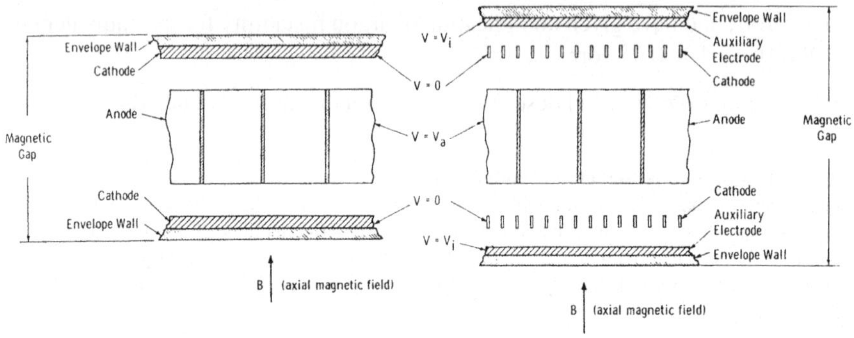

Fig. 2.19 Comparison of the magnetic gaps required for "equivalent" diode and triode pumps (Jepsen et al., 1960) [2-24]

Argon speed and stability behavior of a pump employing slotted cathodes with the outer portions masked were essentially the same as with completely smooth cathodes. With only the central regions masked, behavior appeared to be identical with that for the completely slotted cathodes [2-24].

An example of a diode pump of optimum design is shown schematically in **Fig. 2.19**. Shown also in **Fig. 2.19** is an "equivalent" triode pump in which anode length, cathode-anode spacing, magnetic field strength, and voltage between anode and cathode are the same. In such a triode pump the magnetic gap must be lengthened because of the thickness of the collector electrodes, the spaces required between collector electrodes and cathodes, and the increased thickness of the cathodes themselves.

It was found experimentally that at intermediate pressures the pumping speeds for air of the two pumps are essentially identical. Since the volume of magnetic field occupied by the diode pump is less than that occupied by the "equivalent" triode pump, the pumping speed per unit volume of magnetic field is correspondingly greater for the diode pump [2-24].

Reference

1. S. L. Rutherford, S. L. Mercer, and R. L. Jepsen, *Transactions of the 7th National Vacuum Symposium*, 1960 (Pergamon Press, New York, 1961), p. 380.

Jepsen et al. (1960) [2-24] compared the magnetic field efficiency in diode-type pump and triode pump, as follows.

Andrew (1968) [2-20] reviewed pump configurations (presented in **Fig. 2.20**) which were investigated for improved inert-gas pumping, and discussed their performance.

References

1. W. M. Brubaker 1960, *Trans. 6th AVS Vacuum Symp.*, 1959 (Pergamon Press), pp. 302–6.
2. R. L. Jepsen, A. B. Francis, S. L. Rutherford, and B. E. Kietzmann 1961, *Trans. 7th AVS Vacuum Symp.*, 1960 (Pergamon Press), pp. 45–50.

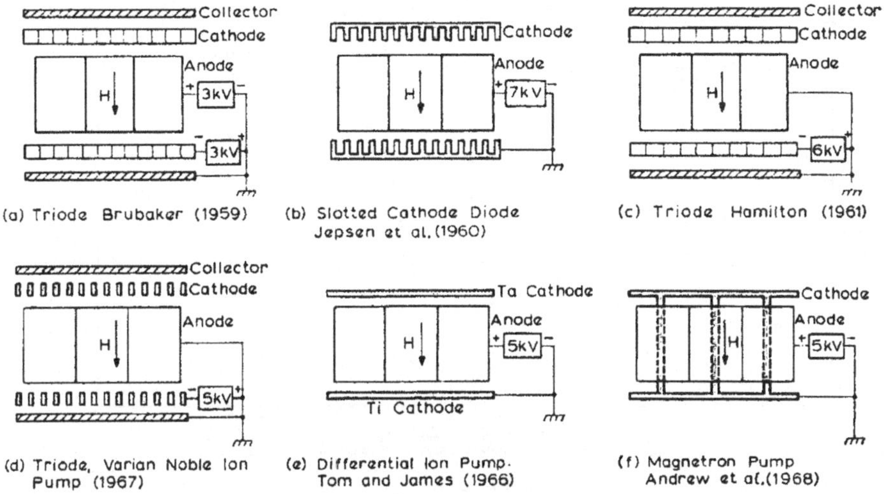

Fig. 2.20 Pump designs for improved inert-gas pumping
(Andrew, 1968) [2-20] **(a)** Brubaker (1959) [1]. **(b)** Jepsen et al. (1960) [2]. **(c)** Hamilton (1961) [3]. **(d)** Triode, Varian Noble Ion Pump (1967). **(e)** Tom and James (1966) [4]. **(f)** Andrew et al. (1968) [5]

3. A. R. Hamilton, 1962, *Trans. 8th AVS Vacuum Symp.*, 1961 ?Pergamon Press), pp. 388–94.
4. T. Tom and B. D. James 1966, *Abs. 13th AVS Vacuum Symp.* (Pittsburgh, Pa.: Herbick and Held), pp. 21–2.
5. Andrew, D., Sethna, D., and Weston, G., 1968, *Proc. 4th Int. Vacuum Congr.*, pp. 337–40.

Jepsen (1968) [2-21] proposed the "energetic-neutrals" hypothesis as the mechanism by which argon is pumped. "The explanation proposed here is that energetic neutral argon atoms are indeed responsible for argon pumping, but the dominant mechanism for the production of these energetic neutrals is through impact of argon ions at the cathodes rather than through charge exchange in gas-phase collisions [2-21]."

Baker and Laurenson (1972) [2-25] presented an article, "Pumping mechanisms for the inert gases in diode Penning pumps."

Abstract [2-25]: The action of a diode Penning pump has been investigated using combinations of the cathode materials aluminum, titanium, and tantalum, and of the gases helium, neon, argon and krypton. The pumping speed S measured in a dynamic system was found to be dependent on the ratio R of the mean of the atomic masses of the cathode pair to the atomic mass of the impinging ion, in fact: $S \propto \log R$. The important factors in an equation for the pumping speed during dynamic and static conditions are discussed.

Figure 2.21 shows the dynamic speeds of the three cathode pairs for each inert gas as a function of the ratio R.

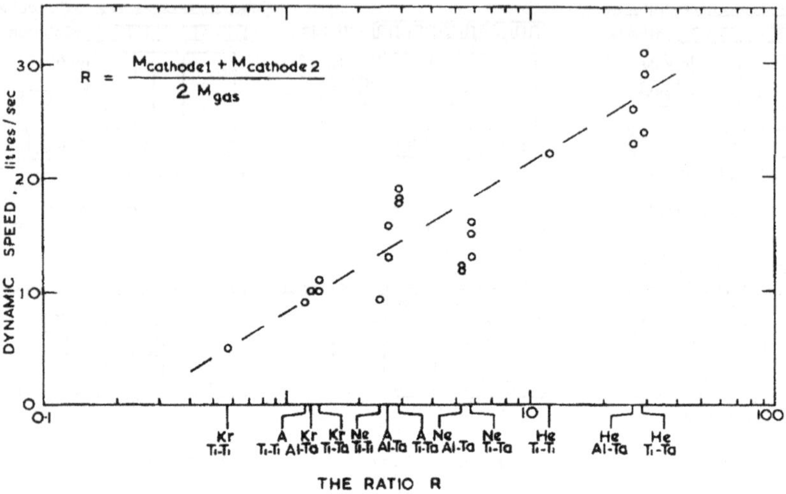

Fig. 2.21 The dynamic speeds of the three cathode pairs for each inert gas as a function of the ratio R (Baker and Laurenson, 1972) [2-25]

Comment: The results suggest that for pumping a high-mass inert gas such as Xe, high-mass cathodes like tantalum are essential to stable pumping.

Denison (1977) [2-26] compared triode and diode SIPs with each other for actual gas pumping. "The triode pump has an advantage at higher pressures while the diode pump exhibits greater pumping speed at low pressures. Accelerated life testing showed the diode pump has more stable pumping characteristics after pumping a large amount of gas [2-26]."

Komiya and Yagi (1969) [2-27] discussed the enhancement of noble-gas pumping for cathodes of different configuration and material. "A multiholed tantalum plate combined with a conventional flat titanium cathode gives a pumping speed for argon about 40% as large as the pumping speed for nitrogen. The multiholed tantalum pump does not show an appreciable pressure pulse (the so-called argon instability) at least for 700 h at 2×10^{-5} Torr with a constant throughput of argon [2-27]."

Yoshimura et al. (1992) [2-28, 2-29] presented an article, "Ar-pumping characteristics of diode-type sputter ion pumps with various shapes of "Ta/Ti" cathode pairs."

Diode-Type Sputter Ion Pumps with Various Shapes of Ti/Ta Cathode Pair [2-28, 2-29]

Pumping characteristics of sputter ion pumps (SIPs) depend on magnetic flux density, anode voltage, and dimensions of anode cells, especially the cell diameter. Further, Ar-pumping characteristics of diode-type pumps would be strongly related to the shape and material of their cathode pairs.

Four pump-element assemblies of different cathode pairs were replaceable with one another for the common pump body. The shape and dimensions of the anode cells composed of 30 cells (24 mm diameter, 28 mm in length, 3 columns × 10 rows) were the same for all the pump-element assemblies. The magnet assembly of 0.15 T of flux density at the center of the discharge area was made, whose dimensions were: gap space, 62 mm; magnet thickness, 30 mm; yoke thickness, 8 mm; magnet-pole area, 100 mm × 285 mm. The distribution of magnet flux density in the pole gap was measured, which is presented in **Fig. 2.22**.

The four pump-element assemblies are presented in **Fig. 2.23 (a)– (d)**. **(a)** represents a "flat Ta/flat Ti" cathode pair, **(b)** a "holed Ta on flat Ti/flat Ti" pair, and **(c)** and **(d)** a "slotted Ta on flat Ti/flat Ti" pair and a "slotted Ta on flat Ti/slotted Ti on flat Ti" pair, respectively. The arrangement of slots of slotted cathodes is seen in the upper part of **Fig. 2.22**. For the slotted cathodes and holed ones, the center axes of all the anode cells face the flat surfaces, not slotted or holed area.

The average flux density over the discharge area was calculated as 0.13 T. The anode voltage was commonly set at 6.5 kV.

Pumping Speed Characteristics [2-28, 2-29]

We measured the pumping speeds by an orifice method.

In measuring Ar-gas pressures the relative sensitivity correction between two Bayard-Alpert gauges (BAGs) (one was a nude type and the other glass-tube type) was conducted.

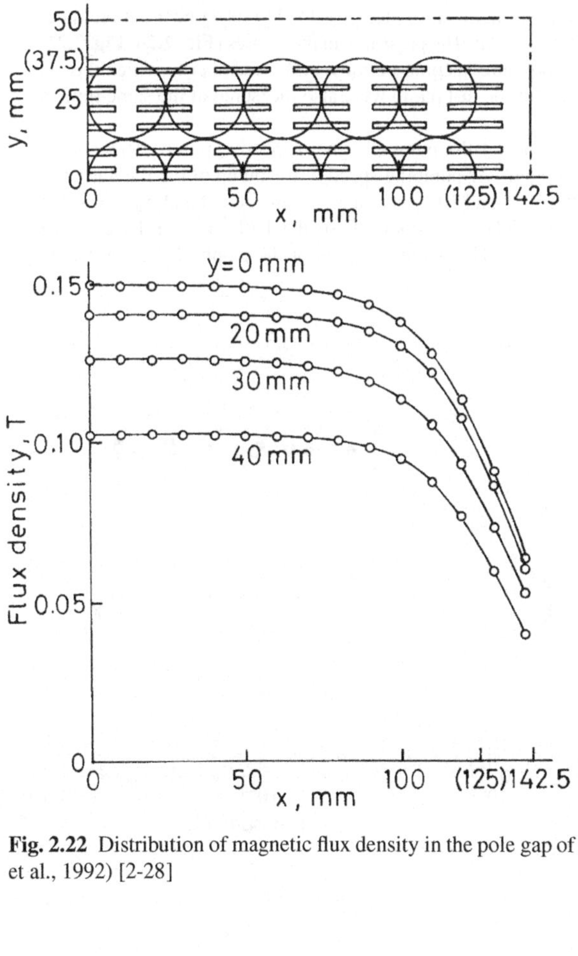

Fig. 2.22 Distribution of magnetic flux density in the pole gap of the magnet assembly (Yoshimura et al., 1992) [2-28]

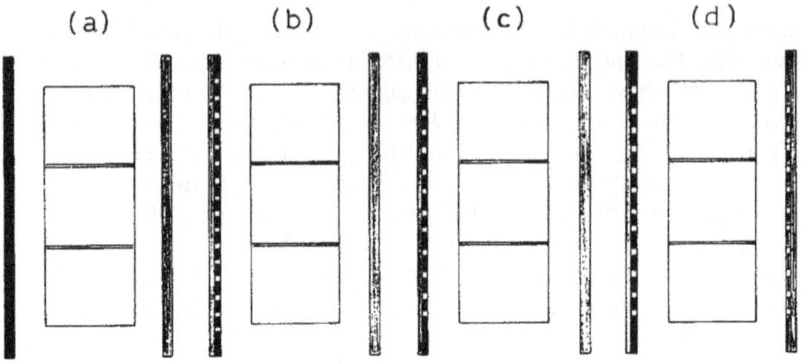

Fig. 2.23 Various shapes of Ta/Ti cathode pairs. (**a**) Flat Ta/flat Ti cathode pair, (**b**) holed Ta on flat Ti/ flat Ti pair, (**c**) slotted Ta on flat Ti/ flat Ti pair, and (**d**) slotted Ta on flat Ti/slotted Ti on flat Ti pair (Yoshimura et al., 1992) [2-28]

Ar pressures were obtained by multiplying a correction factor 0.71 (for Ar to N_2) to pressure indications. In measuring Ar-pumping speeds the pressure was increased from about 2×10^{-5} Pa up to about 4×10^{-4} Pa, step by step, by increasing the introduced gas load through a needle valve. After controlling the introduced gas load to set a one point pressure, the needle valve was not handled for 1–3 h. The pressures in the figures (**Fig. 2.24**, **Fig. 2.25**) were thus obtained at the end of pressure-keeping periods, in the chamber to which the ion pump was connected. Ion currents of the test pump were also measured at the same time in the pumping speed measurement.

Ar-pumping speed characteristics of the pumps of the respective pump elements are presented in **Fig. 2.24**. The "saturated" Ar-pumping speed of the pump of the flat Ta/flat Ti cathode pair was 12–14 L/s in the 10^{-5} and 10^{-4} Pa ranges. On the other hand, the pumps of the "holed Ta on flat Ti/flat Ti" pair and of the "slotted Ta on flat Ti/flat Ti" pair have almost the same pumping speed as 22–23 L/s. The pumping speed of the pump of the "slotted Ta

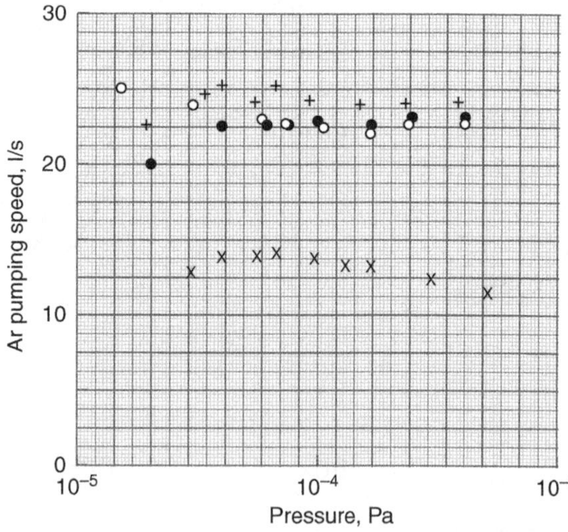

Fig. 2.24 Ar-pumping speeds of the pumps with various cathode pairs as a function of pressure. -×-, "flat Ta/flat Ti cathode" pair; -•-, "holed Ta on flat Ti/flat Ti" pair; -○-, "slotted Ta on flat Ti/flat Ti" pair; -+ -, "slotted Ta on flat Ti/slotted Ti on flat Ti" pair (Yoshimura et al., 1992) [2-29]

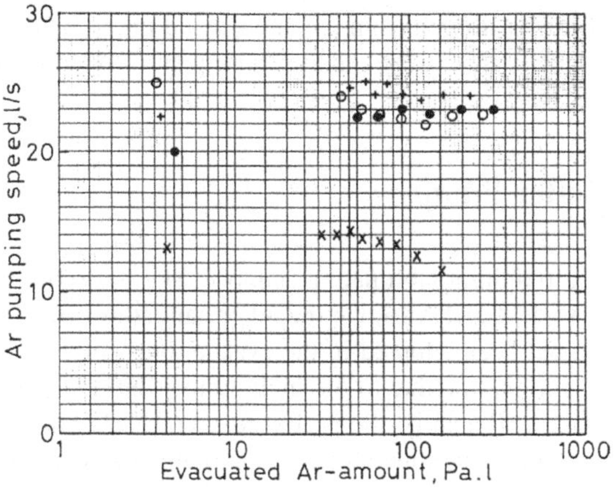

Fig. 2.25 Ar-pumping speeds of the pumps with various cathode pairs as a function of total amount of evacuated Ar. -×-, "flat Ta/flat Ti" cathode pair; -•-, "holed Ta on flat Ti/flat Ti" pair; -○-, "slotted Ta on flat Ti/flat Ti" pair; -+ -, "slotted Ta on flat Ti/slotted Ti on flat Ti" pair (Yoshimura et al., 1992) [2-29]

on flat Ti/slotted Ti on flat Ti" pair was about 25 L/s, which is slightly larger than those of the pump of the "slotted Ta on flat Ti/flat Ti" pair.

The Ar-pumping speeds, as a function of the integrated amount of pumped Ar gas, were calculated using the pumping speed characteristics of **Fig. 2.24**, which are presented in **Fig. 2.25**.

The Ar-pumping speeds for the "flat Ta/flat Ti" cathode pair start decreasing when the pumped Ar-amount reaches 50 Pa · L and decrease rapidly after reaching 100 Pa · L. On the other hand, the pumps with other cathode pairs, the Ar-pumping speed keeps 22–24 L/s when reaching 200–300 Pa · L of the total amount of Ar.

Comments:

The pump having "slotted Ta on a flat Ti/flat Ti" cathode pair is used for JEOL JAMP-series Auger electron spectrometers. In order to pump Xe gas stably, the pump having slotted Ta on flat Ti cathodes is available for the JEOL JAMP (Auger electron micro-probe) series.

Welch et al. (1993) [2-30] presented quantitative data for He speeds and capacities of both noble (Ta/Ti pair) and conventional diode and triode pumps. "The effectiveness of various pump regeneration procedures, subsequent to the pumping of He is reported. These included bakeout and N_2 glow discharge cleaning. The comparative desorption of He with the subsequent pumping of N_2 is reported on."

Welch et al. [2-31] discussed the pumping of hydrogen in the diode pump and triode pump.

Abstract [2-31]: The type of cathode material used in these pumps is shown to have a significant impact on the effectiveness with which hydrogen is pumped. Examples of this include data for pumps with aluminum, titanium, and titanium-alloy cathodes. The use of titanium anodes and titanium shielding of a pump body is also shown to impact measurably the speed of a pump at very low pressures. This stems from the fact that hydrogen is $\times 10^6$ more soluble in titanium than in stainless steel. Hydrogen becomes resident in the anodes because of fast neutral burial. Ions and fast neutrals of hydrogen are also buried in the walls of pump bodies. Outgassing of this hydrogen from the anodes and pump bodies results in a gradual increase in pump base pressure and the consequential decrease in hydrogen pump speed at very low base pressures.

Getter Pumps

The titanium-sublimation pump (TSP) is usually used together with the SIP, for UHV-EM and AES. On the other hand, the non-evaporable getter (NEG) pump has not yet been adopted in the evacuation system of the JEOL electron microscopes, which may be due to the high temperatures of the pump during operation and degassing. Recently, the performance of NEG has been improved in the direction of lowering the temperatures of the pump during operation and degassing.

Titanium-Sublimation Pumps

Titanium sublimates at a high temperature a little lower than its melting point (1660 °C), and deposits on the surfaces of the pump vessel, forming getter films.

Harra (1976) [2-32] reviewed many literatures on the sticking coefficients and sorption capacities of gases on titanium films.

Abstract [2-32]: The pumping speed of a titanium-sublimation pump (TSP) or a sputter ion pump (SIP) using titanium cathodes is a function of many variables including the sticking coefficient and sorption capacity of gases on titanium films. Consequently, in vacuum systems employing such pumping mechanisms, the sticking coefficients and sorption capacities play a major role in determining the partial pressures at any given time of various gas species. The literature on the sticking coefficient data has been reviewed and the data for H_2, D_2, H_2O, CO, N_2, O_2 and CO_2 have been summarized. These results are presented and information related to the sorption mechanism is discussed.

The summary of the sorption data for typical gases on titanium films are presented in **Table 2.6**.

Reference

1. R. Steinberg and D. L. Alger, *J. Vac. Sci. Technol.* **10**, 246 (1973).

Note: If the sticking coefficient is 1, the pumping speed S_M for gas molecules of mass number M at 20 °C is $S_M = 11.6\sqrt{29/M}(\mathrm{L\cdot s^{-1}\cdot cm^{-2}})$.

A TSP is often used together with an SIP as a combination pump. A TSP and an SIP have only a very low pumping speed for methane (CH_4) evolved from a Ti film. Therefore, it is important to identify and if possible reduce the sources of CH_4 outgassing [2-33].

Table 2.6 Summary of the sorption data (Harra, 1976) [2-32]

Gas	Initial sticking coefficient		Quantity sorbed^a ($\times 10^{15}$ molec/cm^2)	
	300 K	78 K	300 K	78 K
H_2	0.06	0.4	$8-230^b$	7-70
D_2	0.1	0.2	$6-11^b$	...
H_2O	0.5	...	30	...
CO	0.7	0.95	5–23	50–160
N_2	0.3	0.7	0.3–12	3–60
O_2	0.8	1.0	24	...
CO_2	0.5	...	4-24	...

^{a} For fresh films thicknesses $> 10^{15}$ Ti atoms/cm^2.
^{b}The quantity of hydrogen or deuterium sorbed at saturation may exceed the number of Ti atoms/cm^2 in the fresh film through diffusion into the underlying deposit at 300 K [1].

Edwards (1980) [2-33] presented an article, "Methane outgassing from a Ti sublimation pump."

Abstract [2-33]: It is well known that the Ti surface will not pump methane and that the ion pump pumping speed for CH_4 is only 1 L/s in the low 10^{-11} Torr range. Thus it is important to identify and if possible reduce the sources of CH_4 outgassing. In this study we report the first quantitative measurements of the methane outgassing rate from a 1500 cm^2 Ti-sublimation chamber both before and after the Ti evaporation. The contrast to the very low CH_4 outgassing rate from the chamber before the Ti is flashed ($< 10^{-12}$ Torr $\cdot$ L/s) the CH_4 outgassing rate after the Ti evaporation is considerable (4×10^{-11} Torr $\cdot$ L/s) with a very slow decrease with respect to time. To essentially eliminate the methane outgassing from the Ti getter chamber, it has been determined sufficient to heat the Ti chamber for four hours at 100 °C after the Ti evaporation has been completed.

Non-Evaporable Getter (NEG) Pumps

The pumping performance of non-evaporable getters (NEGs) has been progressively improved.

Benvenuti and Chiggiato (1996) [2-34] measured the room-temperature pumping speeds of the St707 NEG non-evaporable getter (Zr 70 V 24.6–Fe 5.4 wt%) for individual gases and for gas mixtures as a function of the quantities of gas pumped.

Abstract [2-34]: The interesting feature of this NEG consists in its moderately low activation temperature. It has been found that heating at 400 °C for about 1 h, or at 350 °C for one day, results in pumping speeds of about 1000 L $\cdot$ s^{-1} $\cdot$ m^{-1} for H_2, 2000 L $\cdot$ s^{-1} $\cdot$ m^{-1} for CO, and 450 L $\cdot$ s^{-1} $\cdot$ m^{-1} for N_2, values very close to those obtained after activation at the higher temperature of 740 °C. The St707 NEG is therefore particularly suitable for passive activation during bakeout of stainless-steel vacuum systems, avoiding the need for electrical insulation and feedthroughs which are mandatory when activation is carried out by resistive heating.

Li et al. (1998) [2-35] presented an article, "Design and pumping characteristics of a compact titanium-vanadium non-evaporable getter pump."

Abstract [2-35]: A compact non-evaporable getter (NEG) pump is made by stacking sintered titanium–vanadium blades recently available from SAES Getters as St185. The 50 blades are stacked between copper spacers on a stainless-steel tube, and are activated by a removable heater cartridge inserted into the tube. The room-temperature pumping speeds for CO and N_2 are measured as a function of sorbed gas quantities at various activation temperatures and with different NEG disk spacing, in order to optimize the pump design and operation. The study shows that there is a threshold of activation at a temperature of about 340 °C and optimal pumping speed and pumping capacity can be achieved with an activation temperature around 550 °C. Linear pumping speeds of about $2200 \mathrm{L \cdot s^{-1} \cdot m^{-1}}$ and $1300 \mathrm{L \cdot s^{-1} \cdot m^{-1}}$ can be achieved for CO and N_2, respectively, with activation at temperatures of 500 °C for 1 h. The low activation temperature of the St185 NEG and the compact design make the NEG pump cartridge particularly attractive for situations with tight space and with components sensitive to high temperature.

Methods for Measuring Pumping Speeds

There are two methods for measuring the pumping speeds of sputter ion pumps (SIPs), the throughput method using an orifice and the method using a pipe.

Orifice Method

Yoshimura et al. (1992) [2-28, 2-29] measured the N_2-pumping speeds of an SIP of the slotted Ta on flat Ti/flat Ti pair using an orifice method. The pump has the anode cells composed of 30 cells (24 mm diameter, 28 mm in length, 3 columns × 10 rows). The distribution of magnet flux density in the pole gap is presented in **Fig. 2.22**.

Experimental setup is shown in **Fig. 2.26**. The N_2-pumping speeds and air-pumping speeds measured are presented in **Fig. 2.27**. The shapes of the pumping-speed curves of **Fig. 2.27** are reasonably consistent with the discharge intensity curve of the pump of 0.15 T and ϕ24 mm presented in **Fig. 2.17(b)**.

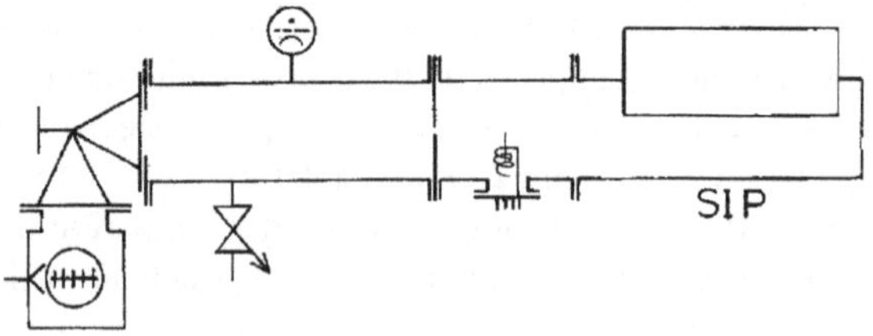

Fig. 2.26 Experimental setup for measuring pumping speed characteristics (Yoshimura et al., 1992) [2-29]

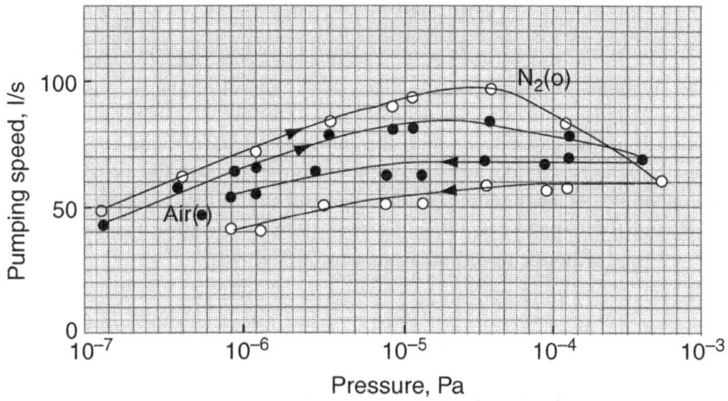

Fig. 2.27 N_2-pumping speeds (○) and air-pumping speeds (•) of the pump of the "slotted Ta on flat Ti/flat Ti" cathode pair. (Yoshimura et al., 1992) [2-29]

Three-Gauge Method (Pipe Method)

Munro and Tom (1965) [2-36] presented an article, "Speed measuring of ion getter pumps by the '**three-gauge' method**."

> **Quoted from Munro and Tom, 1965 [2-36]:**
> The 3-gauge known conductance method can be briefly described with the aid of **Fig. 2.28**. Here the conductance C_{12} between ion gauges No.1 and No.2 is calculated using the long tube formula. The flow rate $Q = C_{12}(P_1 - P_2)$ can now be measured when the pump is operating on a gas leak. Since $Q = SP$, we can obtain the speed at ion gauge No.3 by the relation:

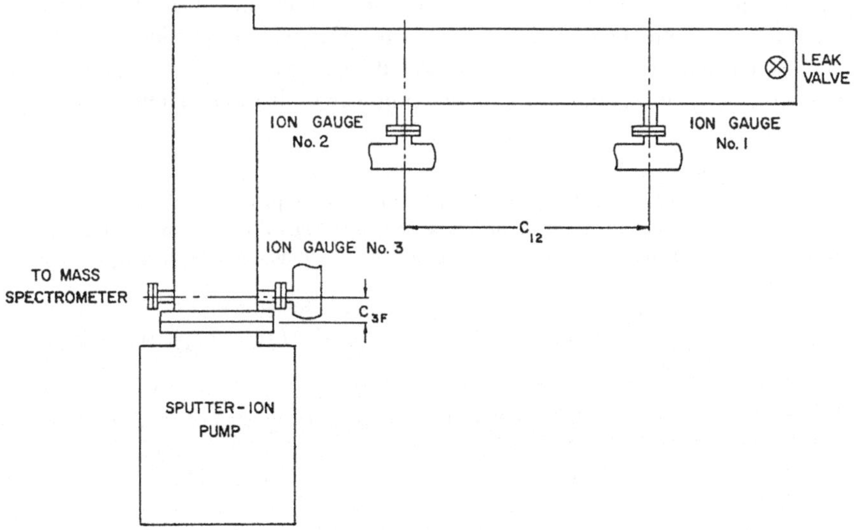

Fig. 2.28 3-gauge speed dome system (Munro and Tom, 1965) [2-36]

$$S_3 = \frac{C_{12}(P_1 - P_2)}{P_3}. \tag{2.3}$$

Since ion gauges are not absolute pressure gauges, there may exist differences in gauge sensitivity among the three gauges. This makes it necessary to calibrate the gauges against each other. The calibration must be carried out with the pump and leak off; the pressure range of calibration must be at least one decade above the base pressure on the system to reduce surface outgassing effects. Under pump off and leak off conditions, the pressure everywhere in the system must be equal or $P_1/P_3 = P_2/P_3 = 1$. Any deviation from unity will be noted and corrected for during the speed measurement as follows:

$$S_3 = C_{12}\frac{K_{31}l_1 - K_{32}l_2}{l_3}, \tag{2.4}$$

where l_1, l_2 and l_3 are, respectively, the ion gauge readings during measurement; $K_{31} = l_3/l_1$ and $K_{32} = l_3/l_2$ are the ratios of the gauge readings during calibration when the pressure is the same everywhere in the system. K_{31} and K_{32} will be referred to as calibration constants.

Note: As seen in Eq. (2.3) the outgassing of the pipe walls is neglected. However, when the introduced leak-gas flow rate becomes small, the outgassing rate from the pipe wall surfaces may affect the flow-gas rate measurement.

Three-Point Pressure Method (3PP Method)

Hirano and Yoshimura (1986) [2-37a] developed the three-point-pressure method (3PP method) for measuring the gas-flow rate through a conducting pipe", in which the pressures at three different points along the pipe are measured to calculate the real gas-flow rate.

Abstract [2-37a]: This method was applied to measuring the flow rate of nitrogen through a pipe. The gas-flow rates calculated using three pressures were compared with those conventionally calculated using two pressures under the same condition. The new method based on three pressures gave real rates which were higher than those calculated conventionally calculated based on two pressures, especially in low pressure regions, thus indicating the presence of distributing outgassing.

Principle [2-37a] [2-37b]

Let us examine the pressure in a portion of a long outgassing pipe where the net gas flow including the leak rate Q_L directs to the left, as shown in **Fig. 2.29**. The conductance C of tube length L is calculated using the long tube formula assuming a linear conductance for a pipe portion.

Consider a very short element with a length Δx and a conductance $CL/\Delta x$. Since the total gas-flow rate at a position x is calculated as $Q_L + (1 - x/L)\,Q_W$, the pressure drop ΔP across the element is

$$\Delta P = \frac{[Q_L + (1 - x/L)\,Q_W]\,\Delta x}{CL}.$$

That is,

$$dP/dx = \frac{Q_L + (1 - x/L)\,Q_W}{CL}. \tag{2.5}$$

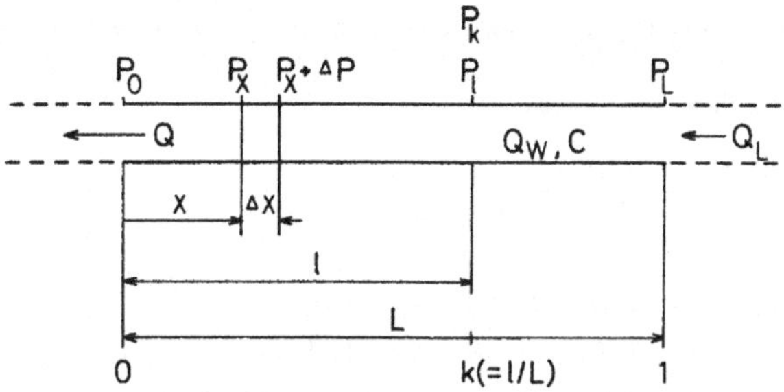

Fig. 2.29 Pipe portion in a long outgassing tube, where gas effectively flows from the right to the left. Gas-flow rate Q is composed of the leak rate Q_L and the outgassing rate Q_W (Hirano and Yoshimura, 1986) [2-37a]

Integrating Eq. (2.5) from 0 to l with respect to x:

$$\int_{P_0}^{P_l} dP = \int_0^l \frac{Q_L + (1 - x/L)\, Q_W}{CL} dx$$

and so

$$P_l = P_0 + \frac{\{Q_L + [1 - l/(2L)]\, Q_W\}\, l}{CL},$$

where P_0 and P_l denote the pressures at the positions $x = 0$ and l, respectively. Substituting a fraction k for l/L and rewriting P_l to P_k, the pressure P_k at a position indicated by the fraction k is given by

$$P_k = P_0 + \frac{k\,[Q_L + (1 - k/2)\, Q_W]}{C}. \tag{2.6}$$

Equation (2.6) shows that P_k varies quadratically [1-3] as a function of k. Equation (2.6) contains three unknown factors Q_L, Q_W, and P_0. These three factors can be calculated from Eq. (2.6) using three pressures P_{k1}, P_{k2}, and P_{k3} measured at three different points in the pipe identified by their fractional positions k_i.

The following equation in a matrix form is obtained for three pressures P_{k1}, P_{k2}, and P_{k3} in the pipe from (2.6):

$$\begin{pmatrix} P_{k1} \\ P_{k2} \\ P_{k3} \end{pmatrix} = \begin{pmatrix} 1 & k_1 & k_1(1 - k_1/2) \\ 1 & k_2 & k_2(1 - k_2/2) \\ 1 & k_3 & k_3(1 - k_3/2) \end{pmatrix} \begin{pmatrix} P_0 \\ Q_L/C \\ Q_W/C \end{pmatrix}.$$

Let $|\alpha|$, $|\beta|$, and $|\gamma|$ represent the following determinates:

$$|\alpha| = \begin{vmatrix} 1 & P_{k1} & k_1(1 - k_1/2) \\ 1 & P_{k2} & k_2(1 - k_2/2) \\ 1 & P_{k3} & k_3(1 - k_3/2) \end{vmatrix},$$

$$|\beta| = \begin{vmatrix} 1 & k_1 & P_{k1} \\ 1 & k_2 & P_{k2} \\ 1 & k_3 & P_{k3} \end{vmatrix},$$

$$|\gamma| = \begin{vmatrix} 1 & k_1 & k_1(1 - k_1/2) \\ 1 & k_2 & k_2(1 - k_2/2) \\ 1 & k_3 & k_3(1 - k_3/2) \end{vmatrix}.$$

Then, the values Q_L and Q_L are derived, using Cramer's formula, as $Q_L = C|\alpha|/|\gamma|$ and $Q_w = C|\beta|/|\gamma|$.

Gas-flow rates Q_L and Q_W are then expressed as

$$Q_L = C\frac{(2-k_2-k_3)(k_2-k_3)P_{k1}+(2-k_3-k_1)(k_3-k_1)P_{k2}+(2-k_1-k_2)(k_1-k_2)P_{k3}}{(k_2-k_3)k_2k_3+(k_3-k_1)k_3k_1+(k_1-k_2)k_1k_2}, \tag{2.7}$$

$$Q_W = -2C\frac{(k_2-k_3)P_{k1}+(k_3-k_1)P_{k2}+(k_1-k_2)P_{k3}}{(k_2-k_3)k_2k_3+(k_3-k_1)k_3k_1+(k_1-k_2)k_1k_2}, \tag{2.8}$$

where $0 \leq k_i \leq 1$. As a result, the total gas-flow rate $Q = Q_L + Q_W$ is expressed as

$$Q = -C\frac{(k_2^2-k_3^2)P_{k1}+(k_3^2-k_1^2)P_{k2}+(k_1^2-k_2^2)P_{k3}}{(k_2-k_3)k_2k_3+(k_3-k_1)k_3k_1+(k_1-k_2)k_1k_2}. \tag{2.9}$$

This is a good solution as long as the gauges are kept sufficiently far from the tube ends in the linear pressure section. We call this gas-flow rate measuring method based on Eq. (2.9), the "three-point pressure method" (or simply the 3PP method).

Measurement of Gas-Flow Rates [2-37a][2-37b]

A conducting pipe with three Bayard-Alpert gauges G1, G2, G3 was connected to a typical test dome with an ion pump (60 L/s, triode type). Three gauges were located at the positions indicated by the fractions k of 0.1, 0.4, and 0.9 as shown in **Fig. 2.30**.

A pipe of 43 mm diam and 800 mm length (corresponding to the length L) with a conductance of 12 L/s was actually selected to make the pressures P_{01}, P_{04}, and P_{09} be in the same pressure range. Substituting 12 for C and rewriting $P_{0.1}$, $P_{0.4}$, and $P_{0.9}$ to $P1$, $P2$, and $P3$, respectively, Eqs. (2.7), (2.8), and (2.9) become

$$Q_L = 5(7P1 - 16P2 + 9P3), \tag{2.10}$$

$$Q_W = -20(5P1 - 8P2 + 3P3), \tag{2.11}$$

$$Q = -5(13P1 - 16P2 + 3P3). \tag{2.12}$$

On the other hand, gas-flow rates are conventionally calculated using two pressures along the pipe. The gas-flow rate Q'_{1-3} calculated using the pressures $P1$ and $P3$ is given by

$$Q'_{1-3} = 15(P3 - P1), \tag{2.13}$$

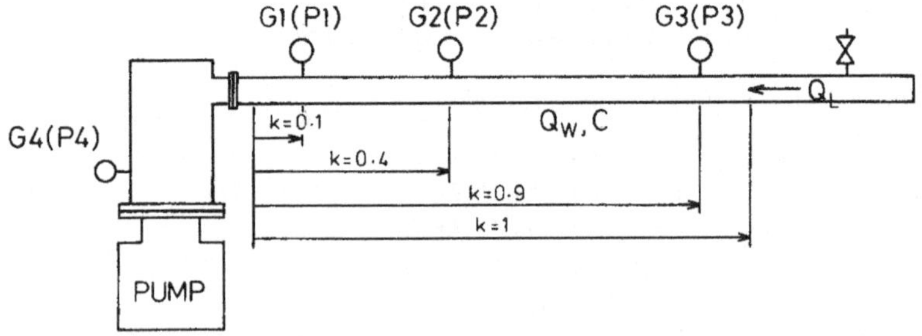

Fig. 2.30 Gas-flow rate measuring system with three gauges installed on a conducting pipe. G1, G2, G3, and G4 are B-A gauges; Q_L is the leak rate of the introduced gas; Q_W is the outgassing rate of the tube portion from $k = 0$ to 1 of the pipe; and C is the conductance of the same portion (Hirano and Yoshimura, 1986) [2-37a]

where 15 is the conductance (L/s) between the gauges G1 and G3. We call this method the conventional pipe method.

The system was pre-degassed by baking (200 °C and 300 °C) under a high vacuum created by a turbomolecular pump.

After the system was evacuated to a pressure less than 1×10^{-6} Pa, nitrogen was leaked into the pipe through a variable leak valve to make the pressure $P1 \sim 1 \times 10^{-5}$ Pa. In the course of reaching equilibrium pressures $P1$, $P2$, and $P3$ were read at elapsed times of 5, 15, 25 min. The leak rate was then increased to obtain higher pressures, and the pressures $P1$, $P2$, and $P3$ were read at elapsed times of 5, 15, 25 min. The process was repeated at various pressures up to $\sim 3 \times 10^{-3}$ Pa.

The pressure distribution in the pipe at 25 min at two different pressure levels are shown in **Fig. 2.31**, where the solid curve shows the pressure distribution in the 10^{-5} Pa range and the broken curve in the 10^{-3} Pa range.

The convex curve in the 10^{-5} Pa range shows that the outgassing rate of the pipe wall is appreciable compared with the leak rate, while the straight line in the 10^{-3} Pa range shows that the effect of the outgassing is negligibly small.

The pumping speed of the ion pump used can be calculated using the flow rate $Q(= Q_L + Q_W)$ and the dome pressure $P4$ measured by the gauge G4 (shown in **Fig. 2.30**) under the following assumptions:

(1) The outgassing rate of the test dome is negligibly small compared with the rate of gases flowing into the test dome.
(2) The ultimate pressure of the pump is negligibly low compared with the test pressure.

The pressure $P0$ and the pumping speed $S0$ at the pump mouth for nitrogen were calculated as

$$P0 = P4 - Q/1200,$$
$$S0 = Q/P0,$$

where 1200 is the conductance (L/s) between the pump mouth and the gauge G4, and Q is the gas-flow rate given by Eq. (2.12).

The pumping speed $S0'$ is also calculated conventionally using Q'_{1-3} of (2.13) (the conventional pipe method).

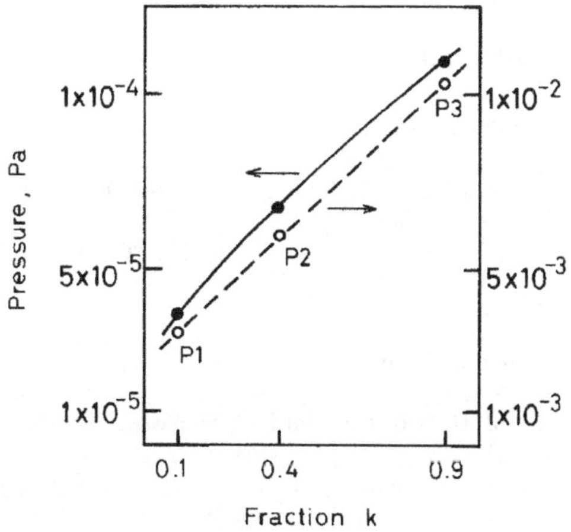

Fig. 2.31 Pressure distributions in the pipe at an elapsed time of 25 min for nitrogen. –•– in the 10^{-5} Pa range, - - o- - in the 10^{-3} Pa range (Hirano and Yoshimura, 1986) [2-37a]

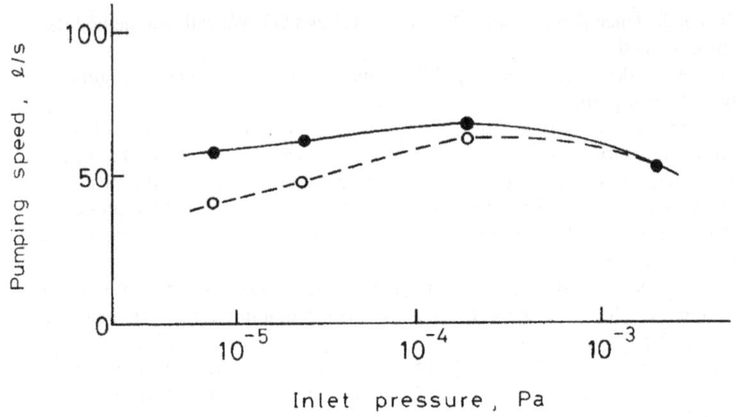

Fig. 2.32 Pumping speeds for nitrogen as a function of pressure at an elapsed time of 25 min. –•– *S*0 by the 3PP method, - - ○- - *S*0′ by the conventional pipe method (Hirano and Yoshimura, 1986) [2-37a]

Pumping speeds for nitrogen as a function of pressure at an elapsed time of 25 min are presented in **Fig. 2.32**, where the solid line curve shows the speed *S*0 derived using the value *Q* calculated by the 3PP method and the broken line curve *S*0′ derived using the value Q'_{1-3} by the conventional pipe method.

The 3PP method can give real gas-flow rate when three pressures are measured accurately.

References

1. H. Hamacher, *Vacuum* **32**, 729 (1982).
2. W. Teubner, *Vak. Tech.* **16**, 69 (1967).
3. K. M. Welch, *Vacuum* **23**, 271 (1973).

References

Mechanical Pump

2-1. N. S. Harris and L. Budgen, "Design and manufacture of modern mechanical vacuum pump", *Vacuum* **26** (12), pp. 525–529 (1976).

2-2. B. S. Ramprasad and T. S. Radha, "On some design aspects of rotary vane pumps", *Vacuum* **23** (7), pp. 245–249 (1973).

Diffusion Pump

2-3. M. H. Hablanian and J. C. Maliakal, "Advances in diffusion pump technology", *J. Vac. Sci. Technol.* **10** (1), pp. 58–64 (1973).

2-4. H.-P. Kabelitz and J. K. Fremerey, "Turbomolecular vacuum pumps with a new magnetic bearing concept", *Vacuum* **38** (8–10), pp. 673–676 (1988).

2-5. G. Levi, "Combination of turbomolecular pumping stages and molecular drag stages", *J. Vac. Sci. Technol. A* **10** (4), pp. 2619–2622 (1992).
2-6. B. Cho, S. Lee, and S. Chung, "Creation of extreme high vacuum with a turbomolecular pumping system: A baking approach", *J. Vac. Sci. Technol. A* **13** (4), pp. 2228–2232 (1995).

Dry Vacuum Pump

2-7. P. A. Lessard, "Dry vacuum pumps for semiconductor processes: Guidelines for primary pump selection", *J. Vac. Sci. Technol. A* **18** (4), pp. 1777–1781 (2000).
2-8. I. Akutsu and T. Ohmi, "Innovation of the fore pump and roughing pump for high-gas-flow semiconductor processing", *J. Vac. Sci. Technol. A* **17** (6), pp. 3505–3508 (1999).
2-9. R.-Y. Jou, H.-P. Cheng, Y.-W. Chang, F.-Z. Chen, and M. Iwane, "Designs, analyses, and tests of a spiral-grooved turbobooster pump", *J. Vac. Sci. Technol. A* **18** (3), pp. 1016–1024 (2000).
2-10. A. Liepert and P. Lessard, "Design and operation of scroll-type primary vacuum pumps", *J. Vac. Sci. Technol. A* **19** (4), pp. 1708–1711 (2001).
2-11. "*Handbook of Vacuum Science and Technology",* edited by D. M. Hoffman, B. Singh, and J. H. Thomas, III, Academic Press. pp. 84–96.
2-12. H.-P. Cheng and M.-T. Chiang, "Pumping performance investigation of a turbobooster vacuum pump equipped with spiral-grooved rotor and inner housing by the computational fluids dynamics method", *J. Vac. Sci. Technol. A* **21** (4), pp. 1458–1463 (2003).
2-13. S. Giors, E. Colombo, F. Inzoli, F. Subba, and R. Zanino, "Computational fluid dynamic model of a tapered Holweck vacuum pump operating in the viscous and transition regimes. I. Vacuum performance", *J. Vac. Sci. Technol. A* **24** (4), pp. 1584–1591 (2006).
2-14. R. P. Davis, R. A. Abreu, and A. D. Chew, "Dry vacuum pump: A method for the evaluation of the degree of dry", *J. Vac. Sci. Technol. A* **28** (4), pp. 1782–1788 (2000).

Cryopump

2-15. P. D. Bentley, "The modern cryopump", *Vacuum* **30** (4/5), pp. 145–158 (1980).
2-16. H.-H. Klein, R. Heisig, and C. M. Augustine, "Use of refrigerator-cooled cryopumps in sputtering plants", *J. Vac. Sci. Technol. A* **2** (2), pp. 187–190 (1984).
2-17. R. W. Dennison and G. R. Gray, "Cryogenic versus turbomolecular pumping in a sputtering application", *J. Vac. Sci. Technol.* **16** (2), pp. 728–730 (1979).
2-18. H.-P. Cheng and Y.-H. Shen, "Effect of heat on the pumping performance of cryopump", *J. Vac. Sci. Technol. A* **24** (4), pp. 1597–1600 (2006).
2-19. R. E. Honig and H. O. Hook, "Vapor pressure data for some common gases", *RCA Review*, September, pp. 360–368 (1960).

Sputter Ion Pump

2-20. D. Andrew, "The development of sputter-ion pumps", *Proc. of the 4th Internl. Vacuum Congress*, 1968 (Manchester, April 1668), pp. 325–331.
2-21. R. L. Jepsen, "The physics of sputter-ion pumps", *Proc. of the 4th Internl. Vacuum Congress*, 1968 (Manchester, April 1968), pp. 317–324.
2-22. S. L. Rutherford, "Sputter-ion pumps for low pressure operation", *Transactions of the 10th National Vacuum Symposium*, 1963 (Macmillan, New York, 1964), pp.185–190.
2-23. K. Ohara, I. Ando, and N. Yoshimura, "Pumping characteristics of sputter ion pumps with high-magnetic-flux densities in an ultrahigh-vacuum range", *J. Vac. Sci. Technol. A* **10** (5), pp. 3340–3343 (1992).
2-24. R. L. Jepsen, A. B. Francis, S. L. Rutherford, and B. E. Kietzmann, "Stabilized air pumping with diode type getter-ion pumps", *Transactions of the 7th National Vacuum Symposium*, 1960 (Pergamon Press, New York, 1961), pp. 45–50.

2-25. P. N. Baker and L. Laurenson, “Pumping mechanisms for the inert gases in diode Penning pumps”, *J. Vac. Sci. Technol.* **9** (1), pp. 375–379 (1972).
2-26. D. R. Denison, “Comparison of diode and triode sputter-ion pumps”, *J. Vac. Sci. Technol.* **14** (1), pp. 633–635 (1977).
2-27. S. Komiya and N. Yagi, “Enhancement of noble gas pumping for a sputter-ion pump”, *J. Vac. Sci. Technol.* **6** (1), pp. 54–57 (1969).
2-28. N. Yoshimura, K. Ohara, I. Ando, and H. Hirano, “Ar-pumping characteristics of diode-type sputter ion pumps with various shapes of ‘Ta/Ti’ cathode pairs”, *J. Vac. Sci. Technol. A* **10** (3), pp. 553–555 (1992).
2-29. N. Yoshimura, K. Ohara, I. Ando, and H. Hirano, “Ar-pumping characteristics of diode-type sputter ion pumps with various shapes of ‘Ta/Ti’ cathode pairs”, *Shinku (J. Vac. Soc. Japan)* **35** (6), pp. 574–578 (1992) (in Japanese).
2-30. K. M. Welch, D. J. Pate, and R. J. Todd, “Pumping of helium and hydrogen by sputter-ion pumps. (1) Helium pumping”, *J. Vac. Sci. Technol. A* **11** (4), pp. 1607–1613 (1993).
2-31. K. M. Welch, D. J. Pate, and R. J. Todd, “Pumping of helium and hydrogen by sputter-ion pumps. (2) Hydrogen pumping”, *J. Vac. Sci. Technol. A* **12** (3), pp. 861–866 (1994).

Getter Pump

2-32. D. J. Harra, “Review of sticking coefficients and sorption capacities of gases on titanium films”, *J. Vac. Sci. Technol.* **13** (1) pp. 471–474 (1976).
2-33. D. Edwards, Jr., “Methane outgassing from a Ti sublimation pump”, *J. Vac. Sci. Technol.* **17** (1), pp. 279–281 (1980).
2-34. C. Benvenuti and P. Chiggiato, “Pumping characteristics of the St707 nonevaporable getter (Zr 70 V 24.6 Fe 5.4 wt%)”, *J. Vac. Sci. Technol. A* **14** (6), pp. 3278–3282 (1996).
2-35. Y. Li, D. Hess, R. Kersevan, and N. Mistry, “Design and pumping characteristics of a compact titanium-vanadium non-evaporable getter pump”, *J. Vac. Sci. Technol. A* **16** (3), pp. 1139–1144 (1998).

Measurement of Pumping Speeds

2-36. D. F. Munro and T. Tom, “Speed measuring of ion getter pumps by the ‘three-gauge’ method”, *1965 Trans. 3rd Internl. Vacuum Congress*, pp. 377–380.
2-37a. H. Hirano and N. Yoshimura, “A three-point-pressure method for measuring the gas-flow rate through a conducting pipe”, *J. Vac. Sci. Technol. A* **4** (6), pp. 2526–2530 (1986).
2-37b. H. Hirano and N. Yoshimura, “A three-point-pressure method for measuring the gas-flow rate through a conducting pipe”, *Shinku (J. Vac. Soc. Japan)* **30** (6), pp. 531–537 (1987).

Other Articles

Mechanical Vacuum Pump

*2-1. M. A. Baker, L. Holland, and D. A. G. Stanton, “The design of rotary pumps and systems to provide clean vacua”, *J. Vac. Sci. Technol.* **9** (1), pp. 412–415 (1972).

Dry Vacuum Pump

*2-2. E. J. Eckle, P. Bickert, R. Lachenman, and B. Wortmann, “Pumping speed of diaphragm pumps for various gases”, *Vacuum* **47** (6–8), pp. 799–801 (1996).

*2-3. J. Y. Tu, Y. Zhu, and Z. Wang, "A new design for the disk-type molecular pump", *J. Vac. Sci. Technol. A* **8** (5), pp. 3870–3889 (1990).

*2-4. I. V. Ioffe, V. A. Koss, M. Gray, and R. G. Livesey, "Modeling of a multistage claw rotor vacuum pump", *J. Vac. Sci. Technol. A* **13** (2), pp. 536–539 (1995).

*2-5. J. C. Helmer, G. Levi, "Transition gas flow in drag pumps and capillary leaks", *J. Vac. Sci. Technol. A* **13** (5), pp. 2592–2599 (1995).

*2-6. H. V. Pitingsrud, "Miniature peristaltic vacuum pump for use in portable instruments", *J. Vac. Sci. Technol. A* **14** (4), pp. 2610–2617 (1996).

*2-7. K. Ando, I. Akutsu, and T. Ohmi, "Gradational lead screw vacuum pump development", *J. Vac. Sci. Technol. A* **17** (5), pp. 3144–3148 (1999).

*2-8. T. Sawada and W. Sugiyama, "Pumping mechanism of helical grooved molecular drag pumps", *J. Vac. Sci. Technol. A* **17** (4), pp. 2069–2074 (1999).

*2-9. R. P. Davis, R. A. Abreu, and A. D. Chew, "Dry vacuum pumps: A method for the evaluation of the degree of dry", *J. Vac. Sci. Technol. A* **18** (4), pp. 1782–1788 (2000).

*2-10. I. Akutsu, T. Matsuoka, M. Ozaki, T. Kyuko, S. Miyashita, T. Ozawa, M. Naka, H. Ohnishi, Y. Narahara, and G. Horikoshi, "A gradational lead screw dry vacuum pump", *J. Vac. Sci. Technol. A* **18** (3), pp. 1045–1047 (2000).

*2-11. J.-S. Heo and Y.-K. Hwang, "Molecular transition and slip flows in the pumping channels of drag pumps", *J. Vac. Sci. Technol. A* **18** (3), pp. 1025–1034 (2000).

*2-12. F. Sharipov, P. Fahrenbach, and A. Zipp, " Numerical modeling of the Holweck pump", *J. Vac. Sci. Technol. A* **23** (5), pp. 1331–1339 (2005).

Diffusion Pump

*2-13. M. H. Hablanian, "Backstreaming measurements above liquid-nitrogen traps", *J. Vac. Sci. Technol.* **6** (1), pp. 265–268 (1969).

*2-14. H. M. Sullivan, "Vacuum pumping equipment and systems", *Rev Sci. Instrum.* **19** (1), pp. 1–15 (1948).

*2-15. T. E. Lucas, "Properties of high vacuum pumps", *Vacuum* **15** (5), pp. 221–229 (1965).

*2-16. N. A. Florescu, "Increase in performance of the vapour vacuum pump", *Vacuum* **12**, pp. 259–265 (1962).

*2-17. D. M. Hoffman, "Operation and maintenance of a diffusion-pumped vacuum system", *J. Vac. Sci. Technol.* **16** (1), pp. 71–74 (1979).

*2-18. D. J. Santeler, "The use of diffusion pumps for obtaining ultraclean vacuum environments", *J. Vac. Sci. Technol.* **8** (1), pp. 299–307 (1971).

*2-19. B. D. Power, N. T. M. Dennis, P. D. Oswald, and B. H. Colwell, "Single structure vapour pumping groups", *Vacuum* **24** (3), pp. 117–122 (1974).

*2-20. E. H. Hirsch and J. Richards, "Pressure fluctuations in a diffusion pump using polyphenyl ether", *Vacuum* **24** (3), pp. 123–124 (1974).

*2-21. K. Watanabe, "Technical developments and examples of modern oil diffusion pumps", *Shinku (J. Vac. Soc. Japan)* **20** (6), pp. 202–212 (1977) (in Japanese).

*2-22. M. H. Hablanian, "Consider diffusion pumps", *Industrial Research/Development* August, pp.84–89 (1979).

*2-23. N. T. M. Dennis, L. Laurenson, A. Devaney, and B. H. Colwell, "Factors influencing the ultimate vacuum of single structure vapor pumping groups", *J. Vac. Sci. Technol.* **20** (4), pp. 996–999 (1982).

*2-24. T. J. Gay, J. A. Brand, M. C. Fritts, J. E. Furst, M. A. Khakoo, E. R. Mell, M. T. Sieger, and W. M. K. P. Wijayaratna, "Clean ultrahigh vacuum system with single-structure diffusion pumps", *J. Vac. Sci. Technol. A* **12** (5), pp. 2903–2910 (1994).

*2-25. L. T. Chu, "Safcty interlock and remote computer monitoring for a diffusion-pumped vacuum system", *J. Vac. Sci. Technol. A* **15** (1), pp. 201–205 (1997).

*2-26. S. Chambreau, M. L. Neuburger, T. Ho, B. Funk, and D. Pullman, "Low cost, mechanically refrigerated diffusion pump baffle for ultrahigh vacuum chambers", *J. Vac. Sci. Technol. A* **18** (5), pp. 2581–2585 (2000).
*2-27. L. Maurice, P. Duval, and G. Gorinas, "Oil backstreaming in turbomolecular and oil diffusion pumps", *J. Vac. Sci. Technol.* **16** (2), pp. 741–745 (1979).
*2-28. N. S. Harris, "Modern diffusion pump vs turbomolecular pump systems", *Vacuum* **30** (4/5), pp. 175–181 (1980).

Turbomolecular Pump

*2-29. M. H. Hablanian, "New concepts in turbomolecular pump design", *J. Vac. Sci. Technol. A* **11** (4), pp. 1614–1619 (1993).
*2-30. C. H. Kruger and A. H. Shapiro, "Vacuum pumping with a bladed axial-flow turbomachine", *1960 7th National Symposium on Vacuum Technology Transactions* (Pergamon Press, New York, 1961), pp. 6–12.
*2-31. W. Becker, "Die Turbomolekularpumpe", *Vakuum-Technik*, **15** (9), pp. 211–218 (1966).
*2-32. W. Becker, "Die Turbomolekularpumpe II", *Vakuum-Technik*, **15** (10), pp. 254–260 (1966).
*2-33. W. Nesseldreher, "The effects of different parameters on the residual gas mass spectrograms of turbo-molecular pumps", *Vacuum* **26** (7), pp. 281–286 (1976).
*2-34. G. E. Osterstrom and T. Knecht, "Grease lublication of turbomolecular vacuum pump bearings", *J. Vac. Sci. Technol.* **16** (2), pp. 746–747 (1979).
*2-35. R. W. Dennison and G. R. Gray, "Cryogenic versus turbomolecular pumping in a sputtering application", *J. Vac. Sci. Technol.* **16** (2), pp. 728–730 (1979).
*2-36. J. A. Basford, "Testing of turbomolecular pumps", *J. Vac. Sci. Technol. A* **10** (4), pp. 2623–2628 (1992).
*2-37. D. Ba, N. Yang, X. Wang, S. Pang, Y. Zhu, and X. Wang, "Pumping performance of a new type of hybrid molecular pump", *J. Vac. Sci. Technol. A* **10** (5), pp. 3352–3355 (1992).
*2-38. M. Iguchi, M. Okamoto, and T. Sawada, "Analysis of TMP's ultimate pressure and development of XHV-TMP and XHV-CMP", *Shinku (J. Vac. Soc. Japan)* **37** (9), pp.742–750 (1994) (in English).
*2-39. N. Ogure, A. Shibata, K. Ono, N. Hayasaka, H. Okano, and K. Okumura, "Enhancement of hydrogen pumping by injecting fluorine into the exhaust system of turbomolecular pumps", *J. Vac. Sci. Technol. A* **13** (3), pp. 524–530 (1995).
*2-40. S. Katsimichas, A. J. H. Goddard, R. Lewington, and C. R. E. de Oliveira, "General geometry calculations of one-stage molecular flow transmission probabilities for turbomolecular pumps", *J. Vac. Sci. Technol. A* **13** (6), pp. 2954–2961 (1995).
*2-41. J. R. Thompson, P. M. Weber, and R. Hellmer, "Extended operation of a wide-range, all-magnetic bearing turbomolecular pump without baking", *J. Vac. Sci. Technol. A* **14** (5), pp. 2965–2967 (1996).
*2-42. N. Konishi, T. Shibata, and T. Ohmi, "Impurity back diffusion through an ultrahigh vacuum turbomolecular pump under large gas throughput", *J. Vac. Sci. Technol. A* **14** (5), pp. 2958–2962 (1996).
*2-43. M. Spagnol, R. Cerruti, and J. Helmer, "Turbomolecular pump design for high pressure operation", *J. Vac. Sci. Technol. A* **16** (3), pp. 1151–1156 (1998).
*2-44. K. Ino, K. Sekine, T. Shibata, T. Ohmi, and Y. Maejima, "Improvement of turbomolecular pumps for ultraclean, low-pressure, and high-gas-flow processing", *J. Vac. Sci. Technol. A* **16** (4), pp. 2703–2710 (1998).
*2-45. H.-P. Cheng, R.-Y. Jou, F.-Z. Chen, and Y.-W. Chang, "Three-dimensional flow analysis of spiral-grooved turbo booster pump in slip and continuum flow", *J. Vac. Sci. Technol. A* **18** (2), pp. 543–551 (2000).
*2-46. M. Harsdorff and R. W. Adam, "On the application of turbomolecular pumps in electron microscopes", *Vacuum* **22** (2), pp. 55–59 (1972).

*2-47. H. Enosawa, C. Urano, T. Kawashima, and M. Yamamoto, "High throughput turbomolecular pump for extreme high vacuum", *J. Vac. Sci. Technol. A* **8** (3), pp. 2768–2771 (1990).
*2-48. M. Heldner and H.-P. Kabelitz, "Reliability of turbomolecular vacuum pumps: A comparison of rolling element and magnetic bearing systems", *J. Vac. Sci. Technol. A* **8** (3), pp. 2772–2777 (1990).

Cryopump

*2-49. R. Dobrozemsky and G. Moraw, "Pumping speed of cryosorption pumps in the pressure range 10^{-6}–10^{-3} Torr", *Vacuum* **21** (12), pp. 587–589 (1971).
*2-50. P. Vijendran and C. V. G. Nair, "Design and performance characteristics of sorption pumps", *Vacuum* **21** (5), pp. 159–164 (1971).
*2-51. O. E. Vilches and J. C. Wheatley, "Techniques for using liquid helium in very low temetature apparatus", *Rev. Sci. Instrum.* **37** (7), pp. 819–831 (1966).
*2-52. J. P. Hobson,"Cryopumping", *J. Vac. Sci. Technol.* **10** (1), pp. 73–79 (1973).
*2-53. C. Benvenuti, "Characteristics, advantages, and possible applications of condensation cryopumping", *J. Vac. Sci. Technol.* **11** (3), pp. 591–598 (1974).
*2-54. H. J. Halama and J. R. Aggus, "Measurements of adsorption isotherms and pumping speed of helium on molecular sieve in the 10^{-11}–10^{-7} Torr range at 4.2 K", *J. Vac. Sci. Technol.* **11** (1), pp. 333–336 (1974).
*2-55. G. Mongodin, V. R. Piacentini, and W. Sajnacki, "Evaluation of a liquid-helimu cryopumping system operating with an electron beam gun", *J. Vac. Sci. Technol.* **11** (1), pp. 340–343 (1974).
*2-56. G. Klipping, "Relations between cryogenics and vacuum technology up to now and in the future", *Proc. 6th Internl. Vacuum Congr. 1974, Japan J. Appl. Phys.* Suppl. 2, Pt. 1, 1974, pp. 81–88.
*2-57. P. Kleber, "Pressure, what does it mean in vacuum chambers with cryosurfaces?", *Vacuum* **25** (5), pp. 191–196 (1975).
*2-58. J. Visser and J. J. Scheer, "Twenty-kelvin cryopumping in magnetron sputtering systems", *J. Vac. Sci. Technol.* **16** (2), pp. 734–737 (1979).
*2-59. C. F. Dillow and J. Palacios, "Cryogenic pumping of helium, hydrogen, and a 90% hydrogen-10% helium mixture", *J. Vac. Sci. Technol.* **16** (2), pp. 731–733 (1979).
*2-60. I. Arakawa, M. Kobayashi, and Y. Tuzi, "Effects of thermal spikes on the characteristics of cryosorption pumps with condensed carbon dioxide layers", *J. Vac. Sci. Technol.* **16** (2), pp. 738–740 (1979).
*2-61. C. Benvenuti and M. Firth, "Improved version of the CERN condensation cryopump", *Vacuum* **29** (11/12), pp. 427–432 (1979).
*2-62. R. A. Haefer, "On the pumping speed of large area cryopumps", *Vacuum* **30** (1), pp. 19–22 (1980).
*2-63. J. L. Hemmerrich and E. B. Deksnis, "Pumping efficiencies of three-dimensional cryopumping structures", *J. Vac. Sci. Technol.* **19** (1), pp. 96–99 (1981).
*2-64. B. A. Hands, "Recent development in cryopumping", *Vacuum* **32** (10/11), pp. 603–612 (1982).
*2-65. R. A. A. Kubiak, W. Y. Leong, R. M. King, and E. H. C. Parker, "On baking a cryopumped UHV system", *J. Vac. Sci. Technol. A* **1** (4), pp. 1872–1873 (1983).
*2-66. M. Kasuya and J. Yuyama, "An ultrahigh vacuum application of a refrigerator-cooled cryopump", *J. Vac. Sci. Technol. A* **8** (4), pp. 3333–3336 (1990).
*2-67. N. Gotoh, T. Momose, H. Ishimaru, and M. P. Larin, "Liquid helium cryopumps with low-emissivity Al film coatings and low helium consumption", *J. Vac. Sci. Technol. A* **13** (5), pp. 2579–2581 (1995).
*2-68. M. M. Menon, G. J. Laughon, R. Maingi, M. R. Wade, D. L. Hillis, and M. A. Mahdavi, "Pumping characteristics of a cryopump with Ar sorbent in He and in a D_2/He mixture", *J. Vac. Sci. Technol. A* **13** (3), pp. 551–555 (1995).

*2-69. M. Xu and Y. Matsui, "Obtaining extremely high vacuum using sintered fine copper powder as a cryosorbent", *J. Vac. Sci. Technol. A* **13** (1), pp. 132–135 (1995).
*2-70. R. Giannantonio, M. Succi, and C. Solcia, "Combination of a cryopump and a non-evaporable getter pump in applications", *J. Vac. Sci. Technol. A* **15** (1), pp. 186–191 (1997).
*2-71. S. Nesterov, J. Vasiliev, L. C. Wagner, and M. Boiarski, "Hydrogen pumping simulation for cryopumps", *J. Vac. Sci. Technol. A* **17** (4), pp. 2099–2103 (1999).
*2-72. E. Wallén, "Adsorption isotherms of He and H_2 at liquid He temperatures", *J. Vac. Sci. Technol. A* **15** (2), pp. 265–274 (1997).
*2-73. J. E. de Rijke, "Factors affecting cryopump base pressure", *J. Vac. Sci. Technol. A* **8** (3), pp. 2778–2781 (1990).
*2-74. G. S. Venuti, "Use of vibration-isolated cryopumps to improve electron microscopes and electron beam lithography units", *J. Vac. Sci. Technol. A* **1** (2), pp. 237–240 (1983).

Sputter Ion Pump

*2-75. S. L. Rutherford, S. L. Mercer, and R. L. Jepsen, "On pumping mechanisms in getter-ion pumps employing cold-cathode gas discharges", *1960 7th National Symposium on Vacuum Technology Transactions* (Pergamon Press, New York, 1961), pp. 380–382.
*2-76. R. L. Jepsen, "Magnetically confined cold-cathode gas discharges at low pressures", *J. Appl. Phys.* **32** (12), pp. 2619–2626 (1961).
*2-77. K. Ohara, I. Ando, and N. Yoshimura, "Pumping characteristics of sputter ion pumps with high magnetic fields in an ultrahigh vacuum range", *Shinku (J. Vac. Soc. Japan)* **35** (6), pp. 567–573 (1992) (in Japanese).
*2-78. H. Hirano and N. Yoshimura, "A three-point-pressure method for measuring the gas-flow rate through a conducting pipe", *Shinku (J. Vac. Soc. Japan)* **30** (6), pp. 531–537 (1987) (in Japanese).
*2-79. W. M. Brubaker, "A method for greatly enhancing the pumping action of a Penning discharge", *1959 6th National Symposium on Vacuum Technology Transactions* (Pergamon Press, New York, 1960), pp. 302–306.
*2-80. E. H. Hirsch, "On the mechanism of the Penning discharge", *Brit. J. Appl. Phys.*, **15**, pp. 1535–1543 (1964).
*2-81. W. Bächler, "Optimale Ausnutzung des Magnetfeldes bei Ionen-Zerstäuberpumpen", *Transactions 3rd Internl. Vacuum Congr.* Vol. 2, Part II (1965), pp. 609–612.
*2-82. A. Dallos and F. Steinrisser, "Pumping speeds of getter-ion pumps at low pressures", *J. Vac. Sci. Technol.* **4** (1), pp. 6–9 (1967).
*2-83. H. Roth und B. Wenzel, "Zur Dimensionierung von Ionenzerstäuderpumpen rotationssymmetrischer Bauart", *Vakuum-Technik* **16** (1/2), pp. 1–8 (1967).
*2-84. W. Schuurman, "Investigation of a low pressure Penning discharge", *Physica* **36**, pp. 136–160 (1967).
*2-85. A. Dallos, " The pressure dependence of the pumping speed of sputter ion pump", *Vacuum* **19**, pp. 79–80 (1969).
*2-86. T. Tom and B. D. James, "Inert gas ion pumping using differential sputter yield cathodes", *J. Vac. Sci. Technol.* **6** (2), pp. 304–307 (1969).
*2-87. V. V. Ryabov and G. L. Saksagansky, "Influence of asymmetry of magnetic and electric fields on the parameters of sputter-ion pumps", *Vacuum* **22** (5), pp. 191–193 (1972).
*2-88. H. Hartwig and J. S. Kouptsidis, "A new approach for computing diode sputter-ion pump characteristics", *J. Vac. Sci. Technol.* **11** (6), pp. 1154–1159 (1974).
*2-89. W. Ho, R. K. Wang, and T. P. Keng, "Calculation of sputtering ion pump speed", *J. Vac. Sci. Technol.* **20** (4), pp. 1010–1013 (1982).
*2-90. M. Pierini and L. Dolcino, "A new sputter ion-pump element", *J. Vac. Sci. Technol. A* **1** (2), pp. 140–142 (1983).
*2-91. M. Pierini, "Use of discharge intensity for evaluation of pumping characteristics of a sputter ion punp", *J. Vac. Sci. Technol. A* **2** (2), pp. 195–197 (1984).

*2-92. M. Audi, "Pumping speed of sputter ion pumps", *Vacuum* **38** (8–10), pp. 669–671 (1988).
*2-93. Y. Yamazaki, M. Miyoshi, T. Nagai, and K. Okumura, "Development of the field emission electron gun integrated in the sputter ion pump", *J. Vac. Sci. Technol. B* **9** (6), pp. 2967–2971 (1991).
*2-94. T. S. Chou, J. Bittner, and J. Schuchman, "Pumping behavior of sputtering ion pump", *J. Vac. Sci. Technol. A* **10** (4), pp. 2639–2641 (1992).
*2-95. H. Kinpara, K. Hirasawa, T. Kotani, M. Nishiyama, K. Hujino, K. Nakajima, N. Takagi, H. Yamakawa, and G. H. Shen, "Generation of extremely high vacuum with a new sputter ion pump", *Shinku (J. Vac. Soc. Japan)* **37** (9), pp. 732–735 (1994) (in English).
*2-96. T. Koizumi, Y. Hayashi, and H. Horiuchi, "Operating characteristics of XHV sputter ion pump", *Shinku (J. Vac. Soc. Japan)* **37** (9) pp. 736–741 (1994) (in English).
*2-97. Y. Suetsugu, "Proposal of a sputter-ion pump structure", *J. Vac. Sci. Technol. A* **12** (1), pp. 248–250 (1994).
*2-98. Y. Suetsugu, "Test results for a sputter-ion pump with a new cell structure", *J. Vac. Sci. Technol. A* **12** (6), pp. 3224–3227 (1994).
*2-99. S. Komiya, N. Takahashi, and M.-S. Xu, "A sputter ion pump to maintain fresh and active cathode surface", *Shinku (J. Vac. Soc. Japan)* **38** (3), pp. 125–127 (1995) (in Japanese).
*2-100. N. Takahashi, M.-S. Xu, and S. Komiya, "Pumping speed of cylindrical magnetron sputter ion pump for Ar, CH_4, He, H_2", *Shinku (J. Vac. Soc. Japan)* **38** (3), pp. 259–261 (1995) (in Japanese).
*2-101. M.-S. Xu, N. Takahashi, and S. Komiya, "Production of extreme high vacuum by using a cylindrical magnetron sputter ion pump", *Shinku (J. Vac. Soc. Japan)* **38** (3), pp. 271–273 (1995) (in Japanese).
*2-102. T. Asamaki, T. Taniguchi, T. Fukaya, A. Kudoh, and K. Yamamoto, "Large-Scale coaxial magnetron discharge containing magnets at extremely high vacuum and its application to sputter ion pump", *Shinku (J. Vac. Soc. Japan)* **38** (1), pp. 17–21 (1995) (in English).
*2-103. H. C. Hseuh, W. S. Jiang, and M. Mapes, "Pumping behavior of ion pump elements at high and misaligned magnetic fields", *J. Vac. Sci. Technol. A* **13** (3), pp. 531–535 (1995).
*2-104. H. C. Hseuh, L. Snydstrup, M. Mapes, and C. Pai, "Beam vacuum system of Brookhaven's muon storage ring", *J. Vac. Sci. Technol. A* **14** (3), pp. 1237–1241 (1996).
*2-105. Y. Suetsugu and M. Nakagawa, "Design study of new distributed ion pumps for the TRISTAN accumulation ring", *Vacuum* **42** (10/11), pp. 625–634 (1991).

Ti Sublimation Pump

*2-106. J. W. Reichardt, "The kinetics of the hydrogen-Titanium reaction", *J. Vac. Sci. Technol.* **9** (1), pp. 548–551 (1972).
*2-107. T. Arai, K. Takeuchi, and Y. Tuzi, "Temperature dependence of the pumping characteristics of titanium getter pump in ultrahigh vacuum", *Shinku (J. Vac. Soc. Japan)* **38** (3), pp. 262–265 (1995) (in Japanese).

Non-evaporable Pump

*2-108. C. Benvenuti and F. Francia, "Room-temperature pumping characteristics of a Zr-Al nonevaporable getter for individual gases", *J. Vac. Sci. Technol. A* **6** (4), pp. 2528–2534 (1988).
*2-109. C. Benvenuti and F. Francia, "Room-temperature pumping characteristics for gas mixtures of a Zr-Al nonevaporable getter", *J. Vac. Sci. Technol. A* **8** (5), pp. 3864–3869 (1990).
*2-110. L. Rosai, B. Ferrario, and P. della Porta, "Behavior of Sorb-ac wafer pumps in plasma machines", *J. Vac. Sci. Technol.* **15** (2), pp. 746–750 (1978).

*2-111. C. Boffito, B. Ferrario, P. della Porta, and L. Rosai, "A nonevaporable low temperature activatable getter material", *J. Vac. Sci. Technol.* **18** (3), pp. 1117–1120 (1981).

*2-112. H. C. Hseuh and C. Lanni, "Evaluation of Zr-V-Fe getter pump for UHV system", *J. Vac. Sci. Technol. A* **1** (2), pp. 1283–1287 (1983).

*2-113. T. A. Giorgi, B. Ferrario, and B. Storey, "An updated review of getters and gettering", *J. Vac. Sci. Technol. A* **3** (2), pp. 417–423 (1985).

*2-114. M. Audi, L. Dolcino, F. Doni, and B. Ferrario, "A new ultrahigh vacuum combination pump", *J. Vac. Sci. Technol. A* **5** (4), pp. 2587–2590 (1987).

*2-115. E. Giorgi, C. Boffito, and M. Bolognesi, "A new Ti-based non-evaporable getter", *Vacuum* **41** (7–9), pp. 1935–1937 (1990).

*2-116. C. Benvenuti and F. Francia, "Room temperature pumping characteristics for gas mixtures of a Zr-Al nonevaporable getter", *J. Vac. Sci. Technol. A* **8** (5), pp. 3864–3869 (1990).

*2-117. R. M. Caloi and C. Carretti, "Getters and gettering in plasma display panels", *J. Vac. Sci. Technol. A* **16** (3), pp. 1991–1997 (1998).

*2-118. F. Watanabe and A. Kasai, "Entrapment pump: Noble gas pump for use in combination with a getter pump", *J. Vac. Sci. Technol. A* **17** (5), pp. 3103–3107 (1999).

*2-119. R. Giannantonio, P. Manini, F. Mazza, D. Dominoni, A. Clozza, and L. Zanin, "Design and characterization of high capacity nonevaporable getter pumps embedded inside the interaction regions of DAΦNE", *J. Vac. Sci. Technol. A* **17** (4), pp. 2093–2098 (1999).

*2-120. S. R. In and S. H. Be, "Pumping characteristics of Zr-V-Fe nonevaporable getter in mixture of H_2 and CO gases", *Shinku (J. Vac. Soc. Japan)* **37** (1), pp. 5–9 (1994).

Measurement of Pumping Speeds

*2-121. B. B. Dayton, "The measured speed of an 'ideal pump'", *Vacuum* **15** (2), pp. 53–57 (1965).

*2-122. W. Steckelmacher, "The measurement of the speed of pumps", *Vacuum* **15** (5), pp. 249–251 (1965).

*2-123. R. Buhl and E. A. Trendelenburg, "Avoiding systematic errors in measuring the pumping speed of high vacuum pumps", *Vacuum* **15** (5), pp. 231–237 (1965).

*2-124. W. Steckelmacher and D. Turner, "Test header arrangements for determining the speed of pumps using an orifice type impact pressure transducer", *Sci. Instrum.*, **43** (5), pp. 893–898 (1966).

*2-125. D. R. Denison and E. S. McKee, "A comparison of pumping speed measurement methods", *J. Vac. Sci. Technol.* **11** (1), pp. 337–339 (1974).

*2-126. D. R. Denison, "Monte Carlo design of a 3-gauge test dome", *J. Vac. Sci. Technol.* **12** (1), pp. 548–550 (1975).

*2-127. F. Yu-guo and Xu. Ting-wei, "The appropriate test domes for pumping speed measurement", *Vacuum* **30** (10), pp. 377–382 (1980).

*2-128. J. K. N. Sharma and D. R. Sharma, "Study of molecular flow inside a test dome connected to the same diameter diffusion pump", *Vacuum* **32** (5), pp. 283–291 (1982).

*2-129. J. K. N. Sharma and D. R. Sharma, "Comparison of two pumping speed measuring methods of oil diffusion pumps", *Vacuum* **32** (5), pp. 253–256 (1982).

*2-130. J. K. N. Sharma and D. R. Sharma, "Gas beaming effect: measurement of pumping speed and determination of effective conductance", *Vacuum* **34** (6), pp. 653–657 (1984).

*2-131. J. M. F. Dos Santos, "Throughput determination in pumping speed measurements: comparison of two methods", *Vacuum* **38** (7), pp. 541–542 (1988).

Chapter 3
Simulation of Pressures in High-Vacuum Systems

In electron microscopes, the pressure at the objective lens pole-piece gap, inside which the specimen is located, cannot be measured. Also, for the electron beam accelerating tube of high voltage, the pressure at the top of the tube cannot be measured, where the electron emitter is located.

In order to design the evacuation system of an electron microscope, one must find the optimum dimensions of evacuation pipes and an orifice that make differential evacuation effective. Also, one must estimate the pressures in the vicinity of the specimen and the electron emitter.

Pressures at various locations depend on many factors such as the outgassing rates of many constituent elements, conductance of evacuation pipes, and pressures at other locations. Pressures in such systems can be analyzed using a vacuum circuit, as presented by Yoshimura et al. (1985) [3A-1].

There is another kind of pressure distribution in a spatial chamber, the gas-flow pattern. The specimen in an electron microscope, located just above the objective-lens (OL) pole-piece, is subjected to the beaming gas molecules which come from the OL-pole-piece bore. The pressure distribution due to beaming gas molecules has been analyzed for typical pipes [3C-1].

Conventional Calculation of System Pressures

Let us consider the simplest evacuation system, where a chamber of outgassing rate Q (Pa·L/s) is evacuated by a pump with speed S (L/s) and intrinsic ultimate pressure P_U (Pa) through a pipe of conductance C (L/s).

The pressure in a chamber is conventionally calculated as follows:

$$P = P_U + \frac{Q}{S_e}, \tag{3.1}$$

where S_e is the effective pumping speed at the evacuation port of the chamber. The effective speed S_e is expressed as

$$\frac{1}{S_e} = \frac{1}{C} + \frac{1}{S} = \frac{C+S}{C \times S}. \tag{3.2}$$

As a result,

$$P = P_U + \frac{(C+S)\,Q}{C \times S}. \tag{3.3}$$

However, the conventional calculation of pressure represented by the above equations contains the following questions.

(1) The net outgassing rate Q of the chamber wall is treated as independent of pressure. However, as Dayton [4-1] (in chapter 4) described, Q is varied depending on the system pressure P.
(2) The vacuum pump is assumed always to pump out the gases in the chamber. However, in some cases, vacuum pumps sometimes act as a gas source for UHV chambers which had been degassed by high-temperature bakeout [3A-5].
(3) The distributed outgassing along an evacuation pipe is ignored in the conventional calculation of pressure.
(4) For a complicated high vacuum system with many evacuation pipes, mathematical calculation becomes very difficult to obtain pressures.

System analysis based on the vacuum circuit is indeed very powerful in system analysis.

3A Vacuum Circuits

For designing a vacuum circuit of a high-vacuum system, the characteristic values of pumping elements and outgassing elements must be clarified.

Yoshimura (1985) [3A-1] presented the basic concept and design of vacuum circuits.

Basic Concept of Vacuum Circuits

Characteristic Values of a Solid Material as an Outgassing Source [3A-1]
The net outgassing rate Q of a chamber wall is expressed as the difference between the free outgassing rate Q_0 and the free sorption rate Q_S as [1]

$$Q = Q_0 - Q_S. \tag{3A.1}$$

When $Q = Q_0 - Q_S < 0$, the chamber wall shows a net pumping function. The effective function of the chamber wall at a pressure P could be represented by the pressure generator with an internal pressure P_X and an internal flow resistance R_X, applied to a field of pressure P, as shown in **Fig. 3A.1** [2]. The internal pressure P_x and the flow resistance R_x are characteristics of the chamber wall in respect of outgassing and pumping. The pressure P_x generally decreases with the elapsed time t after the chamber wall has been evacuated to high vacuum. On the other hand, the flow resistance R_x, representing the porosity of the wall, is considered to be constant, regardless of time t.

The basic concept of the net outgassing rate Q expressed by Eq. (3A.1) can be derived from the pressure-rise curve, as follows.

Let us consider a leak-free chamber of a unit volume consisting of homogeneous walls, which is evacuated by a pump to a sufficiently low pressure P_1. After the chamber has been isolated from the pump, the pressure P gradually rises with the elapsed time and finally reaches a saturated pressure P, as shown in **Fig. 3A.2 (a)**. The curve could be extrapolated to zero pressure, as shown by a broken line. The gradient $(dP/dt)_p$ is the net outgassing rate Q of the chamber wall at P, and $(dP/dt)_0$ at zero pressure is called as the free outgassing rate Q_0 of the wall [1].

The recorded pressure P as a function of the elapsed time t could be assumed to be expressed as

$$P = P_X\left[1 - \exp\{(-Q_0/P_X)t\}\right]. \tag{3A.2}$$

Differentiating Eq. (3A.2) with respect to t,

$$(dP/dt)_p = Q_0 \exp\{(-Q_0/P_X)t\}. \tag{3A.3}$$

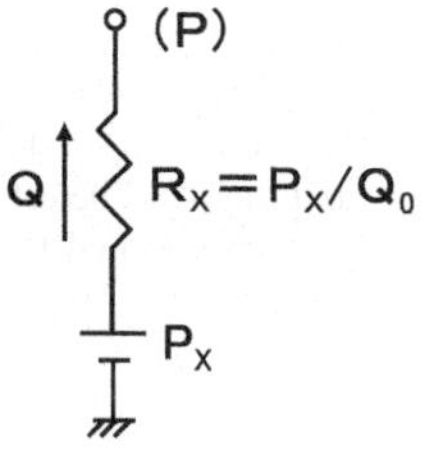

Fig. 3A.1 Pressure generator with P_x and R_x, representing outgassing or pumping for a vacuum field of P

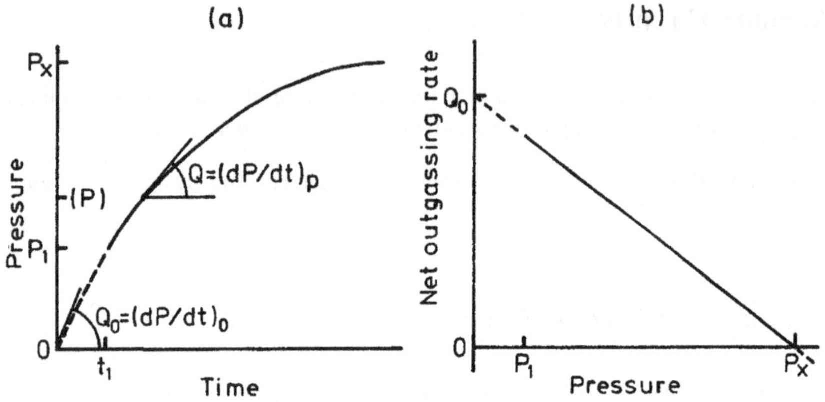

Fig. 3A.2 (**a**) Typical pressure-rise curve, and (**b**) net outgassing rate Q depending on pressure P (Yoshimura, 198) [3A-1]

From (3A.2) and (3A.3), Q defined as $(dP/dt)_p$ could be expressed as

$$Q = Q_0 (1 - P/P_X). \tag{3A.4}$$

The net outgassing rate Q as a function of pressure P is represented by a straight line, as shown in **Fig. 3A.2 (b)**. The extrapolated line cuts the horizontal P axis and the vertical Q axis at P_X and Q_0, respectively. The function of the chamber wall as a gas source could be characterized by P_X and Q_0. The wall effectively evolves gases in the pressure range lower than P_X, where the rate Q reduces with the increase of P. On the other hand, it effectively sorbs (the word "sorb" includes both "adsorb" and "absorb") gases in the range of pressure higher than P_X.
Equation (3A.4) can be rewritten as

$$Q = P_x/Z_x - P/Z_x, \tag{3A.5}$$

if Q_0 is replaced by P_X/Z_X.Z_X with a dimension of flow impedance is also a characteristic value of the wall. The first term of Eq. (3A.5) represents the free outgassing rate Q_0, whereas the second term means the free sorption rate Q_s at pressure P. The effective function of the chamber wall as a gas source at a pressure P could be represented by that of a pressure generator with internal pressure P_X and internal flow impedance Z_X for a field of P_X . The pressure generator introduces gases into the field of P at the net rate Q represented by the difference between the free outgassing rate Q_0 and the free sorption rate Q_s, as expressed by Eq. (3A.1).

Solid materials as a gas sources placed inside a vacuum chamber could be treated as a kind of walls, and so the net outgassing rate K per unit surface area of a solid material could be expressed by the following equation analogous to Eq. (3A.4):

$$K = K_0 (1 - P/P_X). \tag{3A.6}$$

The characteristic values, internal pressure P_x and free outgassing rate K_0 per unit surface area of a solid material depend on the history of the solid under vacuum. The values P_X and K_0 could be estimated by two different K values, K_1 and K_2, measured at two different pressures P_1 and P_2 as

$$P_X = (K_1 P_2 - K_2 P_1)/(K_1 - K_2) \tag{3A.7}$$

$$K_0 = (P_1 K_2 - P_2 K_1)/(P_1 - P_2). \tag{3A.8}$$

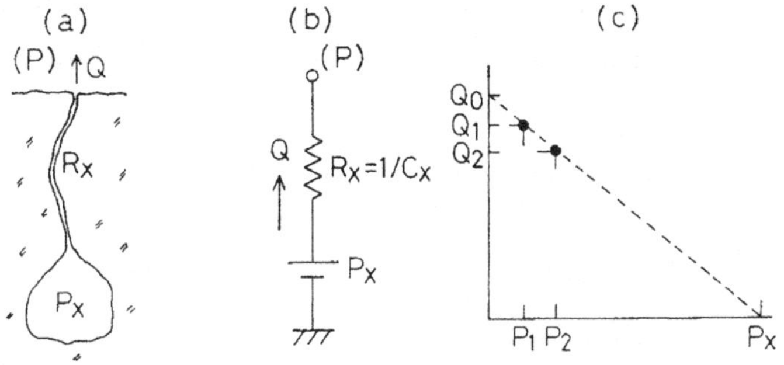

Fig. 3A.3 Concept of outgassing source. (**a**) Illustration of outgassing source, (**b**) Pressure generator as an outgassing source, and (**c**) The characteristic values P_X and Q_0 (Yoshimura, 1990) [3A-2]

References

1. B. B. Dayton, *AVS 6th Vacuum Symposium Transactions* (1959), p. 101.
2. S. Ohta, N. Yoshimura, and H. Hirano, *J. Vac. Sci. Technol. A* **1**, 84 (1983).

The concept of outgassing source described above is illustrated in **Fig. 3A.3** [3A-2], where the characteristic values of an outgassing source, P_X and Q_0 are defined graphically as (**c**) of the figure. Equation (3A.4), $Q = Q_0(1 - P/P_X)$, can be easily derived from **Fig. 3A.3**.

Designing of Vacuum Circuits

Vacuum Circuit Composed of the Characteristics Values of Constituent Elements [3A-1]

Now, net gas flows and pressures in a high-vacuum system under a steady state of pressure could be analyzed using a vacuum circuit composed of the characteristic values of constituent elements, as described below.

Let us consider the simple system of **Fig. 3A.4 (a)**, where a chamber with P_X and Q_0 is evacuated by a pump with ultimate pressure P_u and speed S through a pipe with P'_X, Q'_0 and conductance C. The vacuum circuit corresponding to the system (a) is obtained by converting the system elements into the circuit elements, as shown in (**b**).

Both P_X and P'_X of the walls are usually much higher than P_u of the pump and pressure P in the chamber. Therefore, the vacuum circuit can be simplified without an appreciable error under several conditions as follows.

When P'_X can be considered to be the same as P_X , the vacuum circuit (**b**) can be simplified to the circuit (**c**). If P_u is negligibly low, the circuit (**c**) can be further simplified to the circuit (**d**). If Q'_0 can be treated as zero, the circuit (**c**) can be simplified to the circuit (**e**). Finally, the circuit (**f**) is the simplest one acceptable under all the requirements described above. The circuits (**b**)–(**f**) can be analyzed by Kirchhoff's laws.

It should be noted that the vacuum circuit naturally involves the relationship among the effective functions of constituent elements as gas sources in an original systems. In a high vacuum system, the element with the lowest internal pressure P_X always works as an effective pump, whereas the element with the highest P_X always works as an effective gas

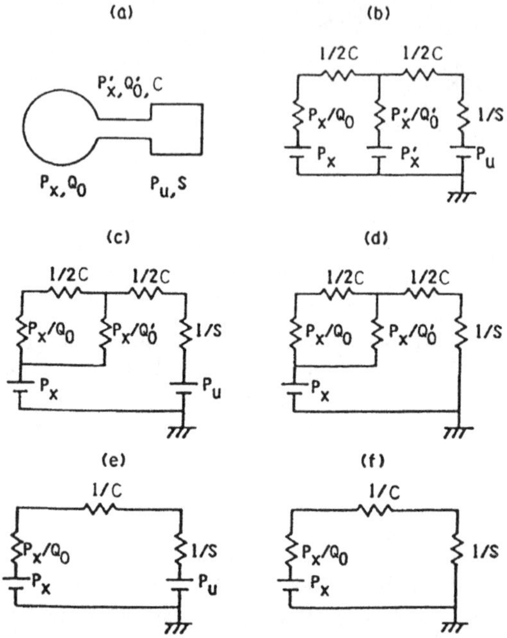

Fig. 3A.4 High-vacuum system and vacuum circuits. (**a**) Original system, (**b**) vacuum circuit corresponding to (**a**), and (**c**), (**d**), (**e**), and (**f**) simplified vacuum circuits under several conditions (Yoshimura, 198) [A-1]

source. The element with intermediate P_X works as a pump when its P_X is lower than the pressure P in the vacuum field, and works as a gas source when P_X is higher than P in the field. Pressures including internal pressures P_X of elements could be regarded as kinds of potential governing net gas flows in the entire high vacuum system.

Analyses on net gas flows and pressures must be conducted on the individual gas species separately since both the conductance of a pipe and pumping speed of a pump depend on the gas species. Partial pressures P_i in a high vacuum system are analyzed by a vacuum circuit for individual gas species "i". Total pressures P in a molecular flow region are obtained as $P = \sum_i P_i$.

References

1. B. B. Dayton, *AVS 6th Vacuum Symposium Transactions* (1959), p.101.
2. N. Yoshimura, *Applied Surface Science* (VASSCAA-1, *Proceedings of the First Vacuum and Surface Science Conference of Asia and Australia*) Vols. 169/170, 685 (2001).
3. S. Ohta, N. Yoshimura, and H. Hirano, *J. Vac. Sci. Technol. A* **1**, 84 (1983).

As seen above, the vacuum circuit of a high-vacuum system can be designed by replacing every outgassing element with the corresponding pressure generator with internal pressure P_X and internal flow resistance R_x, and by connecting the output terminals of pressure generators with gas-flow-route resistances. However, almost all available outgassing data are net outgassing rates Q measured in high vacuum. Therefore, one must estimate the characteristic values P_X and R_X from the equation $R_X \cong P_X/Q$ by assuming the P_X value. For a high-vacuum system, one can assume that the P_X of all outgassing elements is, for instance, 10^5 Pa, and the R_X of individual element becomes $R_X \cong 10^5/Q$ (in s/L) [3A-3]. For an ultrahigh-vacuum system, one can assume the P_X of all outgassing elements is, for instance, 1 Pa.

Note: There is another king of vapor sources, that is, "vapor pressure" of material elements. Its function as a gas source can be represented by the pressure generator with an internal pressure P_X (vapor pressure) and zero internal flow resistance, applied to a field of pressure P.

Simulation of Pressures

Ohta et al. (1983) [3A-3] presented a practical pressure-simulation technique using an equivalent electric circuit, and simulated the pressures in an electron microscope by measuring voltages at the corresponding positions of a simulator circuit, an equivalent electric-resistor network.

Resistor-Network Simulation Method

Simulation with Resistor-Network Simulator [3A-3]
Figure 3A.5 shows a typical example of an electron microscope with two pumps and five evacuation pipes. The simulation method was applied to the vacuum system of **Fig. 3A.5** to obtain a reasonable pressure distribution in the whole system. At the start of simulation, the conductance of every pipe and the outgassing rate of every part of the whole system were estimated. The outgassing rates of component materials were measured by the differential pressure-rise method [3A-1]. Typical rates measured under several specific conditions are tabulated in **Table 3.1**. The conditions in the table were selected so as to be consistent with

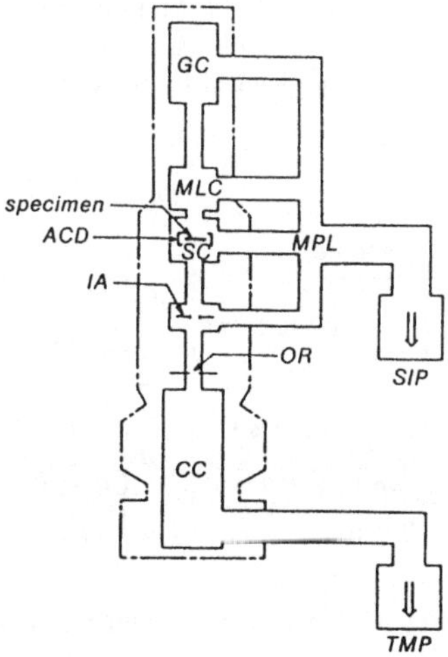

Fig. 3A.5 The high-vacuum system of an electron microscope. GC, gun chamber; MLC, minilab chamber; ACD, anti-contamination device; SC, specimen chamber; IA, intermediate aperture; OR, orifice; CC, camera chamber; MPL, main pumping line; SIP, sputter ion pump; TMP, turbomolecular pump (Ohto et al., 1983) [3A-3]

Table 3.1 Outgassing rates ($Pa \cdot L \cdot s^{-1} \cdot cm^{-2}$) of the component materials (Ohta et al., 1983) [3A-3]

Evacuation time	30 h following bakeout[a]	200 h	24 h
Stainless steel	3.3×10^{-8}	6.7×10^{-8}	
Steel, Ni coated	3.3×10^{-8}	6.7×10^{-8}	
Casting materials of camera chamber			1.3×10^{-6}
Alumina (fine in structure)		1.3×10^{-10}	
Viton-A O-ring[b]	1.3×10^{-7}	1.3×10^{-6}	1.3×10^{-5}
EM film			2.7×10^{-7}

[a] One-week vacuum bakeout at about 60 °C.
[b] One-week vacuum bakeout at about 100 °C before assembly.

the practical conditions in the vacuum system of this microscope. The amount of gas load at every part of the system was calculated using the outgassing-rate data in the table.

The typical procedure for simulation of the simplest system is shown in **(a)** to **(e)** of **Fig. 3A.6**, instead of the procedure for the electron microscope.

(a) First, let us estimate the outgassing rate Q, conductance C, and pumping speed S of the system as 10^{-3} Pa · L/s, $10L/s$, and 100L/s, respectively.

(b) Draw the corresponding vacuum circuit **(b)**. Note that the value P_Q is set to 10^5 Pa which is large enough as compared with the vacuum pressure.

(c) Convert the vacuum circuit into the corresponding electric circuit **(c)**, that is, Pa into V (volts) and s/L into Ω (Ohms).

In this circuit, the voltage generator of 10^5 V corresponds to the pressure generator of 10^5 Pa, and the electric resistors correspond to the vacuum resistors.

In the circuit **(c)** 10^5 V of the voltage generator is too large and 10^{-2} and $10^{-1}\Omega$ of the resistors are too small to assemble a practical simulator. Therefore, the following modifications are made as shown in **(d)** and **(e)**.

(d) Modify the electric circuit as follows: (1) Multiply the voltage of the generator 10^5 V by 10^{-4}. As a result, the reconverting factor k_1 from voltage V to pressure Pa becomes 10^4 Pa/V. (2) Multiply the resistances of all resistors in circuit **(c)** by 10^3.

It should be noted that the relative distribution of the voltage in the circuit **(d)** is not changed by these modifications. In this case the reconverting factor k_1 remains

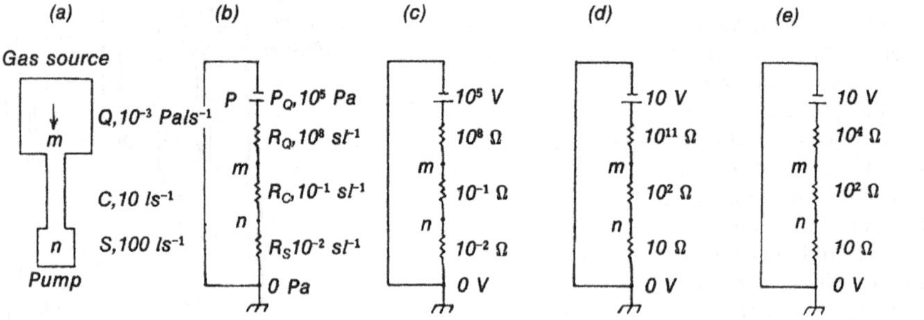

Fig. 3A.6 Steps for designing a simulator circuit of a vacuum system (Ohta et al., 1983) [3A-3]

10^4 Pa/V. In this circuit, however, 10^{11} Ω for the resistor is too large to assemble a simulator.

(e) Further modify the circuit into the simulator circuit **(e)** in which 10^4 Ω is used in place of $10^{11}\,\Omega$. The resistor of 10^4 Ω is easily obtainable.

Note that 10^4 Ω is still large enough compared with other resistances, 10^2 and 10Ω. So, the relative distribution of voltage in the circuit is kept constant with an error of one percent.

As a result, the reconverting factor k_2 from V to Pa in (e) is given by

$$k_2 = 10^{-7} \times k_1 = 10^{-3}\,\text{Pa/V}.$$

The procedure for designing the simulator circuit of the high-vacuum system of an electron microscope is the same as the one described above, although the actual high-vacuum system has many elements. **Figure 3A.7** shows the simulator circuit of the electron-microscope high-vacuum system, where the reconverting factor is 10^{-3} Pa/V.

In the simulator circuit of **Fig. 3A.7**, the outlined part corresponds to the whole column of the electron microscope. The resistors along the center line between the gun chamber (GC) and the camera chamber (CC) correspond to the resistance of an orifice, conducting pipes, apertures, etc. The resistors located on the left side of the outlined part correspond to

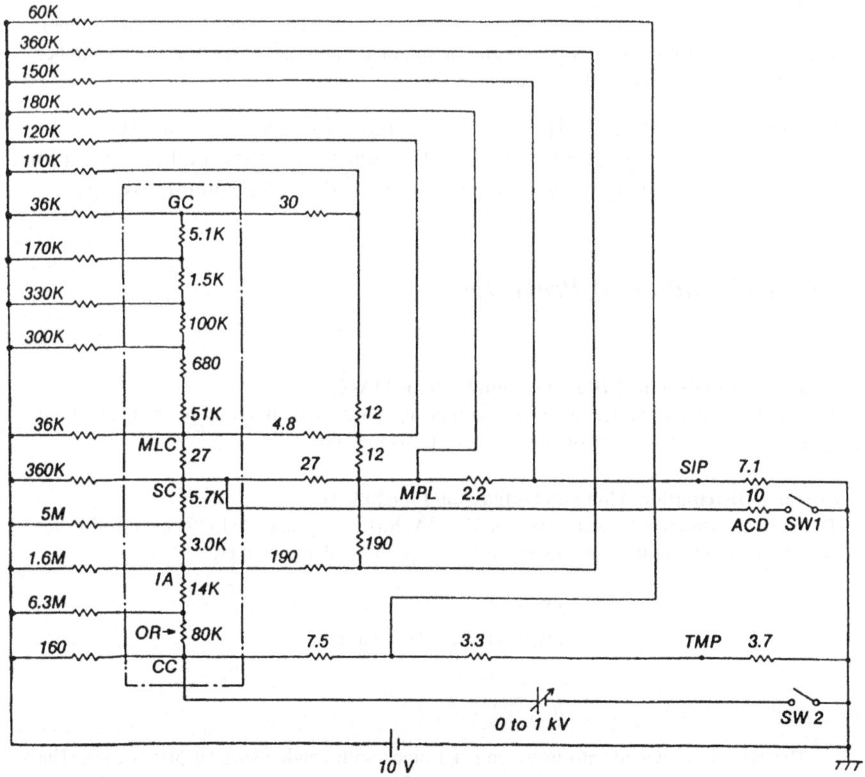

Fig. 3A.7 Circuit of the simulator for the electron-microscope high-vacuum system. The outlined part corresponds to the whole column. The resistors are represented in Ω and the symbols K and M used for resistance mean $\times 10^3$ and $\times 10^6$, respectively (Ohta et al., 1983) [3A-3]

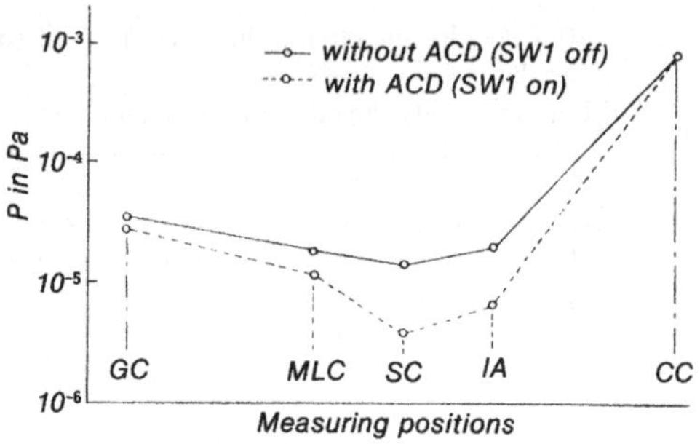

Fig. 3A.8 Pressure distribution in the electron-microscope column (Ohta et al., 1983) [3A-3]

the outgassing rates of each part of the whole column. The resistors located above the outlined part correspond to the outgassing rates of the evacuation pipes connecting the pumps to the column. These resistors related to the outgassing rates are to be connected to the voltage generator which supplies 10 V as a whole.

The 10 Ω resistor connected to switch 1 (SW1) corresponds to the cryo-pumping speed of an anti-contamination device (ACD) consisting of liquid-nitrogen-cooled fins. The pumping speed of this device is estimated to be 100 L/s. The voltage generator with a variable voltage range of 0–1 kV, connected to CC (camera chamber) through SW2, corresponds to variable pressures of 0–1 Pa.

Figure 3A.8 shows the simulated pressures at some positions in the microscope column. The pressure-distribution curves indicate the minimum pressure at the specimen chamber SC. This proves that the ACD is effective for the specimen chamber.

Hirano et al. (1988) [3A-4] described an algorithm for matrix analysis using a computer for analyzing vacuum circuits. The matrix-analysis method was applied to two high-vacuum systems, an outgassing pipe and an electron microscope.

Matrix Calculation of Pressures

Application to Practical High-Vacuum Systems [3A-4]
Two high-vacuum systems, an outgassing pipe and an electron microscope, were analyzed by a digital computer, based on the matrix analysis method.

Pressure Distribution Along an Outgassing Pipe [3A-4]
The pressures in an outgassing pipe of **Fig. 3A.9(a)** were calculated under the following conditions, assuming the same P_X regardless of position along the pipe:

$$P_X = 1 \times 10^{-4}\,\mathrm{Pa},$$

$$Q_0(\mathrm{or}Q) = 1 \times 10^{-4}\,\mathrm{Pa \cdot L/s},$$

$$C = 10\,\mathrm{L/s},$$

$$S = 100\,\mathrm{L/s}.$$

The pipe was divided into five elemental pipes with conductance of 50 L/s each. The vacuum circuit corresponding to the system of **Fig. 3A.9 (a)** is presented in **(b)**. The gas sources are represented by pressure sources with a conductance. Another vacuum circuit, assuming the same net outgassing rate regardless of the position along the pipe, is presented in **(c)**. The gas sources in this case are represented by the current sources.

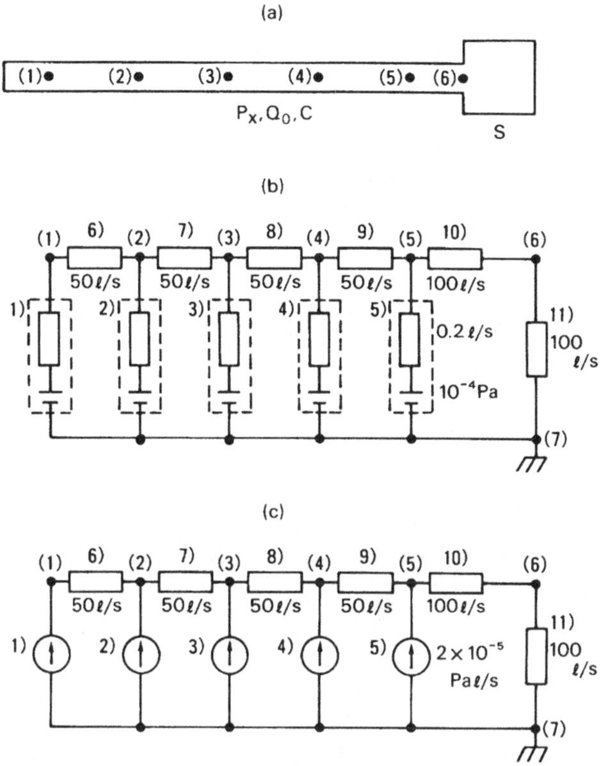

Fig. 3A.9 Outgassing pipe system (**a**), and the corresponding vacuum circuits (**b**) and (**c**). The circuit (**b**) is composed of pressure sources and flow resistance, and (**c**) of current sources and flow resistance (Hirano et al., 1988) [3A-4]

The pressures at six positions of the circuits of **Fig. 3A.9 (b)** and **(c)** were calculated, and the results are presented in **Fig. 3A.10**.

A pressure distribution along an outgassing pipe has been mathematically analyzed by Yoshimura and Hirano [3A-5] under the assumption that the net outgassing rate per unit surface area of the pipe is same regardless of position. The pressure P_k in the pipe is expressed by the equation

$$P_k = Q/S + k(1-k/2)\,Q/C, \tag{3A.9}$$

where k is the fraction showing the position, and Q is the net outgassing rate of the whole pipe. Substituting $1\times10^{-4}\,\text{Pa}\cdot\text{L/s}$, 100 L/s, and 10 L/s for Q, S and C, respectively, the pressure P_k for the system of **Fig. 3A.9 (a)** is expressed as

$$P_k = 1\times10^{-6} + k(1-k/2)\times10^{-5}\,\text{Pa}. \tag{3A.10}$$

The pressure distribution calculated using Eq. (3A.10) is also presented in **Fig. 3A.10** by a solid line.

Substituting 1 for k in the Eq. (3A.9) results in

$$P_1 = Q/S + Q/2C, \tag{3A.11}$$

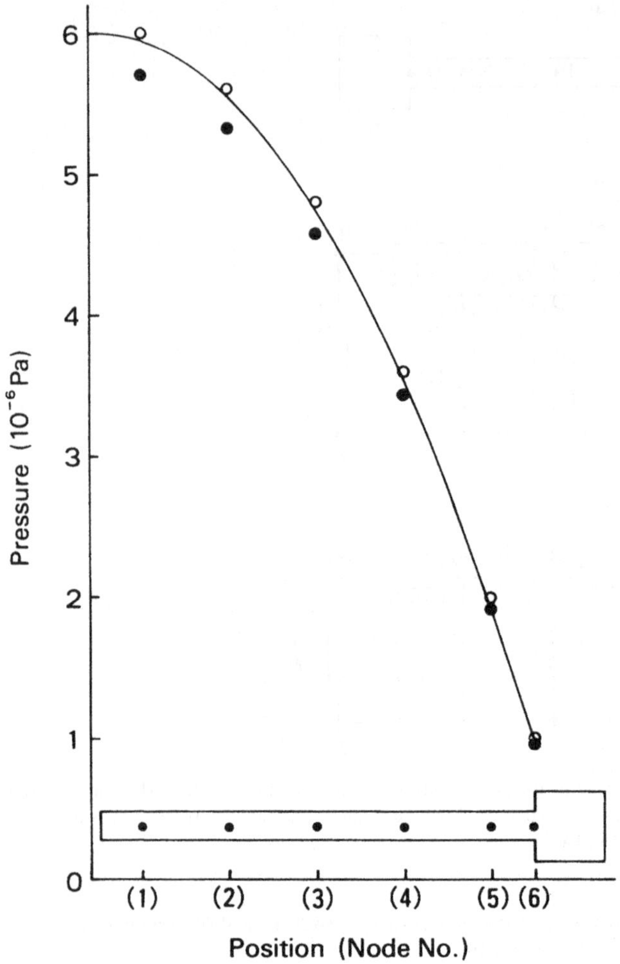

Fig. 3A.10 Pressure distribution along the outgassing pipe of **Fig. 3A.9(a)**. The points (•) correspond to the circuit **(b)**, and the points (∘) to the circuit **(c)**. The solid line shows the pressure distribution given by Eq. (3A.10) (Hirano et al., 1988) [3A-4]

One can see that the pressure drop across the pipe is half of the quotient, Q/C.

Comparing the three values (•, ∘, calculated line) at each position in **Fig. 3A.10**, one can find the following:

(1) There is a difference of about 5% between the pressures at each position in the circuits of **Fig. 3A.9 (b)** and **(c)**. This difference is due to whether or not the pressure dependence of outgassing is reflected in circuit analysis.
(2) The output node pressures in the circuit **(c)** show good agreement with the corresponding mathematically calculated pressures P_k (solid line) using Eq. (3A.10). Quite small difference will vanish if the number of elements into which the pipe is divided is increased sufficiently.

Pressures in an Electron-Microscope High-Vacuum System [3A-4]
A typical high-vacuum system composed of a pressure source P_0, outgassing sources Q and Q', conductance C_0 and C, and pumping speed S, which is analogous to an electron-microscope high-vacuum system, is presented in **Fig. 3A.11 (a)**. The chamber with gas load Q is connected to a field of pressure P_0 through an orifice with conductance C_0 . The vacuum circuit corresponding to the system of **(a)** is presented in **(b)**, where outgassing sources of Q and Q' are directly given as current sources, not pressure sources.

The vacuum circuit for the electron microscope of **Fig. 3A.5** has been designed using the outgassing rates, conductance of the pipes and pumping speeds of the pumps, as shown in **Fig. 3A.12**. Pressures in the vacuum circuit of the electron microscope have been successfully calculated using a personal computer.

Yoshimura (2000) [3A-6] presented an article, "Modeling of outgassing or pumping functions of the constituent elements such as chamber walls and high-vacuum pumps."

Abstract [3A-6]: Chamber walls, subjected to "in situ" baking, sometimes show a pumping function for a high vacuum, while a high-vacuum pump sometimes shows outgassing in an ultrahigh vacuum. Such functions of the system elements can be represented by the internal pressure P_X. All the system elements, such as chamber walls, pumps, and pinholes through a pipe wall, can be replaced by a pressure generator with the internal pressure P_X and the internal flow impedance R_X in the equivalent vacuum circuit. The internal pressure P_X of the chamber wall varies depending on the wall history under high vacuum. The equivalent vacuum circuit composed of many characteristic values (P_X, R_X) and flow impedances R can represent the gas flows in the original high-vacuum system.

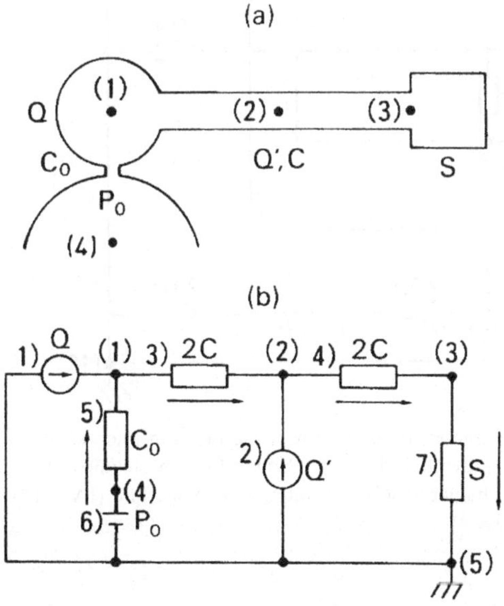

Fig. 3A.11 **(a)** High-vacuum system and **(b)** the corresponding vacuum circuit (Hirano et al., 1988) [3A-4]

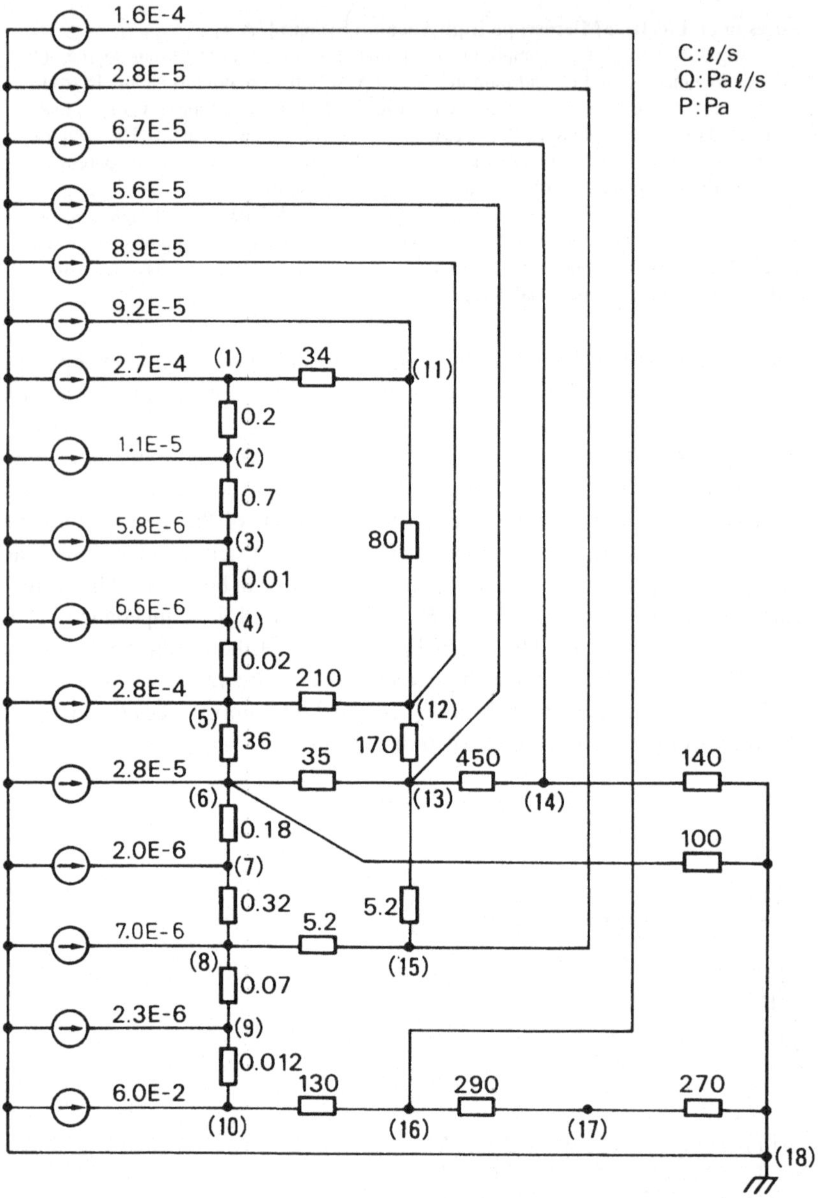

Fig. 3A.12 Vacuum circuit representing the high-vacuum system of the electron microscope of **Fig. 3A.5**. The input data are seen for the individual elements. Node (1), gun chamber (GC); (5), minilab chamber (MLC); (6), specimen chamber (SC); (8), intermediate aperture (IA); (10), camera chamber (CC) (Hirano et al., 1988) [3A-4]

Pressure Analysis Using Equivalent Vacuum Circuits [3A-6]

In an ultrahigh-vacuum system, the chamber wall, subjected to "in situ" baking, sometimes shows a pumping effect, and the vacuum pump sometimes shows an outgassing phenomenon. Alpert [1] reported on an interesting ultrahigh-vacuum system, where 2×10^{-10} Torr was kept in the "sealed-off" portion for a long period of time after the portion is isolated from the diffusion pump (DP), while the pressure in the portion connected to the DP, rose gradually and reached a saturated pressure of 1×10^{-7} Torr, as shown in **Fig. 3A.13**. Alpert [1] attributed the pressure rise with time to the gradual deterioration of the pumping function of the ion gauge, IG-B.

Narushima and Ishimaru [2] reported the "inverted pressure distribution" in an aluminum alloy pipe system, as shown in **Fig. 3A.14**. That is, after "in situ" baking, the

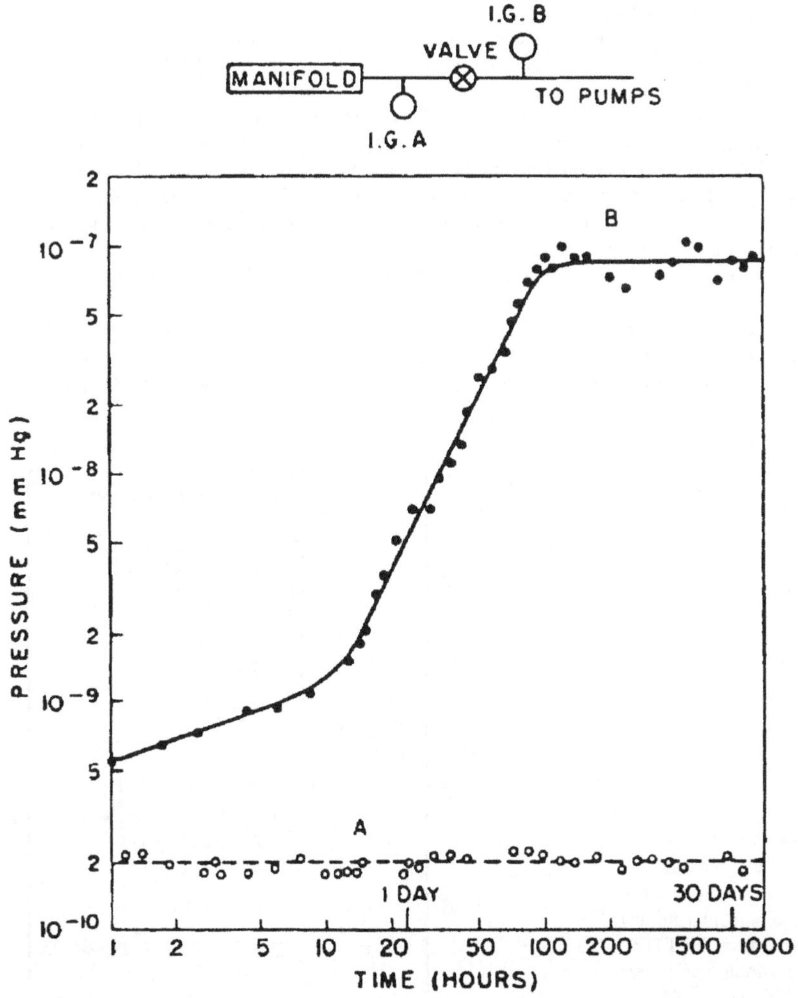

Fig. 3A.13 A: pressure vs. time in the sealed-off portion of vacuum system; B: pressure vs. time in the ion gauge in contact with the DP (diffusion pump), from Alpert [1] (Yoshimura, 2000) [3A-6]

pressure at the position far from the turbomolecular pump (TMP) became lower than the pressure near the TMP. They attributed the "inverted pressure distribution" to the pumping function of the pipe wall [2]. We shall here discuss on the "footprint" of pressure in **Fig. 3A.13** [1] and the "inverted pressure distribution" in **Fig. 3A.14** [2] by using the characteristic values of the system elements.

The pressure distributions shown in **Fig. 3A.13** and **Fig. 3A.14** can be described by the internal pressures of the system elements, which vary according to the history of the elements under vacuum. Consider the "sealed-off" portion isolated from the diffusion pump in **Fig. 3A.13** [1]. The manifold wall and the pipe wall were made of the same kind of glass and were both treated by in situ baking at 420 °C, and so the internal pressures of the respective walls would be likewise 2×10^{-10} Torr after in situ baking. The internal pressure P_{IG} of the ion gauge scarcely varied with time in the "sealed-off" portion, which might be a little lower than 2×10^{-10} Torr. The equivalent vacuum circuit for the "sealed-off" portion is presented in **Fig. 3A.15 (a)**.

Next, consider the portion connected to the diffusion pump. At a short elapsed time after in situ baking (420 °C), the internal pressure P_X of the pipe wall must be as low as 2×10^{-10} Torr. However, residual gas molecules, mainly coming from the DP, were sorbed into the pipe wall, and the P_X of the pipe wall gradually rose to a saturated pressure of 1×10^{-7} Torr, i.g. the ultimate pressure P_U of the DP used. The equivalent vacuum circuit corresponding to the portion connected to the diffusion pump, under the saturated condition, is presented in **Fig. 3A.15 (b)**, where the internal pressure P_X of the pipe and IG is the same as the P_U of the DP. There is not a gas flow in (b), which is recognized merely as a vacuum box of pressure P_U of 1×10^{-7} Torr.

Now, consider the "inverted pressure distribution" of **Fig. 3A.14** [2]. Assume that the pumping function of the Bayard-Alpert gauge was negligibly small. According to the proposed model of outgassing, the internal pressure of the pipe wall, during in situ baking, was higher than the ultimate pressure (the same as the internal pressure) of the TMP, and so the pressure at the position far from the TMP was higher than the pressure near the TMP.

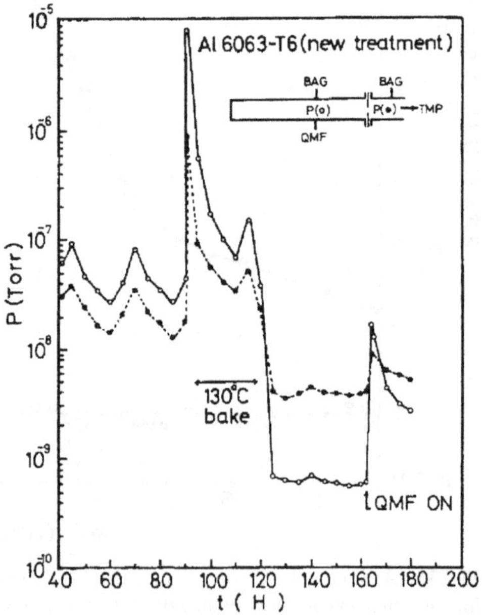

Fig. 3A.14 Pressure variation at 130 °C, 24 h bake. (○): pressures far from the TMP (turbomolecular pump); (•) pressures near the TMP; QMF: quadrupole mass filter, from Narushima and Ishimaru [2] (Yoshimura, 2000) [3A-6]

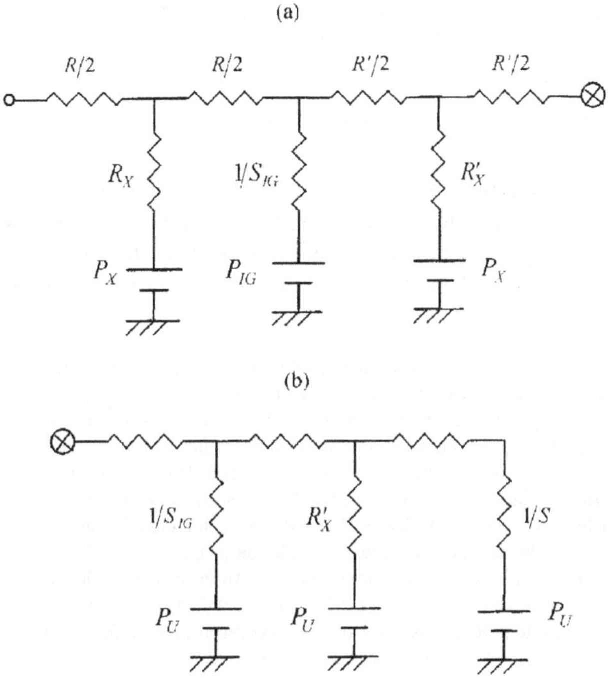

Fig. 3A.15 Equivalent vacuum circuits corresponding to the systems of Fig. 3A.13 at the saturated condition. In **(a)** $R = 1/F$, where F is the flow conductance of the manifold, and $R' = 1/F'$, where F' is the flow conductance of the pipe. $R_{IG} = 1/S_{IG}$, where S_{IG} is the pumping speed of IG-A. In **(b)** P_U and $1/S$ correspond to the characteristic values P_X and R_X of the pump, respectively (Yoshimura, 2000) [3A-6]

On the other hand, after in situ baking, the internal pressure of the pipe wall become lower than the ultimate pressure of the TMP, resulting in the pressure far from the TMP becoming lower than that near the TMP. The equivalent vacuum circuit corresponding to the system of **Fig. 3A.14** is presented in **Fig. 3A.16**.

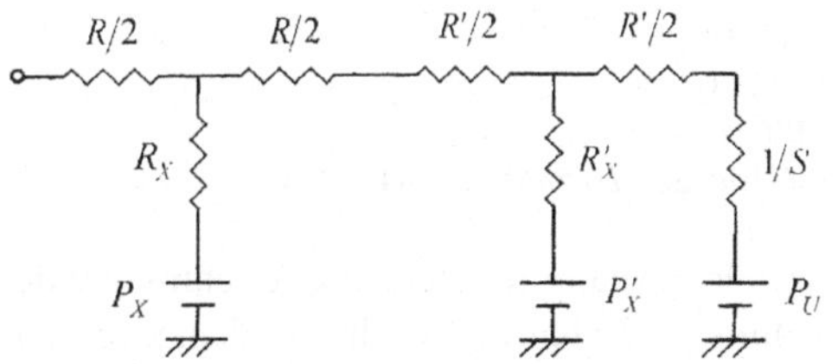

Fig. 3A.16 Vacuum circuit corresponding to the system of **Fig. 3A.14.** **(a)** $P_X, P'_X > P_U$, during "in situ" baking. **(b)** $P_X < P'_X < P_U$, variable with time, after in situ baking (Yoshimura, 2000) [3A-6]

References

1. D. Alpert, *J. Appl. Phys.* **24** (7) 860 (1953).
2. K. Narushima, H. Ishimaru, *Shinku (J. Vac. Soc. Japan)* **25** (4) 172 (1982).

Comment: Zajec and Nemani (2005) [4–27] recently presented an article, "Hydrogen pumping by austenitic stainless steel." The pumping function of the stainless steel surface can be explained by the relationship between its characteristic value P_X and the field pressure P. That is, when $P_X < P$, the surface shows the pumping function.

Kendall (1983) [3A-7] reviewed some works on pressure simulation by resistor networks, which have been developed during the past several decades as follows:

Molecular-flow network theory is based on the similarities between the equations describing current flow in electrical networks and the equations describing gas flow under pure molecular-flow conditions. The long history of the subject can be traced back to Dushman (1922) [1], who discussed the analogies between molecular-flow resistance, conductance, flow rate, and pressure on the one hand, and electrical resistance, conductance, current, and potential on the other. These ideas received limited circulation as early as 1922, and appeared again in 1949 in Dushman's well known textbook [2]. An early application of molecular-flow network theory was given by Aitken (1953) [3] in 1953. He described an electrical analog of the vacuum system of the Oxford 140 MeV synchrotron, complete with current sources which could be switched to various points to simulate real and virtual leaks. This was probably the first application of a large, complex vacuum system. A brief note by Stops (1953) [4] and a much more detailed paper by Teubner (1962) [5] extended the theory to include chambers interconnected by flow routes and to non-equilibrium flow conditions. The volumes of the chambers were represented by capacitances. Both writers used the concept of a pump as a resistive sink for gas molecules. Stops included the ultimate pressure of the pump by introducing a series electromotive force. These concepts reappear in the recent work by Ohta et al. [6] [3A-7].

References

1. S. Dushman, *Production and Measurement of High Vacuum* (General Electric, New York, 1922).
2. S. Dushman, *Scientific Foundations of Vacuum Technique*, 1st ed. (Wiley, New York, 1949). See also 2nd ed., 1962.
3. M. J. Aitken, *Brit. J. Appl. Phys.* **4**, 188 (1953).
4. D. W. Stops, *Brit. J. Appl. Phys.* **4**, 350 (1953).
5. W. Teubner, *Exp. Tech. Phys.* **10**, 279 (1962).
6. S. Ohta, N. Yoshimura, and H. Hirano, *J. Vac. Sci. Technol. A* **1**, 84 (1983).

Degras (1956) [3A-8] presented an article, "Analogies élémentaires entre le vide et l'électricité." Degras converted the volume of a chamber V, the conductance of a conduit F, the pumping speed of a pump S, the pressure P, and the ultimate pressure limit of a pump P_u into the capacitance of a condenser C, the conductance (the reciprocal of the resistance R) of a flow route $1/R$, the conductance of a pressure generator $1/r$, the voltage E, and the generating pressure of a generator E_1, respectively. That is,

V (volume) $\rightarrow$ C (capacitance)

F (conductance) $\rightarrow$ $1/R$ (R : resistance)

S (pumping speed) $\rightarrow$ $1/r$ (conductance of pressure generator)

P (pressure) $\rightarrow$ E (voltage)

P_u (ultimate pressure of pump) $\rightarrow$ E_u (generating pressure of generator).

Aitken (1953) [3A-9] made a short correspondence, "An electrical analogue to a high vacuum system."

An Electrical Analogue to a High Vacuum System [3A-9]

In a complex vacuum system where there are several pumping lines in parallel, or, where, as in synchrotrons and betatrons, the vacuum chamber is a toroidal tube of low pumping speed, the calculation of the effect on the pressure of altering various parts of the pumping system (e.g. speed or number of pumps, length of pumping lines) is rather tedious. So tool is the estimation of the whereabouts of a leak from observed pressure gauge readings on the system. An electrical analogue can be used to obtain the required answers very quickly by simple voltage measurements. The analogue is only applicable when the pressure is small enough for the mean free path to be larger than the dimensions of the system; the rate of mass flow through a tube is then given by,

$$Q = S(p_1 - p_2) \tag{3.12}$$

where S is the speed of the tube and depends only on its dimensions, p_1 and p_2 are the pressures at either end of the tube.

Similarly for a diffusion pump of speed S_0, at the throat of which the pressure is p_0, the rate of mass flow through the pump is,

$$Q = S_0 p_0. \tag{3.13}$$

Both equations (3.12) and (3.13) are analogous to Ohm's law, so that an electrical analogue can be constructed in which voltage corresponds to pressure, current to mass flow, and resistance to the reciprocal of speed. **Figure 3A.17** shows the analogue constructed for the vacuum system of the Oxford 140 MeV synchrotron.

The procedure for finding the locality of the leak is first to measure the pressures on the five gauges on the vacuum system. The probe leak on the analogue is then connected on to various points of the electrical network until one is found for which the ratios of the voltages at the points corresponding to the gauges is the same as the ratios of the actual gauge readings.

The permanent leak into the system is due to diffusion of air through the eight neoprene sleeves (see **Fig. 3A.17**) which join the eight sectors of the orbit tube together and the residual pressure due to this leak for various different pumping arrangements is found by connecting eight equal current sources (not shown) to the points corresponding to the sleeves; the electrical circuit is then altered to corresponding to the various arrangements and the voltages observed in each case. Pressure due to evolution of vapor (e.g. from the gun heater) which is condensed on the cold trap, can be dealt with by closing the three switches shown, connecting on the probe leak at the point corresponding to the gun heater and proceeding as before.

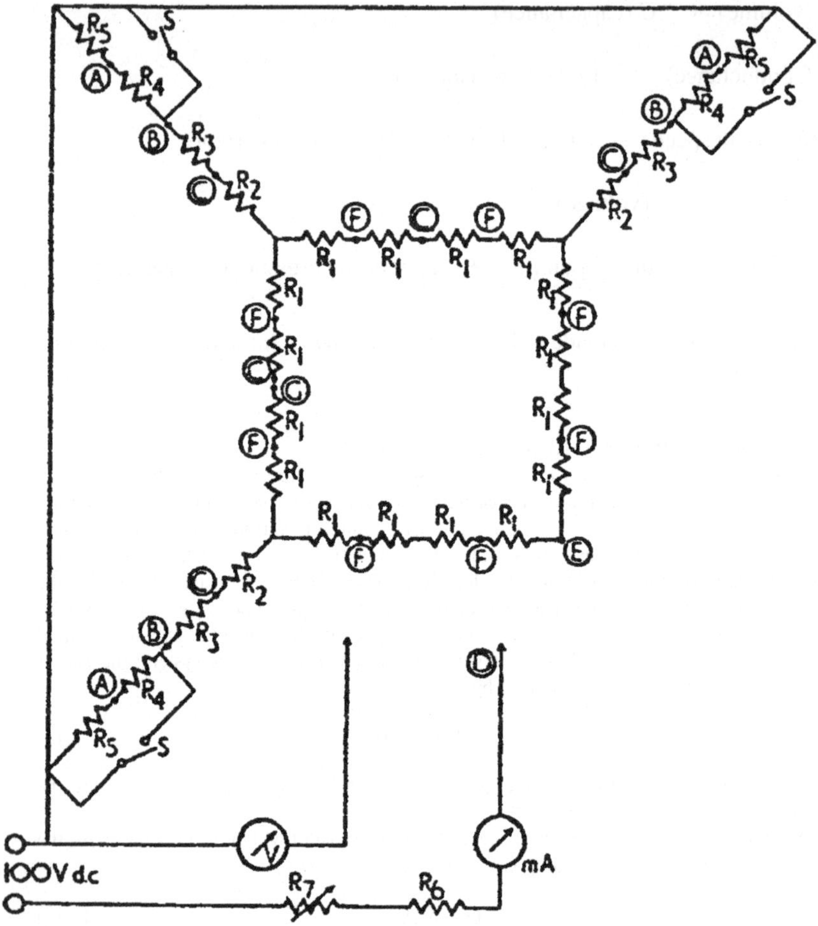

Fig. 3A.17 Electrical analogue R1 = 90 Ω, R2 = 170 Ω, R3 = 95 Ω, *R*4 = 6 Ω, R5 = 850 Ω, R6 = 25 *k*Ω, *R*7 = 100 *k*Ω V = voltmeter (f.s.d. 2.5 V), mA = milliameter (f.s.d. 4 mA), S = switches, A, pump throat; B, cold trap; C, gauge; D, probe leak; E, electron gun; F, neoprene sleeves; G, target (Aitken, 1953) [3A-9]

The analogue has been in use for the past year and has been found invaluable, particularly in the localization of leaks. There are fifteen places around the orbit tube at which leaks may occur and, by use of the analogue, suspicion can in a few minutes be centered on one of these places. With this localization procedure a straightforward method of leak finding (a jet of coal gas or butane in conjunction with a normal ionization gauge) has been found quite adequate [3A-9].

Stops (1953) [3A-10] made a short correspondence, "Further applications of the electrical analogue to vacuum systems."

Further Applications of the Electrical Analogue to Vacuum Systems [3A-10]
The analogous aspects of the vacuum systems and certain electrical circuits have long been recognized, but apparently like the application of analogue techniques to heat and mass flow

problems, the practical usefulness has taken some time to be appreciated. The letter from M. J. Aitken [1], has indicated that the electrical analogue may play an increasingly important part in the study and design of large vacuum systems, and to these ends, a transient response type would often prove useful.

Elementary considerations show that the ordinary type of vacuum system operating under free molecular gas flow conditions may be represented by an R-C electrical circuit with various sources of e.m.f.'s or currents. The related equations are:

$Q = \delta P/W = VdP/dt$ at constant volume.

$i = \delta v/R = Cdv/dt$ at constant capacity.

The analogous quantities are apparent from comparison of these two sets of equations.

A simple vacuum system, shown diagrammatically by the upper part of **Fig. 3A.18**, has a pump of speed S, lumped elements of volume V and tabulation resistances W with leaks Q' at various points. These leaks may be viscous in character and therefore not satisfy the linear pressure relation, but the changes in the gas pressure in the system will usually be so small relative to the external pressure P_A, that there will be little error in writing,

$Q_1 = (P_A - P_1)/W_1'$, etc.,

with W_1' as the effective resistance of the leak.

The lower part of **Fig. 3A.18** shows the analogous electrical circuit, with resistors R replacing the tubes, condensers C replacing the volumes and voltages v corresponding to the pressures. Measurements of voltages with time will give the corresponding variations of pressures with time, and likewise the various currents i will give the appropriate values of throughputs Q.

The scaling factors introduced into the values of the parameters will determine the relative time scales, which in turn will usually decide the mode of presentation of the information provided by the analogue. This subject has been discussed at length by Pashkis and Baker [2], Lawson and McGuire [3] and others. Lawson and McGuire have also examined mathematically the accuracy attainable with a lumped element analogue of a continuously distributed system, which will be relevant in the vacuum case when the system tabulation is also the main volume.

With the analogue envisaged, the effects of inserting pumps or pumping devices at various points in a vacuum line may be easily determined. By switching a metered quantity of charge into a point, the sudden evolution of known quantities of gas or vapor, may be simulated. Similarly, the response of a system to mixtures of gas and/or vapors may be found by varying the values of the parameters appropriate to each component.

References

1. M. J. Aitken, *Brit. J. Appl. Phys.*, **4**, p. 188, 1953.
2. V. Pashkis and H. O. Baker, *Trans. Amer. Soc. Mech. Engrs.*, **64**, p. 105, 1942.
3. D. I. Lawson and J. H. McGuire, Private communication.

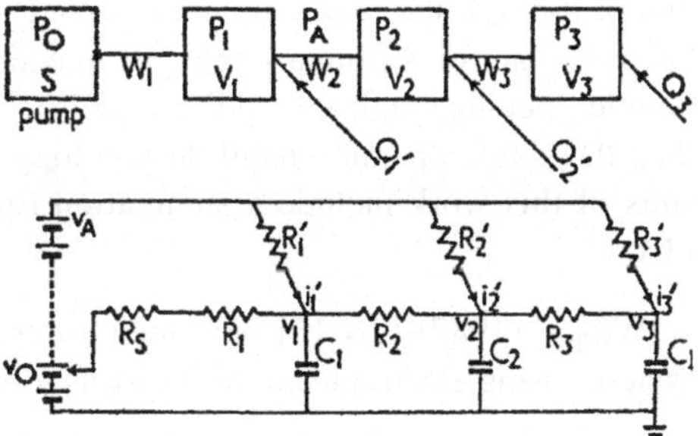

Fig. 3A.18 Analogous vacuum and electrical systems (Stops, 1953) [3A-10]

Kendall (1968) [3A-11] analyzed a vacuum system with two pumps by the corresponding vacuum circuit where one vacuum pump was treated as a gas source.

Abstract [3A-11]: Factors affecting the pressure in a vacuum chamber pumped by two dissimilar pumps are investigated with the aid of a simplified theoretical model. The possibility of one of the pumps acting as a significant source of gas or vapor is taken into account. It is shown that minimum chamber pressure may be obtainable only with finite pumping-line impedance in series with this pump. An experimental illustration of the phenomenon is given. Applications of the theory to the design of vacuum systems are discussed.

Kendall and Pulfrey (1969) [3A-12] presented an article, "Theory of pulsed molecular-flow networks."

Abstract [3A-12]: A general molecular-flow network theory applicable to volumes interconnected by small orifices or porous membranes is discussed. Parameters which describe the properties of such orifices and membranes are first defined. Basic theory is then extended to cover operation with mixtures of gases having different molecular weights. Characteristics of several networks with pulsed pressure inputs are then derived. It is shown that the mass-dependent resistance to flow of an orifice or porous membrane makes it possible to use certain molecular-flow networks as gas analyzers. Other networks are described which have potential applications in improving gauge response to rapid pressure fluctuations.

Ohta (1962) [3A-13] presented a vacuum circuit corresponding to a multiple-pump system, where vacuum pumps were converted to negative pressure generators of -760 Torr. He also presented the respective vacuum circuits corresponding to a long orbit accelerator and an instrument to make a clean, pure plasma.

Horikoshi et al. (1990) [3A-14] presented an article, "An analysis of a complex network of vacuum components and its application."

Abstract [3A-14]: An analysis of a complex network composed of many vacuum components was made by using an analogy between pipe conductance and electric conductance of circuit components. In the analysis, we have applied concepts of Kirchhoff's law and Green function. The Green function matrix introduced in the analysis gives effective pumping speeds for all connection points in the network. As an example, we applied the analytic method to a network, with a satisfactory result.

Theil (1995) [3A-15] applied a matrix method to a gas manifold system to acquire the pressure distribution. "The methodology for determining the flow distribution through the two simplest topologies of gas manifold, single- and double-opening manifolds from a single gas-injection point, is derived. It is shown that the double-opening manifold topology tends to provide more uniform flow distribution than the single-opening manifold topology for similar conductance ratios. The results of this work include a summation formula for the single-opening manifold [3A-15]."

Wilson (1987) [3A-16] presented an article, "Numerical modeling of vacuum systems using electronic circuit analysis tools."

Abstract [3A-16]: The analysis of large and complex high-vacuum systems is a tedious process, especially when distributed loads and pumps are involved. It has been suggested that by using the appropriate transformations, a vacuum system can be modeled as an electrical network. Well-developed numerical techniques exist for the computational characterization of electrical networks, often implemented in the form of a general purpose circuit simulation code. In this paper, a method is presented for the modeling of high-vacuum systems using a circuit analysis code. The method includes the capability to model outgassing, and to obtain the response of a system to transient and time-varying loads. An example is given of the analysis of a vacuum system with distributed pumps and distributed loads using the commonly available SPICE II program. The use of this technique is expected to significantly reduce the effort required to analyze such systems as accelerators, storage rings, and process lines.

Santeler (1987) [3A-17] presented an article, "Computer design and analysis of vacuum systems."

Abstract [3A-17]: A computer program has been developed for an IBM compatible personal computer to assist in the design and analysis of vacuum systems. The program has a selection of 12 major schematics with several thousand minor variants incorporating diffusion, turbomolecular, cryogenic, ion, mechanical, and sorption pumps as well as circular tubes, bends, valves, traps, and purge-gas connections. The gas throughput versus the inlet pressure of the pump is presented on a log-log graphical display. The conductance of each series component is sequentially added to the graph to obtain the net system behavior $Q_{(P)}$. The conductances component may be calculated either from the inlet area and the transmission probability or from the tube length and the diameter. The gas-flow calculations are valid for orifices, short tubes, and long tubes throughout the entire pressure range from molecular through viscous to choked and nonchoked exit flows. The roughing-pump and high-vacuum-pump characteristic curves are numerically integrated to provide a graphical presentation of the system pumpdown. Outgassing data for different materials is then combined to produce a graph of the net system "outgassing pressure." Computer routines are provided for differentiating a real pumpdown curve for system analysis. The computer program is included with the American Vacuum Society course, "Advanced Vacuum System Design and Analysis," or it may be purchased from Process Applications, Inc.

Reid (1992) [3A-18] presented an article, "Vacuum calculations for large systems."

Abstract [3A-18]: In designing large installations such as particle accelerators, it is essential to be able to predict vacuum conditions with some confidence. In particular an ultimate vacuum pressure distribution will often be required. To calculate this analytically is very difficult, but a number of approximation techniques using desk-top computers are available. Some of these techniques, particularly Monte Carlo techniques, finite-element models using matrix inversion techniques and the use of electrical circuit analysis packages such as SPICE will be discussed. Their application to some simple model structures will be illustrated.

3B Molecular-Flow Conductance

Molecular-flow conductance of various evacuation pipes must be calculated to design the vacuum circuits for high-vacuum systems.

DP-evacuation systems come with an opaque water-cooled baffle and liquid-nitrogen cold trap, both of which show gas-flow resistance. An ion shield constructed with opaque step plates may be inserted into an ion-pump evacuation pipe for electron microscopes, showing gas-flow resistance. The conductance of such items can be calculated using the diagrams of transmission probability presented by Davis (1960) [3B-1] and Levenson et al. (1960) [3B-2].

Conductance

Conductance of Orifice, C_O

R_0, gas constant per mole; T, temperature in K; M, molecular weight

$C_O = A\sqrt{\frac{R_0 T}{2\pi M}}$ where A is the area of the aperture.

$= 3.64A\sqrt{T/M}$ (L/s) where A is in cm^2.

$= 2.86d^2\sqrt{T/M}$ (L/s) where d is diameter in cm for circular cross-section.

$C_{O-air} = 11.6A$ (L/s) for air at 20 °C.

$= 9.16d^2$ (L/s) where d (diameter) is in cm for circular cross-section.

Conductance of Long Tube of Constant Circular Cross-Section, C_{LT}

$C_{LT} = 3.81\frac{d^3}{L}\sqrt{\frac{T}{M}}$ (L/s) where d is in cm and L (length) in cm.

$C_{LT-air} = 12.1d^3/L$ (L/s) for air at 20 °C

Conductance of Short Tube of Constant Circular Cross-Section, C_{ST}

$$\frac{1}{C_{ST}} = \frac{1}{C_{LT}} + \frac{1}{C_O} = \frac{1}{C_{LT}}\left(1 + \frac{C_{LT}}{C_O}\right).$$

$$C_{ST} = \frac{C_{LT}}{(1 + C_{LT}/C_O)} = C_{LT}\frac{1}{(1 + C_{LT}/C_O)}.$$

$C_{ST} = 3.81\frac{d^3}{L}\sqrt{\frac{T}{M}}\frac{1}{(1+\frac{4}{3}\frac{d}{L})}$ (L/s) where d is in cm and L in cm.

$C_{ST-air} = 12.1\frac{d^3}{L}\frac{1}{(1+\frac{4}{3}\frac{d}{L})}$ (L/s) for air at 20 °C.

Transmission Probability

The conductance F is derived by $F = F_O \times P$, where F_O is the conductance of the pipe opening.

Davis (1960) [3B-1] presented an article, "Monte Carlo calculation of molecular flow rates through a cylindrical elbow and pipes of other shapes."

Abstract [3B-1]: A method is devised for the calculation of molecular flow rates through pipes where the mean free path for intermolecular collisions is large compared to the dimensions of the pipes. Results of the calculation are given (in terms of the transmission probability P) for a straight cylindrical pipe, a cylindrical elbow, an annulus between two concentric cylinders, a straight cylindrical pipe with restricted openings, and a straight cylindrical pipe with restricted openings and a plate to block the direct beam between the openings (**Figs. 3B.1, 3B.2, 3B.3, 3B.4**).

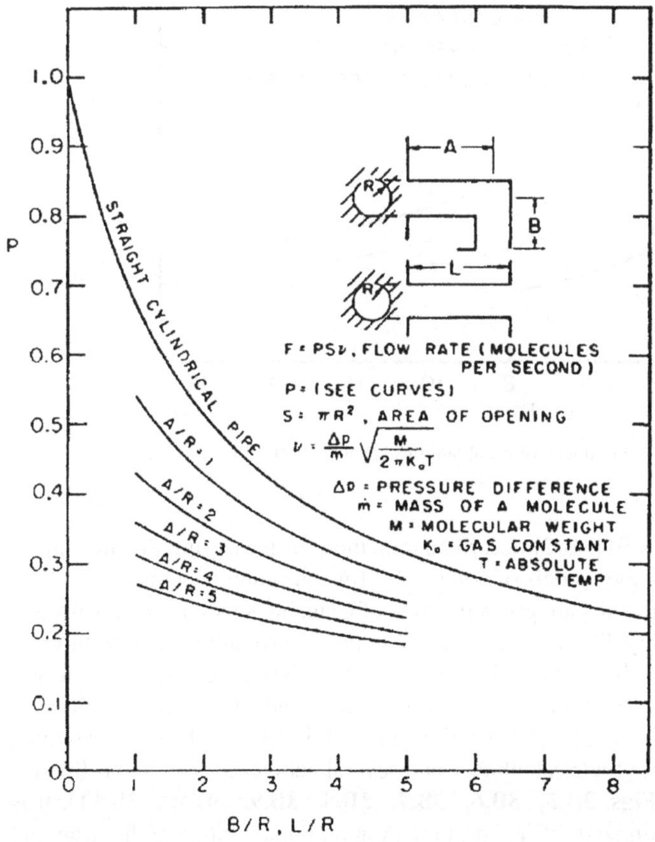

Fig. 3B.1 Molecular flow rate in a 90° cylindrical elbow calculated by the Monte Carlo method, and in a straight cylindrical pipe calculated by Clausing [P. Clausing, *Ann Physik* **12**, 404 (1932), p. 961] (Davis, 1960) [3B-1]

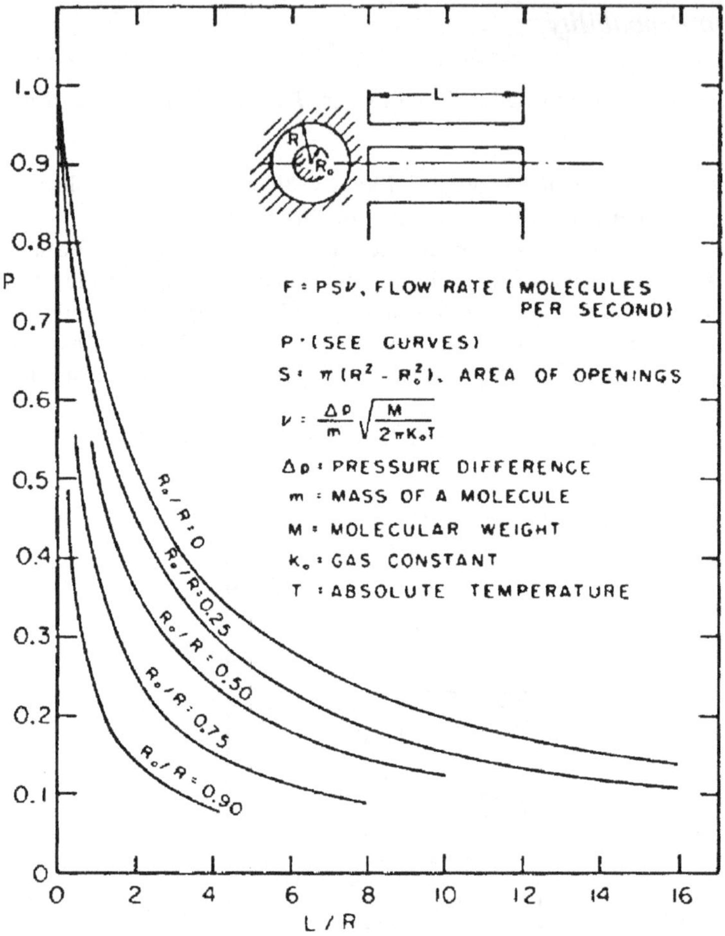

Fig. 3B.2 Molecular flow rate in a cylindrical annulus (Davis, 1960) [3B-1]

Levenson et al. (1960) [3B-2] described a method for measuring the molecular-flow conductance of scale models of complete diffusion-pump systems.

Abstract [3B-2]: Results are given in terms of Clausing's factor P (transmission probability), defined by $P = F/F_0$, where F is the net volume-flow rate through a geometry and F_0 is the volume-flow rate of gas striking the orifice of the geometry. Use of the factor P allows prediction of the conductance of the full-scale system represented by the model. Experimental and theoretical data are presented for straight cylinders, cylindrical elbows, and several geometries having no-line-of-sight through them (**Figs. 3B.5, 3B.6, 3B.7, 3B.8, 3B.9, 3B.10, 3B.11**). It is shown that for an optimized diffusion-pump system, comprising a baffle, trap and valve, $0.3 < P < 0.5$ can be obtained.

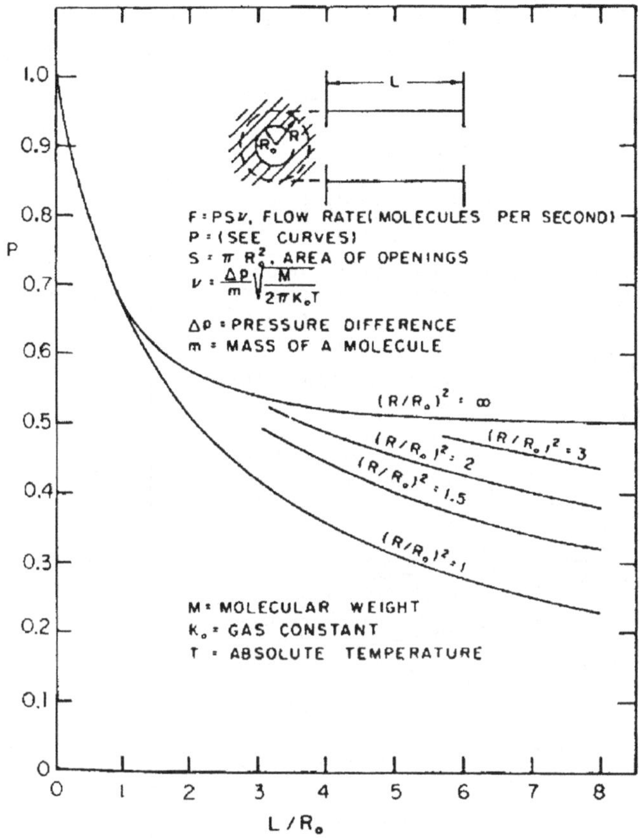

Fig. 3B.3 Molecular flow rate in a cylindrical pipe with restricted openings (Davis, 1960) [3B-1]

The transmission probability P , called Clausing's factor, is defined as $P = F/F_0$, where F is the net volume-flow rate (pipe conductance) through the geometry, and F_0 is the net volume-flow rate of gas striking the orifice of the geometry (orifice conductance).

Fan et al. (2002) [3B-3] presented an article, "Conductance calculation of a long tube with equilateral triangle cross section."

Abstract [3B-3]: Using the stratified gas flow model for calculating the conductance of long tubes with constant cross section, an analytical expression for calculating the conductance of a long tube with equilateral triagle cross section has been derived. The formula given is applicable to the full pressure range. A minimum in the conductance in the intermediate flow state is shown.

Steckelmacher (1966) [3B-4] presented an article, "A review of the molccular flow conductance for systems of tubes and components and the measurement of pumping speed."

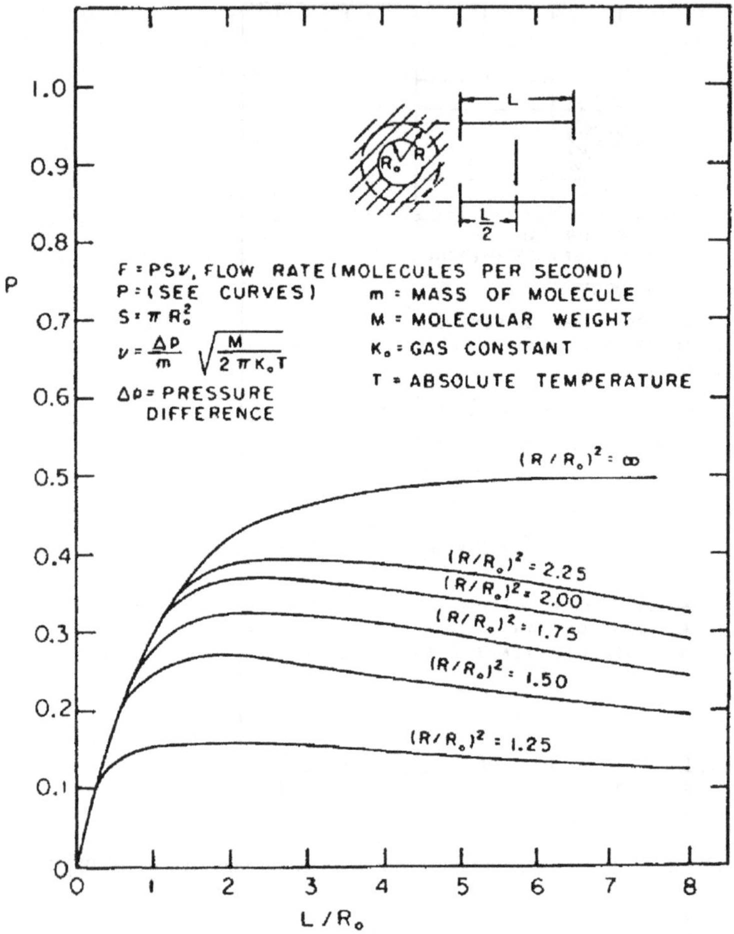

Fig. 3B.4 Molecular flow rate in a cylindrical pipe with restricted openings and a central blocking plate (Davis, 1960) [3B-1]

Abstract [3B-4]: The definition of molecular flow conductance for tubes and components is considered and the calculation of the conductance of long tubes with differently shaped cross-sections reviewed. The conductance and transmission probability for short tubes is discussed in relation to Clausing type integral equations and their approximate solutions. A review is given of theoretical and experimental investigations of molecular gas flow emission patterns for tubes and components. An estimation for the effective transmission probability and conductance of more complex systems, particularly of tubes, diaphragms, baffles etc. connected in series, is considered as well as their experimental determination using scale models and statistical Monte Carlo type methods for their more accurate calculation.

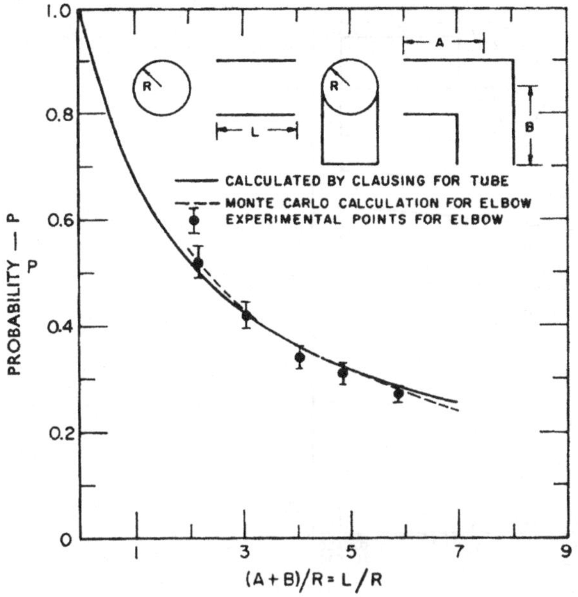

Fig. 3B.5 *P* for 90° elbow (Levenson et al., 1960) [3B-2]

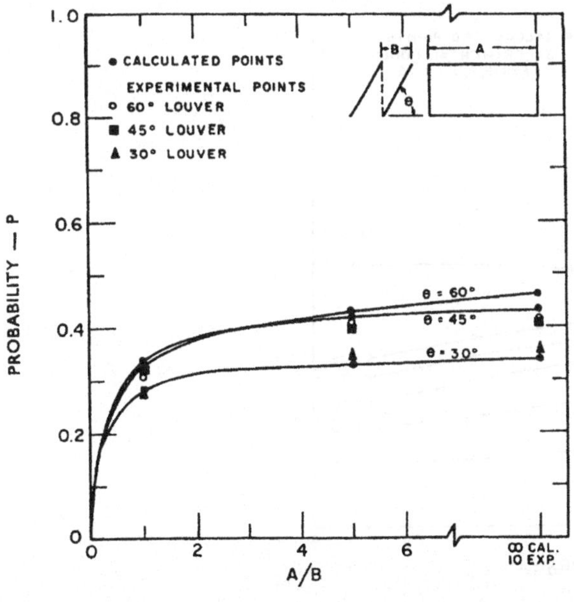

Fig. 3B.6 *P* for louver geometries (Levenson et al., 1960) [3B-2]

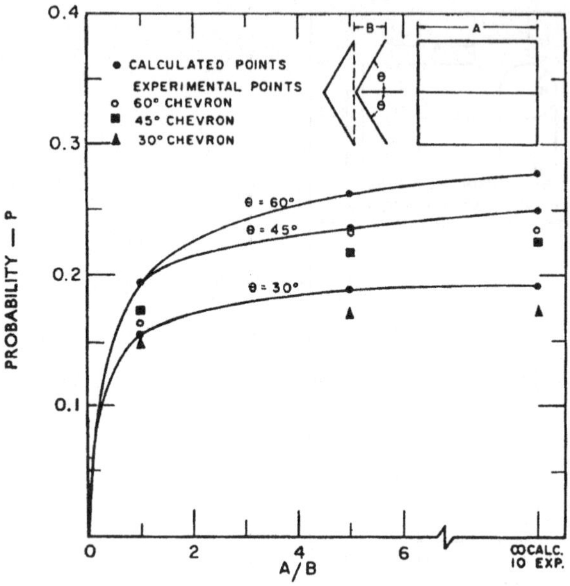

Fig. 3B.7 *P* for chevron geometries (Levenson et al., 1960) [3B-2]

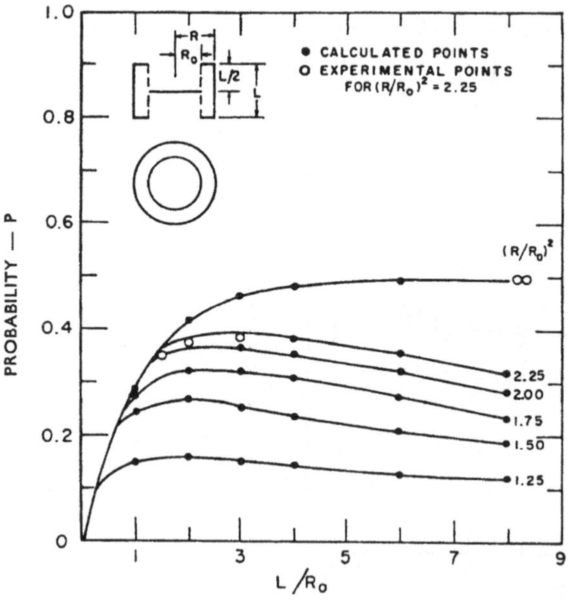

Fig. 3B.8 *P* for straight cylinder with two restricted ends and circular blocking plate (Levenson et al., 1960) [3B-2]

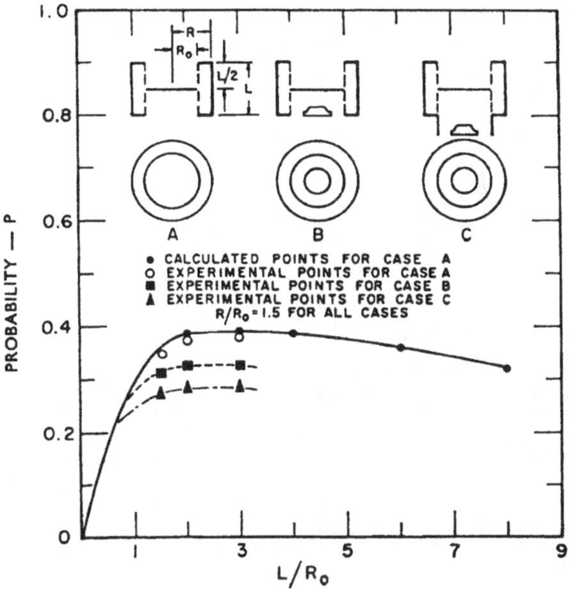

Fig. 3B.9 *P* for straight cylinder with two restricted ends and circular blocking plate in diffusion-pump system (Levenson et al., 1960) [3B-2]

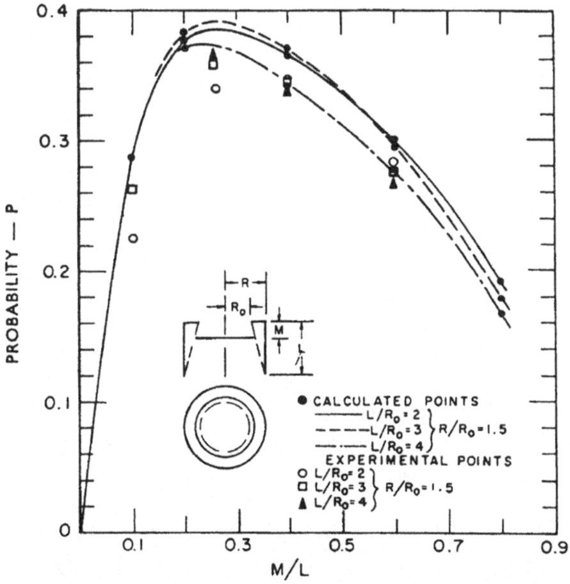

Fig. 3B.10 *P* for small end of straight cylinder with one restricted end and circular blocking plate (Levenson et al., 1960) [3B-2]

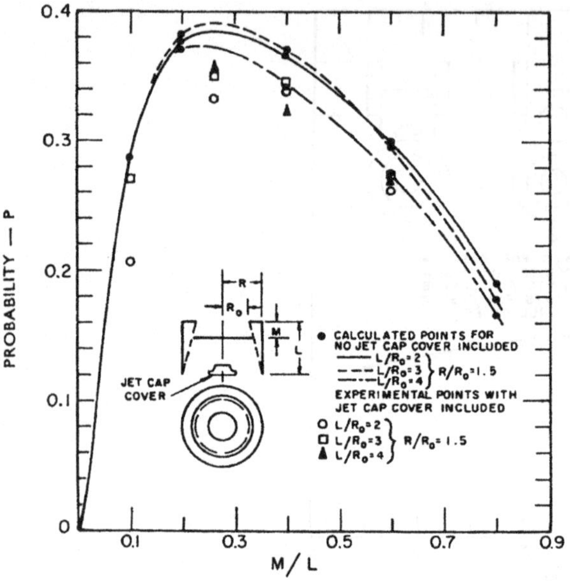

Fig. 3B.11 *P* for small end of straight cylinder with one restricted end and circular blocking plate in diffusion-pump system (Levenson et al., 1960) [3B-2]

3C Gas-Flow Patterns

In an electron-microscope column, gas flow patterns exist due to complicated narrow beam paths such as various lens pole-pieces and apertures. Though it is very difficult to calculate actual gas flow patterns in an electron-microscope column, one can roughly estimate gas flow patterns from the reported patterns of simple pipes with various ratios of length to diameter [3C-1].

Dayton (1956) [3C-1] presented an article, "Gas flow patterns at entrance and exit of cylindrical tubes."

Abstract [3C-1]: The molecular flow of gas through a straight cylindrical tube produces a beam pattern at the exit whose departure from the cosine law distribution depends on the ratio of length to radius for the tube. A complementary angular distribution pattern is formed at the entrance by those molecules which return without passing through. Equations and diagrams are presented for the beaming patterns. The flow of gas across the junction of tubes of different diameter is discussed.

Dayton wrote as follows: "In 1930 P. Clausing (*Z. f. Physik* **66**, 471–76, 1930)" derived a formula for the angular distribution pattern of gas molecules emerging from the exit of a short cylindrical tube under conditions of molecular flow. Clausing showed that the gas emerges in the form of a jet, and that the departure of the dispersion pattern from the cosine law distribution depends only on the ratio of length to radius for the tube. Clausing applied his formula to calculate the angular distribution for the case in which the ratio of length to radius equals two and presented a diagram of his results as shown in **Fig. 3C.1** [3C-1].

The cosine law distribution, which is shown by the dashed line, is obtained when the ratio of length to radius is zero, corresponding to a hole in a thin plate. As Clausing points out, this circular pattern for the limiting case of a hole in a thin

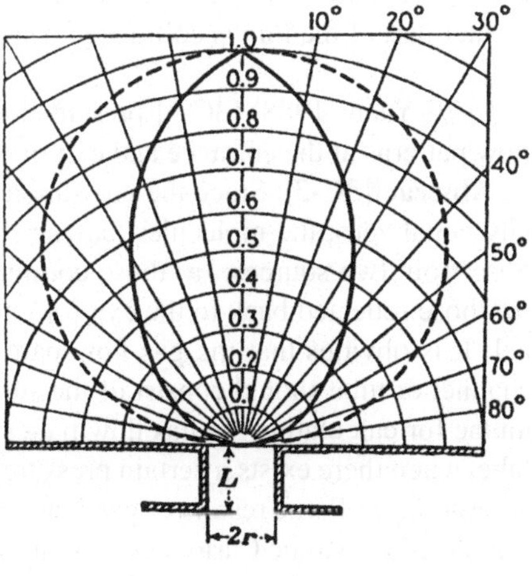

Fig. 3C.1 Clausing's diagram of the exit pattern for $L = 2r$. [P. Clausing, *Z. f. Physik* **66**, 471–76 (1930)] (Dayton, 1956) [3C-1]

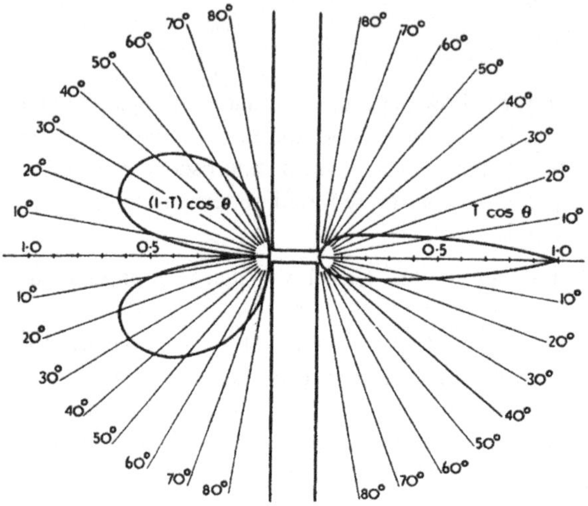

Fig. 3C.2 Entrance and exit patterns for L=10 r (Dayton, 1956) [3C-1]

plate must also be considered as a jet because more molecules are beamed in the direction of the axis of the hole than in any other direction [3C-1].

Equations and diagrams have been presented for flow patterns for cylindrical tubes with various ratios of length L to radius r. **Figure 3C.2** shows the entrance and exit patterns for the case $L/r = 10$ [3C-1].

Nanbu (1985) [3C-2] calculated the angular distribution of molecular flux from a circular orifice for thickness-diameter ratios less than 2 by use of the test-particle Monte Carlo method.

Abstract [3C-2]: The best fit equations are made from the obtained numerical data. Even for a small thickness-diameter ratio such as 0.3 the distribution shows a marked deviation from the ideal cosine law. For large thickness-diameter ratios the greater part of the flux is collimated near the axis of the orifice.

Tu Ji-Yuan (1988) [3C-3] presented an article, "A further discussion about gas flow patterns at the entrance and exit of vacuum channels."

Abstract[3C-3]: Since the pressure at the exit of vacuum channel does not usually equal zero, the molecules leaving the exit (or entrance) of a vacuum channel arise from two sources: (a) those coming through from the entrance (or exit), and (b) those reflected back to the exit (or entrance) without having traversed the channel. It is obvious that the gas flow patterns at the exit (or entrance) of a vacuum channel or tube should consist of the sum of the above two parts. The general formulae for calculating the gas flow patterns at the exit and entrance of a cylindrical tube, when there exists a certain pressure at the exit of the tube, are presented in the current paper. Some results of gas flow patterns in the inclined vacuum channels, as calculated by Monte Carlo methods, are also discussed.

References

Vacuum Circuits

3A-1. N. Yoshimura, "A differential pressure-rise method for measuring the net outgassing rates of a solid material and for estimating its characteristic values as a gas source", *J. Vac. Sci. Technol. A* **3** (6), pp. 2177–2183 (1985).

3A-2. N. Yoshimura, "Discussion on methods for measuring the outgassing rate", *Shinku (J. Vac. Soc. Japan)* **33** (5), pp. 475–481 (1990) (in Japanese).

3A-3. S. Ohta, N. Yoshimura, and H. Hirano, "Resistor network simulation method for a vacuum system in a molecular flow region", *J. Vac. Sci. Technol. A* **1** (1), pp. 84–89 (1983).

3A-4. H. Hirano, Y. Kondo, and N. Yoshimura, "Matrix calculation of pressures in high-vacuum systems", *J. Vac. Sci. Technol. A* **6** (5), pp. 2865–2869 (1988).

3A-5. N. Yoshimura and H. Hirano. "Two point pressure method for measuring the outgassing rate", *J. Vac. Sci. Technol. A* 7 (6), pp. 3351–3355 (1989).

3A-6. N. Yoshimura, "Modeling of outgassing or pumping functions of the constituent elements such as chamber walls and high-vacuum pumps", *Applied Surface Science* (VASSCAA-1, *Proceedings of the First Vacuum and Surface Science Conference of Asia and Australia*) Vols. **169/170** (2001) pp. 685–688.

3A-7. B. R. F. Kendall, "Comments on: Resistor network simulation method for a vacuum system in a molecular flow region" [*J. Vac. Sci. Technol. A* 1, 84 (1983)], *J. Vac. Sci. Technol. A* **1** (4), pp. 1881–1882 (1983).

3A-8. A. D. Degras, "Analogies élémentaires entre le vide et l'électricité", *Le Vide*, No 64, Juillet-Août, pp. 15–162 (1956) (in French).

3A-9. M. J. Aitken, "An electrical analogue to a high vacuum system", *Brit. J. Appl. Phys.* **4**, p. 188 (1953).

3A-10. D. W. Stops, "Further applications of the electrical analogue to vacuum systems", *Brit. J. Appl. Phys.* **4**, p. 350 (1953).

3A-11. B. R. F. Kendall, "Theoretical analysis of a two-pump vacuum system", *J. Vac. Sci. Technol.* **5** (2), pp. 45–48 (1968).

3A-12. B. R. F. Kendall and R. E. Pulfrey, "Theory of pulsed molecular-flow networks", *J. Vac. Sci. Technol.* **6** (2), pp. 326–332 (1969).

3A-13. S. Ohta, "On parallel evacuation", *Proceedings of 3rd Meeting on Vacuum Technology, Shinku (J. Vac. Soc. Japan)* pp. 21a–3 (1962) (in Japanese).

3A-14. G. Horikoshi, Y. Saito, and K. Kakihara, "An analysis of a complex network of vacuum components and its application", *Vacuum* **41** (7–9), pp. 2132–2134 (1990).

3A-15. J. A. Theil, "Gas distribution through injection manifolds in vacuum systems", *J. Vac. Sci. Technol. A* **13** (2), pp. 442–447 (1995).

3A-16. S. R. Wilson, "Numerical modeling of vacuum systems using electronic circuit analysis tools", *J. Vac. Sci. Technol. A* **5** (4), pp. 2479–2483 (1987).

3A-17. D. J. Santeler, "Computer design and analysis of vacuum systems", *J. Vac. Sci. Technol. A* **5** (4), pp. 2472–2478 (1987).

3A-18. R. J. Reid, "Vacuum calculations for large systems", *Vacuum* **43** (1/2), pp. 31–33 (1992).

Molecular Flow Conductance

3B-1. D. H. Davis, "Monte Carlo calculation of molecular flow rates through a cylindrical elbow and pipes of other shapes", *J. Appl. Phys.* **31** (7), pp. 1169–1176 (1960).

3B-2. L. L. Levenson, N. Milleron, and D. H. Davis, "Optimization of molecular flow conductance", *Transactions of the 7th National Vacuum Symposium* 1960 (Pergamon Press, New York, 1961), pp. 372–377.

3B-3. P. Fan, Ji-Guo Chu, and Jian-Da Shao, "Conductance calculation of a long tube with equilateral triangle cross section", *J. Vac. Sci. Technol. A* **20** (6), pp. 2119–2122 (2002).
3B-4. W. Steckelmacher, "A review of the molecular flow conductance for systems of tubes and components and the measurement of pumping speed", *Vacuum* **16** (11), pp.561–584 (1966).

Gas Flow Patterns

3C-1. B. B. Dayton, "Gas flow patterns at entrance and exit of cylindrical tubes", *1956 National Symposium on Vacuum Technology Transactions* (Pergamon Press, New York, 1957), pp. 5–11.
3C-2. K. Nanbu, "Angular distributions of molecular flux from orifices of various thicknesses", *Vacuum* **35** (12), pp. 573–576 (1985).
3C-3. Tu Ji-Yuan, "A further discussion about gas flow patterns at the entrance and exit of vacuum channels", *Vacuum* **38** (7), pp. 555–559 (1988).

Other Articles

Vacuum Circuits

*3-1. N. Yoshimura, "Vacuum circuit composed of elements with characteristic values corresponding to those of high vacuum system", *IONICS* **23** Feb. 1985, pp. 23–28 (in Japanese).

Gas Flow/Conductance

*3-2. W. Teubner, "Molecular flow of gases through tubes with regard to various particle-wall interaction processes. Part 1", *Vakuum-Technik* **16** (4), pp. 69–80 (1967) (in German).
*3-3. W. Teubner, "Molecular flow of gases through tubes with regard to various particle-wall interaction processes. Part 2–Experimental evaluation of deviations from Knudsen's law for the stationary flow", *Vakuum-Technik* **16** (5), pp. 95–101 (1967) (in German).
*3-4. K. M. Welch, "The pressure profile in a long outgassing vacuum tube", *Vacuum* **23** (8), pp. 271–276 (1973).
*3-5. Jay D. Pinson, and A. W. Peck, "Monte Carlo analysis of high speed pumping system", *AVS Transactions of the 9th Vacuum Symposium* pp. 406–410 (1962).
*3-6. W. Dong and L. A. Bromley, "Vacuum flow of gases through channels with circular, annular, and rectangular cross sections", *1961, Trans. 8th National Vacuum Symposium combined with the 2nd International Congress on Vacuum Science and Technology*, pp. 1116–1132.
*3-7. V. E. Stubblefield, "Net molecular flow conductance of series elements", *J. Vac. Sci. Technol. A* **1** (3), pp. 1549–1552 (1983).
*3-8. A. W. Ross, M. Fink, and H. F. Wellenstein, "Monte Carlo simulation of chevron baffle performance", *J. Vac. Sci. Technol. A* **11** (3), pp. 723–725 (1993).
*3-9. O. Boulon, R. Mathes, and J.-P. Thibault, "Direct simulation Monte Carlo method for molecular and transitional flow regimes in vacuum components with static and moving surfaces", *J. Vac. Sci. Technol. A* **17** (4), pp. 2080–2085 (1999).
*3-10. W. Steckelmacher, "Comment on 'Practical way to calculate molecular conductances of pipes with constant annular cross section'", *J. Vac. Sci. Technol.* **18** (1), pp. 54–55 (1981).
*3-11. G. Horikoshi, "An analysis and a consideration of molecular flow in a long conduit pipe", *Shinku (J. Vac. Soc. Japan)* **30** (8), pp. 658–665 (1987) (in Japanese).
*3-12. W. Harries, "Zur Bemessung von Vakuumleitungen", *Chem.-Ing.-Technik* **21** (7/8), pp. 139–142 (1949) (in German).

*3-13. J. A. Perry, Jr., “Critical flow through sharp-edged orifices”, *Transactions of the ASME*, October, pp. 757–764 (1949).
*3-14. R. A. Haefer, “Addition theorem for the resistance to flow of composite systems and its relationship with the pumping speed in the region of molecular flow”, *Vacuum* **30** (6), pp. 217–223 (1980).
*3-15. L. Füstöss, “Evaluation and calculation of gas flow through axially symmetric tubes”, *Vacuum* **33** (1/2), pp. 13–16 (1983).
*3-16. D. J. Santeler and M. D. Boeckmann, “Molecular flow transmission probabilities of rectangular tubes”, *J. Vac. Sci. Technol. A* **9** (4), pp. 2378–2383 (1991).
*3-17. A. Pace and A. Poncet, “Monte Carlo simulations of molecular gas flow: some applications in accelerator vacuum technology using a versatile personal computer program”, *Vacuum* **41** (7–9), pp. 1910–1913 (1990).
*3-18. P. Clausing, “The flow of highly rarefied gases through tubes of arbitrary length”, *J. Vac. Sci. Technol.* **8** (5), pp. 636–646 (1971).
*3-19. G. Melin, “Dynamic pumping and pulsed outgassing in waveguides and vacuum systems for plasma physics experiments”, *J. Vac. Sci. Technol. A* **5** (5), pp. 2945–2953 (1987).
*3-20. M. J. Ernst and H. F. Hemond, “Vacuum pumpdown model for long outgassing tubes”, *J. Vac. Sci. Technol. A* **12** (2), pp. 554–563 (1994).
*3-21. R. G. Livesey, “Method for calculation of gas flow in the whole pressure regime through ducts of any length”, *J. Vac. Sci. Technol. A* **19** (4), pp. 1674–1678 (2001).
*3-22. H. Hamacher, “An analytical approach for calculating vacuum properties of outgassing tubing systems”, *Vacuum* **32** (12), pp. 729–733 (1982).

Gas-flow Patterns

*3-23. J. W. Ward, M. V. Fraser, and R. L. Bivins, “Monte Carlo analysis of the behavior of divergent conical effusion orifices”, *J. Vac. Sci. Technol.* **9** (3), pp. 1056–1061 (1972).
*3-24. P. Krasuski, “Angular distribution of flux at the exit of cylindrical tubes”, *J. Vac. Sci. Technol. A* **5** (4), pp. 248–2492 (1987).

Other

*3-25. Y. Suetsugu, “Application of the Monte Carlo method to pressure calculation”, *J. Vac. Sci. Technol. A* **14** (1), pp. 245–250 (1996).
*3-26. D. Santeler and D. Warren, “Computer solution to chamber-gas fractionation during pumpdown”, *J. Vac. Sci. Technol. A* **11** (1), pp. 231–239 (1993).
*3-27. A. K. Rebrov, “Free jets in vacuum technologies”, *J. Vac. Sci. Technol. A* **19** (4), pp. 1679–1986 (2001).
*3-28. T. Takiya, F. Higashino, Y. Terada, and A. Komura, “Pressure wave propagation by gas expansion in a high vacuum tube”, *J. Vac. Sci. Technol. A* **17** (4), pp. 2059–2063 (1999).
*3-29. V. Singh, B. Berney, and A. Krishnan, “Designing low pressure systems with continuum models”, *J. Vac. Sci. Technol. A* **14** (3), pp. 1252–1257 (1996).
*3-30. J. G. Chu, “Model of differential electrical network for the blades of turbomolecular pumps”, *J. Vac. Sci. Technol. A* **11** (2), pp. 412–416 (1993).
*3-31. M. H. Hablanian, “Concept of compatibility in conductance calculations”, *J. Vac. Sci. Technol.* **7** (1), pp. 237–240 (1970).

Chapter 4
Outgassing

Outgassing occurs when the surfaces of the chamber walls are exposed to vacuum, affecting the pressures in vacuum systems. Therefore, outgassing rates of various construction materials must be estimated to simulate the pressures in high-vacuum systems.

Stainless-steel evacuation pipes are exclusively used in the high-vacuum evacuation systems of scientific instruments because they have apparent advantages in outgassing and in machining. The surfaces of the stainless-steel chambers of UHV electron microscopes and UHV analysis instruments (JEOL Ltd.) are mirror-polished by electrolytic polishing (EP) or chemical polishing (CP). Electrodes of the electron guns and the accelerating tubes of electron microscopes are made of stainless steel, titanium alloy or Inconel, whose surfaces are mirror-polished and treated by high-temperature (about 1000 °C) pre-degassing under vacuum. Evacuation pipes for electron microscopes (JEOL Ltd.) are vacuum-brazed at 1000 °C, resulting in very low outgassing.

Specimen stages and specimen holders for electron microscopes are made of non-magnetic materials such as aluminum alloys. Machined metal parts are cleaned by ultrasonic cleaning using detergent solution and alcohol. Some units for ultra-high vacuum are cleaned by rf discharge using oxygen and/or hydrogen in the final cleaning stage.

Process of Outgassing

Almost all metal surfaces have porous oxide layers which contain a large amount of water vapor and air. Such gas molecules diffuse out gradually when such surfaces are exposed to high vacuum. **Figure 4.1** shows an anodic oxidized film of aluminum, which is amazingly thick and porous.

Outgassing from an elastomer or polymer placed inside the vacuum chamber also occurs through the mechanism of diffusion.

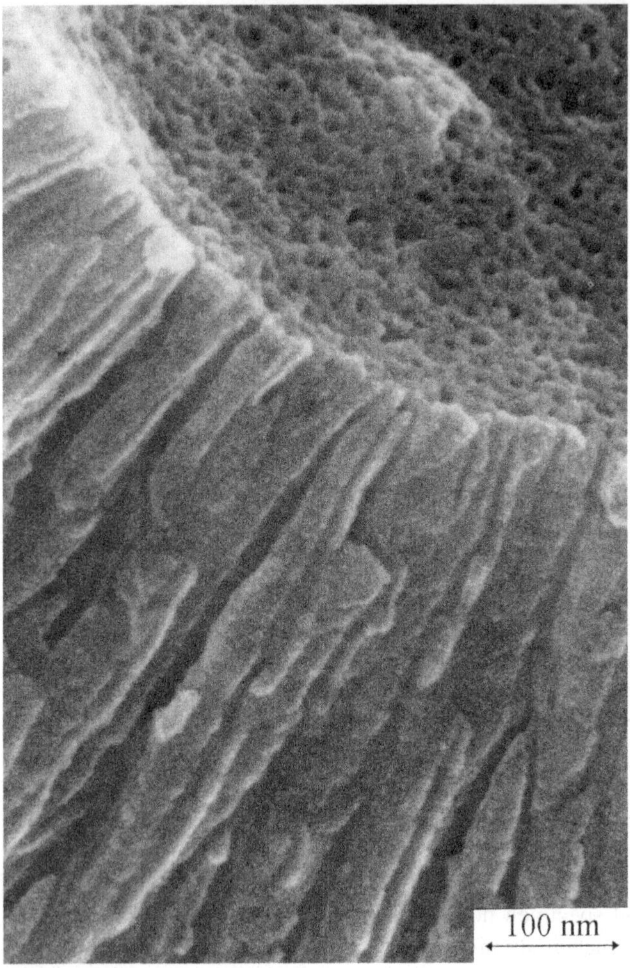

Fig. 4.1 SEM image of anodic oxide layer of aluminum (JEOL Ltd.)

Diffusion

Dayton (1959) [4-1] presented an article, "Relations between size of vacuum chamber, outgassing rate, and required pumping speed," where semi-empirical formulas on the outgassing rate depending on the pumping time are discussed.

Semi-empirical Formulas [4-1]

The outgassing rate is preferably expressed in $\text{Torr} \cdot \text{L} \cdot \text{s}^{-1} \cdot \text{cm}^{-2}$ at a specified temperature and a specified time after the beginning of pumping. When outgassing rate is plotted against time on log–log graph paper, the resulting curve frequently has a nearly constant slope for the first few hours. As shown by the data of Santeler [1], Geller [2] and others, this initial part of the curve can often be presented by

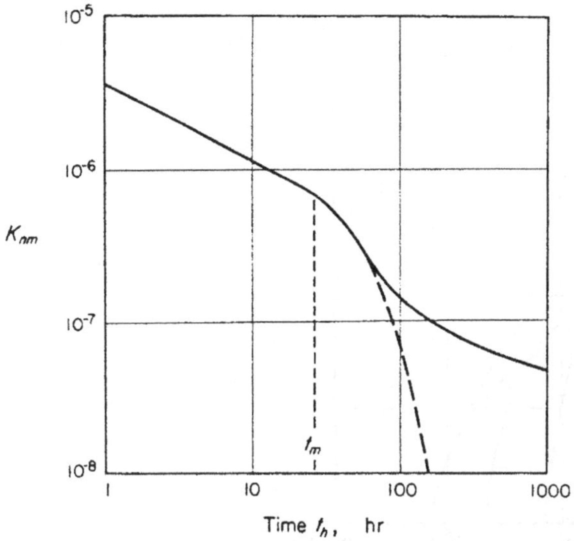

Fig. 4.2 Typical outgassing curve for an elastomer (Dayton, 1959)[4-1]

$$\log(K_m - K_u) = \log K_1 - \alpha \log t \tag{4.1}$$

where K_m is the outgassing rate at the time t, K_u and K_1 are constants, and the slope of the curve, α, frequently has values between 0.5 and 0.8. When experimental data on the outgassing rate is not available, K_u, K_1 and α can sometimes be estimated from other known physical properties of the material by formulas given in the following.

Dayton (1959) [4-1] presented the typical outgassing curve for an elastomer in **Fig. 4.2**.

References

1. D. J. Santeler, *Vacuum Symposium Transactions*, 1958, Pergamon Press, London. General Electric Co. Report No. 58GL303 (1958).
2. R. Geller, *Le Vide* **13** (74), 71 (1958).

Dayton (1961) [4-2] presented an article, "Outgassing rate of contaminated metal surfaces."

Abstract [4-2]: Successive evaporation of various molecules from a heterogeneous layer of surface contamination can result in an outgassing rate which is inversely proportional to the time of pumping. A relatively uniform distribution of the activation energies of desorption will result in a sum of the individual outgassing curves for constant desorption energy which appears as a straight line with a slope of minus one on the log–log plot of outgassing rate vs. time of pumping. This line forms an "outgassing barrier" which limits the ultimate pressure in unbaked metal vacuum systems.

Comment: "A heterogeneous layer of surface contamination" means "an oxide layer" of metal surface.

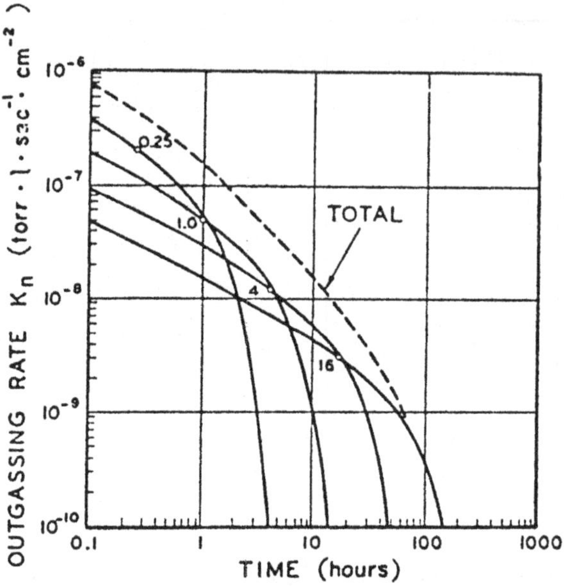

Fig. 4.3 Total outgassing rate as a sum of curves for constant values of τ (Dayton, 1961) [4-2]

Dayton (1961) [4-2] presented the relationship that the outgassing rates are inversely proportional to the time of pumping as shown in **Fig. 4.3**.

Comments: Diffusion time constants τ_i have various values depending on the oxide layer. In this graph, Dayton has chosen $\tau_1 = 0.25\,\text{h}$, $\tau_2 = 1.0\,\text{h}$, $\tau_3 = 4\,\text{h}$ and $\tau_4 = 16\,\text{h}$ in anticipation of the requirement that the outgassing rate K_n decays as t^{-1} in the region from τ_0 to τ_m.

Dayton (1961) [4-2] described the persistency of chemisorbed water as follows.

"As the temperature is raised above 100 °C the chemisorbed water begins to desorb at an appreciable rate, but a temperature of more than 500 °C is required to remove all the chemisorbed water from oxides according to the data by Garner [1]. Chemisorption of H_2O on oxide is believed to be associated with the formation of hydroxide groups (OH^- ions). Data is given in the book edited by Garner [1] which indicates that most of the chemisorbed water is located on the surface of the oxide (including the walls of the pores).

Reference

1. W. E. Garner (editor), *Chemisorption,* Academic press, New York (1957) loc. cit. pp. 59–75.

Rogers (1963) [4-3] presented an article, "The variation in outgassing rate with the time of exposure and pumping."

Abstract [4-3]: Past experiments have shown that the pumpdown time of a vacuum chamber varies with the exposure to moist air. This paper presents a theoretical analysis of the variation. It is shown that when the vapor outgassing dominates the pumpdown, the slope of the log pressure versus log pumping time curve will vary from 1/2 to 3/2 for a semi-infinite plate. This is in contrast to the theoretical slope of 1/2 that is found for uniformly distributed gas in a semi-infinite plate.

Rogers (1963) [4-3] described the variation in outgassing rate with parameters of prior pumping times in second pumpdown.

Second Pumpdown [4-3]
If the initial moist atmosphere exposure time was very long compared to the time of interest, the concentration can be considered uniform, and the first pumpdown will exhibit the square root relationship between outgassing rate and pumping time. If the system is now exposed to a moist atmosphere, and again pumped down, the resulting outgassing variation will depend upon both the previous pumpdown time and the following moist atmosphere exposure time. Examples of the resulting outgassing variations are shown in **Figs. 4.4** and **4.5**.

Notes: In **Figs. 4.4** and **4.5**, C, concentration in molecules/length^3; C_0, initial concentration; D, diffusion coefficient in length^2/time.

Dayton (1962) [4-4] presented an article, "The effect of bake-out on the degassing of metals."

Abstract [4-4]: Equations are derived for the effects of a bake-out period on the rate of outgassing of hydrogen and other gases dissolved in metals. It is shown that van Liempt's formula for the time to remove 95% of the gas is incorrect, and a corrected formula is derived. The equations are compared with available experimental data.

Dayton (1962) presented the theoretical curve for effect of bake-out in **Fig. 4.6**.

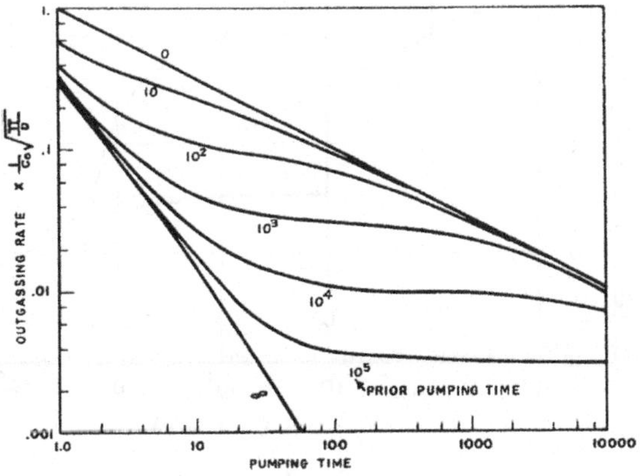

Fig. 4.4 Variation in outgassing rate with prior pumping time for a system with a uniform initial concentration (Any consistent time units) (Rogers, 1963) [4-3]

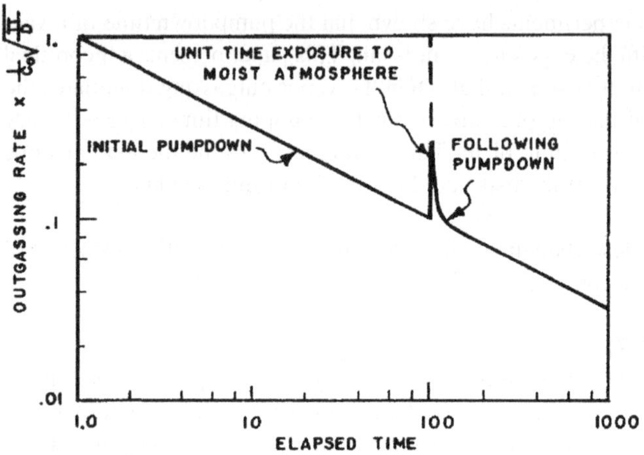

Fig. 4.5 Variation in outgassing rate with total elapsed time (Any consistent time units) (Rogers, 1963) [4-3]

Yoshimura et al. (1991) [4-5] measured the pressure-rise rates repeatedly using an extractor ionization gauge (EG) (degassed by electron bombardment) by isolating an electropolished (EP) type 304 stainless-steel (SS304) pipe (in situ baked). The isolation test was repeated four times in sequence. For the first two tests, the valve was closed for 60 minutes each, while for the last two tests, the valve was closed for 90 minutes each. The evacuation period between each successive isolation test was 10 minutes. The resulting pressure-rise curves are presented in **Fig. 4.7**.

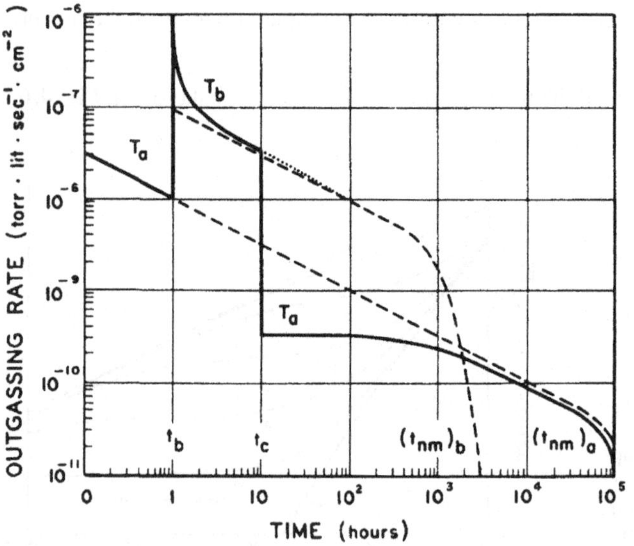

Fig. 4.6 Theoretical curve for effect of bake-out. T_a, ambient temperature and T_b, bake-out temperature (Dayton, 1962) [4-4]

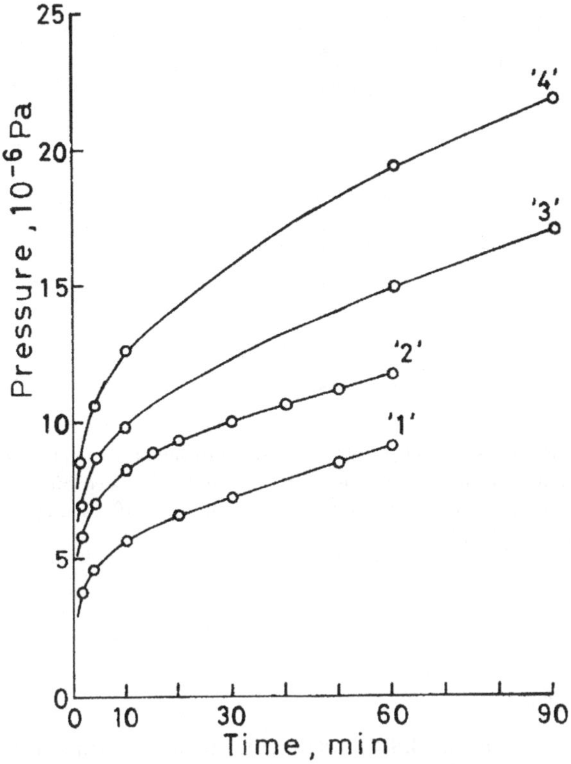

Fig. 4.7 Pressure-rise curves for the isolated pipe which had been in situ baked (150 °C, 20 h). Pressures were measured with the EG (Yoshimura et al., 1991) [4-5]

The pressure-rise rates got faster and the attainable pump-down pressures (the starting pressures for the successive build-up tests) got higher when the build-up tests (for 90 minutes or 60 minutes) and the evacuation (10 minutes each) were repeated, as clearly seen in **Fig. 4.7**. This experimental result clearly shows that the gas molecules adsorbed and absorbed on the pipe wall increased with the intermittent evacuation, resulting in increasing outgassing rates. That is, gas molecules evolved from the pipe wall and the EG were accumulated on the wall surface during the isolation period, resulting in increased outgassing rates of the pipe wall in the successive isolation period. The source of the outgassing gases must be the bulk, not the surface of the pipe wall.

Redhead (1996) [4-6] introduced the pressure-rise curve measured by Jousten [1]. "Jousten (1994) [1], using a system designed for calibrating gauges, measured the pressure rise in a type 316LN stainless steel chamber that has been vacuum fired and then baked for 48 h at 250 °C. This system contained an extractor gauge, a spinning rotor gauge, and a quadrupole mass spectrometer. After pumping down to 1.5×10^{-10} Torr(H_2) it was sealed off and the $p(t)$ curve was measured with the spinning rotor gauge for 4 weeks; the curve was linear and the pressure increased by a factor of about 10^6 (see **Fig. 4.8**) [4-6]."

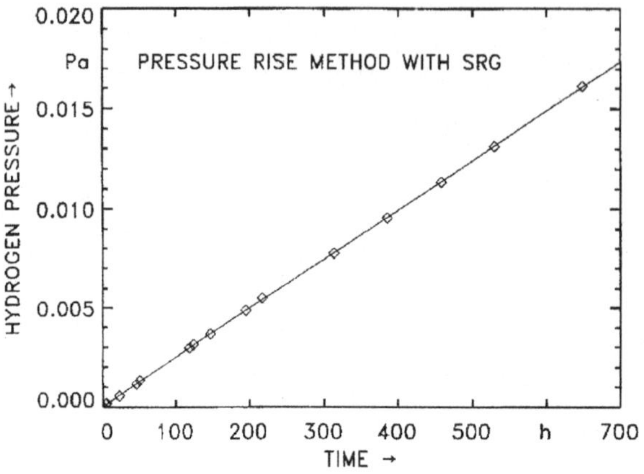

Fig. 4.8 Pressure rise after sealoff of a stainless steel chamber, previously vacuum fired and baked at 250 °C for 48 h, measured with the spinning rotor gauge. The chamber was sealed off at a pressure of 2×10^{-8} Pa, the first point taken at 33 minutes at a pressure of 1.52×10^{-5} Pa. From Jousten [1] with permission (Redhead, 1996) [4-6]

Reference

1. K. Jousten, Vacuum **47**, 325 (1996).

The pressure-rise characteristics of **Fig. 4.8** shows that the outgassing from the a type 316LN stainless steel chamber, which has been vacuum fired and then baked for 48 h at 250 °C, is a bulk phenomemon, not a surface one. Let us consider the situation that the chamber has been isolated for a long time, 4 weeks. The amount of gas molecules adsorbed and absorbed on the chamber walls must be much increased compared with the amount on the surface before isolation, because many, many gas molecules impinge on the surface frequently under high pressure of residual gases during a long period of isolation. The gas molecules in the bulk wall must be reduced through diffusion during a long isolation period.

Schram (1963) [4-7] presented an article, "La desorption sous vide."

Abstract [4-7]: A general survey is given of our present knowledge in this field. Important discrepancies between the experimental results and the theoretical explanations put forward by the different authors are pointed out. The complexity of the phenomena is emphasized and the relative importance of the different elementary process are discussed. A limited choice of research is proposed for desorption in vacuum. Theoretical calculations based on simplified models is compared with some experimental results. The knowledge of the true surface area is found necessary and a simple experimental method used to measure this surface is described. The experimental curves of desorption are discussed taking in account the first results of true surface area measurements.

In unbaked systems the absorbed gas layers at the surface are proved to yield the most important part of the total desorption rate. The residual desorption of baked systems in ultrahigh vacuum techniques seems more likely to be a bulk (diffusion, permeation), rather than a surface phenomenon.

Calder and Lewin (1967) [4-8] calculated the outgassing rate of a stainless-steel plate, based on diffusion mechanism. "Calculations show that the outgassing rate should be reduced by several orders of magnitude by a high-temperature treatment. The greater the thickness of the metal, the higher the temperature has to be. Measurements are in reasonable agreement with the calculations [4-8]."

Effect of Temperature on Degassing (Theory) [4-8]
The one-dimensional diffusion equation is

$$D\frac{\partial^2 c}{\partial x^2} = \frac{\partial c}{\partial t} \tag{4.2}$$

where D is the diffusion coefficient and c the concentration.

Equation (4.2) is solved for a slab of unit cross-section and thickness d. Initially the concentration is c_0 and constant throughout. At time $t = 0$ vacuum is applied to both faces. The initial and boundary conditions are

$$c = c_0 \quad \text{for} \quad 0 \le x \le d \quad \text{at} \quad t = 0$$

and

$$c = 0 \quad \text{for} \quad x = 0 \quad \text{and} \quad x = d \quad \text{at} \quad t > 0.$$

The solution is [1] (Levin, 1965, p. 32)

$$c(x,t) = c_0\frac{4}{\pi}\sum_0^\infty (2n+1)^{-1}\sin\frac{\pi(2n+1)x}{d}\exp\left\{-\left(\frac{\pi(2n+1)}{d}\right)^2 Dt\right\}. \tag{4.3}$$

The instantaneous gas flow from one face of the slab is

$$\dot{Q} = D\left(\frac{\partial c}{\partial x}\right)_{x=0} = \frac{4c_0 D}{d}\sum_0^\infty \exp\left\{-\left(\frac{\pi(2n+1)}{d}\right)^2 Dt\right\}. \tag{4.4}$$

Values of

$$\sum_0^\infty \exp\left\{-\left(\frac{\pi(2n+1)}{d}\right)^2 Dt\right\}$$

as a function of Dt/d^2 are given in **Fig. 4.9**. For $Dt/d^2 > 0.025$, as is the case in many practical situations, we can write, to a sufficient approximation,

$$c = c_0\frac{4}{\pi}\sin\frac{\pi x}{d}\exp(-\pi^2 d^{-2} Dt). \tag{4.5}$$

$$\dot{Q} = 4c_0 D d^{-1}\exp(-\pi^2 d^{-2} Dt). \tag{4.6}$$

If a sheet has been degassed for t_1 seconds at a temperature T_1 the outgassing rate (immediately) afterwards at room temperature T_r is

$$\dot{Q}_r = 4c_0 D_r d^{-1}\exp(\;\pi^2 d^{-2} D_1 t_1). \tag{4.7}$$

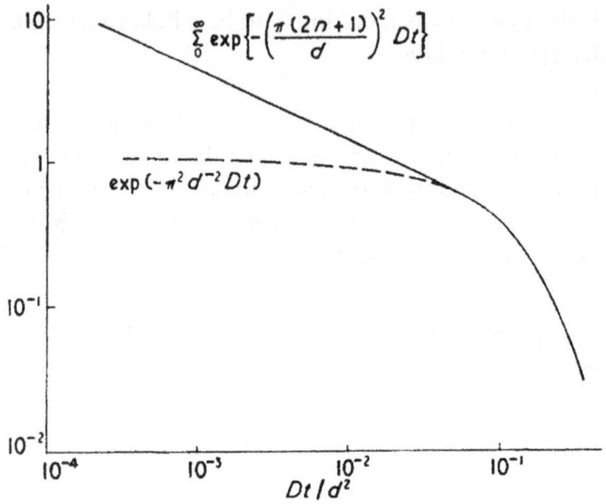

Fig. 4.9 Plot of the infinite series of Eq. (4.4) and its first term (Calder and Lewin, 1967) [4-8]

Reference

1. G. Lewin, 1965, *Fundamentals of Vacuum Science and Technology* (New York: McGraw-Hill).

Santeler (1992) [4-9] presented an article, "Estimating the gas partial pressure due to diffusive outgassing."

Abstract [4-9]: Hydrogen outgassing from iron-based metals such as stainless steel, and water outgassing from plastic or polymeric materials are typical examples where diffusive outgassing processes have been demonstrated to be major process. The diffusive transport of gas through materials is governed by the second order differential equation $D\delta^2C/\delta X^2 = \delta C/\delta t$, known as Fick's Law. Two alternate solutions of this equation are available which provide for simplified approximations for both the short-term and the long-term behavior, as well as for the exact numerical answer. An important aspect of true diffusive outgassing is its excellent predictability as compared to experiment. This permits calculating the outgassing rate following a known time-temperature-environment history. A simple graphical solution is demonstrated to accomplish this for the example of hydrogen outgassing from a thick walled stainless steel vessel operating at an elevated temperature following various bakeout schedules.

Diffusive Outgassing—Fick's Law [4-9]

The outgassing rate decays according to Fick's Law of diffusion [1, 2, 3] given by the second order differential equation

$$D\frac{\delta^2 C}{\delta x^2} = \frac{\delta C}{\delta t}. \tag{4.8}$$

One standard solution to this typical vacuum-wall boundary-value problem is

$$q = C_0 \left(\frac{D}{\pi t}\right)^{0.5} \left[1 + 2\sum_{n=1}^{n=\infty} (-1)^n \times e^{-n^2 l^2/Dt}\right] \text{ Torr} \cdot \text{L}/(\text{s} \cdot \text{cm}^2), \tag{4.9}$$

where C_0 is the initial concentration in Torr · L/cm^3(Torr · L/cc), D is the temperature-dependent diffusion coefficient in cm^2/s, both C_0 and D depend on the material and the gas type), t is the time in seconds, and l is the material thickness in cm. Equation (4.9) as applied to vacuum systems is discussed in [4] and [5]. An alternate solution of Eq. (4.8) for a vacuum wall is discussed in [6] and [7] where the time-temperature outgassing is given as

$$q = \left(\frac{4C_0 D}{l}\right) \left[\sum_{n-0}^{\infty} - \exp\left\{-\left[\pi (2n+1)/l\right]^2 Dt\right\}\right] \text{ Torr} \cdot \text{L}/(\text{s} \cdot \text{cm}^2). \tag{4.10}$$

The two solutions given by Eqs. (4.9) and (4.10) appear completely different but both give the same numerical answers. Note that while both equations are summations of exponential terms, the solutions are effectively summed from opposite directions. As a result, the two solutions have different single term approximations and for different time periods. For outgassing times which are short relative to Dt/l^2, the summation of the exponential terms in the bracketed term in Eq. (4.9) is negligible relative to the unity term and the entire bracketed term is approximately equal to 1.0. Equation (4.9) can then be approximately by

$$q = C_0 \left(\frac{D}{\pi t}\right)^{0.5} \text{ Torr} \cdot \text{L}/(\text{s} \cdot \text{cm}^2). \tag{4.11}$$

Conversely, for times which are long compared to 0.0025 l^2 Eq. (4.10) can be approximated by the first exponential term

$$q = \left[4C_0 D/l\right] \left\{\exp\left[-\left(\pi/l\right)^2 Dt\right]\right\} \text{ Torr} \cdot \text{L}/(\text{s} \cdot \text{cm}^2). \tag{4.12}$$

It is important to note that these outgassing processes are partial pressure dependent; thus, with a low partial pressure of hydrogen on both sides of the wall, the material outgasses in both direction. For this condition, the half thickness should be used for l. The concentration gradient which is moving into the material from both sides will meet at the center. Each half side may be treated as a separate material which is outgassing through one face and sealed on the other face. Since the two sides are symmetric, no gas flows across the interface.

Diffusive outgassing is extremely temperature dependent through the diffusivity D which is given by

$$D = D_0 e^{-Ed/RT} \text{cm}^2/\text{s}, \tag{4.13}$$

where D_0 is the diffusivity at infinite temperature, in cm^2/s, Ed is the activation energy for diffusion in cal/(g · mol), R is the gas constant, (1.987 cal/(g · mol · K), and T is the absolute temperature in K.

Typical values of these parameters for hydrogen in stainless steel are $D_0 = 0.012\,\text{cm}^2$ and $E_d = 13100\,\text{cal/ gmol}$.

References

1. W. Jost, *Diffusion in Solids, Liquids, Gases* (Academic, New York, 1960).
2. R. M. Barrier, *Diffusion In and Through Solids* (Cambridge University, Cambridge, 1941).
3. H. S. Carslaw and J. C. Jaeger, *Conduction of Heat in Solids*, 2nd ed. (Clarendon, Oxford, 1959).
4. F. Pagano, *13th National Vacuum Symposium*, 1966, p. 103.

5. D. J. Santeler et al., *Vacuum Technology and Space Simulation*, NASA SP-105, p. 188, p. 204, 1966.
6. R. Calder and G. Lewin, *J. Appl. Phys.* **18**, 1459 (1967).
7. G. Lewin, *Fundamentals of Vacuum Science and Technology* (McGraw-Hill, New York, 1965), p. 25.

Li and Dylla (1993) [4-10] presented an article, "Model for the outgassing of water from metal surfaces."

Abstract [4-10]: An analytic expression for the outgassing rate is derived based on the assumption that the rate of water diffusing through the passivation oxide layer to the surface governs the rate of its release into the vacuum. The source distribution function for the desorbing water is assumed to be a combination of a Gaussian distribution centered at the interior surface driven by atmospheric exposure, and a uniform concentration throughout the bulk. We have measured the outgassing rate from a clean stainless-steel (type 304) chamber as a function of water exposure to the chamber surface from <1 to 600 monolayers. The measured outgassing rate data show that α (in a power law of the form $Q = Q_{10}t^{-\alpha}$) tends to 0.5 for low H_2O exposures and tends to 1.5 for high H_2O exposures as predicted by the model.

Recombination-limited Outgassing

Moore (1995) [4-11] presented an article, "Recombination limited outgassing of stainless steel."

Abstract [4-11]: It was observed two generations ago that at low pressures the permeation of hydrogen through metals is no longer a function of the square root of pressure, but approaches zero. This has been attributed to surface effects, specifically including the recombination limit; the atomic hydrogen outgassed to the surface must recombine into molecular hydrogen before it can escape into vacuum. The recombination rate is a function of the square of the surface coverage by atomic hydrogen. The concentration at the surface cannot be zero as assumed in conventional diffusion theory, but must remain at finite levels consistent with the outgassing rate. In this article a finite-difference analysis is made of a reported 2 h bake at 950 °C; the diffusion of hydrogen to the surface is modified to include a recombination limit. The outgassing rates with time are found together with the concentration profiles in the steel. The recombination coefficient is adjusted to match the observed postbake outgassing rate. It is concluded that the addition of the recombination limit to diffusion theory may be a viable analysis method to predict the results of specific time and temperature bake patterns.

Recombination Limited Concentration Profiles Versus Bake Time [4-11]
The recombination limited concentration of hydrogen atoms through the thickness of the metal is shown in **Fig. 4.10** for increasing bake times. The finite concentrations at the surfaces are a major departure from diffusion limited analyses which assume the surface concentration to be zero. As time increases, the distributions approach a sine function plus a constant, instead of the sine function found with diffusion limited outgassing [1].

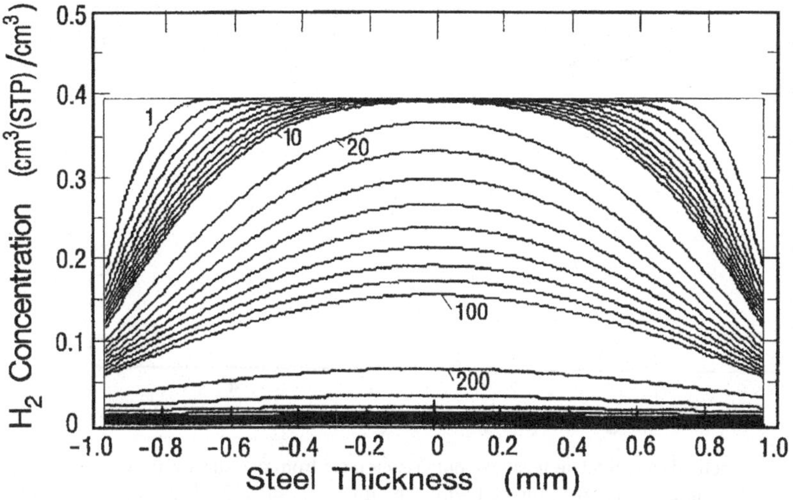

Fig. 4.10 Concentration profiles of atomic hydrogen calculated by recombination limited outgassing, within a 1.9 mm-thick 304 LN stainless-steel sheet, vacuum furnace baked at 950 °C. The concentration is shown as a function of the cross-sectional position measured from the center of the thickness of the steel sheet. The number of bake seconds are labeled on the profiles. The recombination coefficient assumed is $6 \times 10^{-22}\,\mathrm{cm}^4/(\mathrm{atom}\cdot\mathrm{s})$. Initial concentration assumed is 0.3 Torr · L at $0\,°C/\mathrm{cm}^3$ (Moore, 1995) [4-11]

Outgassing Rate Versus Time [4-11]

Diffusion limited outgassing as a function of time is compared with recombination limited in **Fig. 4.11**. The outgassing rate is shown in Torr · L at $0\,°\mathrm{C}/(\mathrm{s}\cdot\mathrm{cm}^2)$. For the first few seconds of the bake the diffusion limited outgassing is slightly greater, but within a few hundred seconds the recombination limited outgassing is far higher. At the end of the bake, the slope of the recombination limited outgassing approaches -2 on the log–log plot; increasing the time by a factor of 10 would reduce the outgassing rate by a factor of 100.

The cooldown time after the completion of the bake is neglected, so the outgassing drops vertically by a factor of 114 000, the ratio scaled from Hsueh and Cui [2]. Beyond this time there was no change reported in outgassing rate.

References

1. R. Calder and G. Lewin, *Br. J. Appl. Phys.* **18**, 1459 (1967).
2. H. C. Hseuh and Xiahua Cui, *J. Vac. Sci. Technol. A* **7**, 2418 (1989).

Akaishi et al. (2001) [4-12] presented an article, "True and measured outgassing rates of a vacuum chamber with a reversibly adsorbed phase."

Abstract [4-12]: The modeling for the pumpdown of a vacuum chamber with a reversibly adsorbed phase is carried out and the outgassing equation which predicts variations of surface coverage and gas density with time is derived. When it is assumed that the wall surface of a vacuum chamber consists of oxide layer and this layer involves weakly chemisorbed (nondissociate) H_2O with desorption energies

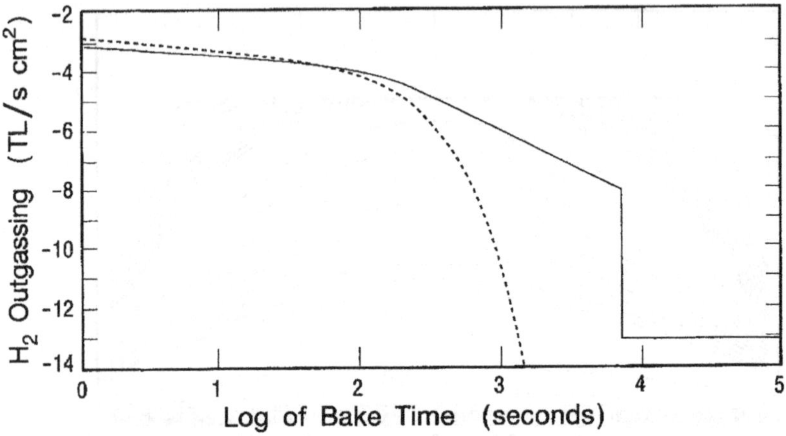

Fig. 4.11 Two methods to calculate postbake outgassing rates from one surface of a 1.9 mm-thick sheet of 304LN stainless steel are compared with experimental results [2]. The widely used "diffusion limited" method [1] (*dashed line*) gives rates even during the bake which become far less than the postbake room-temperature measurement of about 10^{-13} Torr · L at 0 °$C/($ s · cm$^2)$. The solid line is a combination of calculated and measured data. For the duration of the 2 h vacuum furnace bake at 950 °C (the 4 h warmup ramp is neglected), the "recombination limited" method is used with a recombination coefficient assumed to be 3×10^{-22} cm$^4/($atom · s$)$. After the end of the bake, the solid line represents experimental data. The cooldown from 950 °C to 25 °C causes a reduction in outgassing rate by a factor of 114 000; this number is extrapolated from the measured postbake changes up to 300 °C [2]. After this, at room temperature, there were no further changes in rate [2]. It appears that this recombination coefficient predicts a postbake outgassing rate near that observed. A number of uncertainties limit the utility of this result: (1) The initial concentration in the experimental beam tubes is not known, it is only assumed to be 0.3 Torr · L at 0 °$C/$cm^3 to be consistent with Calder and Lewin [1]. (2) The calculations are one dimensional, they assume that the sheet is a flat sheet, semi-infinite in extent so that the only variations in concentration are normal to the surface. However, the experimental samples were finite beam tubes, cylindrical in shape. (3) The calculation could be refined. The outgassing during warmup and cooldown could be included (Moore, 1995) [4-11]

from 20 to 30 kcal/mol, it is shown that the calculated outgassing rate from the outgassing equation well explains the experimental result that the outgassing rate measured by the orifice pumping method in the unbaked stainless steel chamber is dependent on pumping speed.

Data of Outgassing

Stainless Steel

Adams (1983) [4-13] presented an article, "A review of stainless steel surface."

Abstract [4-13]: The characteristics of the surface of various stainless steels are reviewed. The property of these alloys that makes them stainless is the formation of Cr_2O_3 layer on the surface. It has been found that this protective layer can be modified relatively easily by heating, abrading, chemical treatment, or ion bombardment. Modification can be changes in the chemical composition of the surface layer or the formation of a layer of segregated material on the surface. These changes may alter the protective nature of the surface films. The outgassing characteristics of stainless steel surfaces also vary depending upon the treatment these surfaces receive.

Fujita (1976) [4-14] presented an article, "Stainless steel as a vacuum industrial material," where the characteristics of stainless steel (SS304, 304L) such as work hardening after machining, corrosion resistance, and forming defects after welding are described in detail.

Ishimori et al. (1971) [4-15] measured the outgassing rates of stainless steel, mild steel and chromium plated mild steel by an orifice method after different pretreatments.

"The pretreatments were (1) drying by a sirocco fan after degreasing with carbon tetrachloride; (2) baking stainless steel at 100–450 °C and mild steel and chromium plated mild steel at 100–300 °C; and (3) exposing to argon and atmosphere after baking.

After pretreatment (1), the outgassing rates of stainless steel and chromium plated mild steel were of the order of $10^{-9}\,\mathrm{Torr \cdot L/(cm^2 \cdot s)}$, and the outgassing rates of mild steel were one order of magnitude higher than those of the first two. The effect of baking is the highest for stainless steel and the lowest for chromium plated mild steel. They discussed that the low effect of baking for chromium plated mild steel must be due to many pin-holes existing on the surface of chromium plated film [4-15]."

Electro-polishing and Vacuum Firing

Young (1969) [4-16] measured the outgassing characteristics of stainless steel and aluminum with different surface treatments by an orifice method.

Experimental Procedure and Results [4-16]
The apparatus used to measure the outgassing rate is shown in **Fig. 4.12**. The sample tanks were made of 1/8 in. thick 304 stainless steel or aluminum, 6 in. in diameter and 18 in. long.

The outgassing rate Q (shown in **Table 4.1**) was determined by measuring the difference in pressure between gauges No.1 and No.2, $(P_1 - P_2)$, and knowing the conductance C of the aperture, then $Q = (P_1 - P_2)C$. Corrections were made to the pressure determinations $(P_1 - P_2)$ and the conductance C to account for the outgassing gas being H_2.

Nuvolone (1977) [4-17] evaluated heat-treatment procedures for obtaining reproducible low degassing rates in 316 L stainless-steel high-vacuum systems.

Experimental Technique [4-17]
The measurement system is a standard type of apparatus, using the throughput method. The pressures are measured with Bayard-Alpert ionization gauges, and in addition is connected

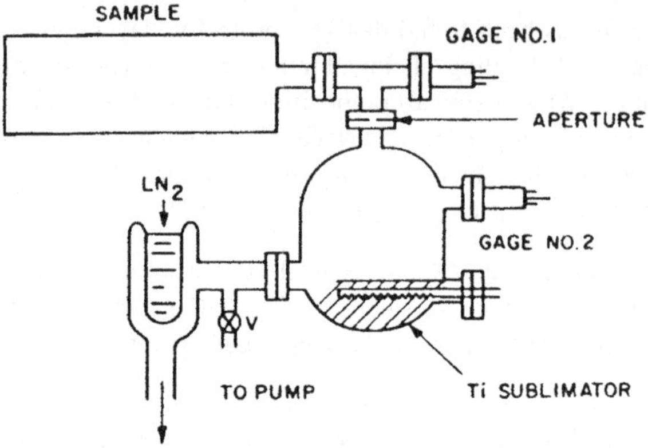

Fig. 4.12 Schematic of experimental apparatus for measuring the outgassing rate of stainless steel and aluminum samples (Young, 1969) [4-16]

a quadrupole residual gas analyzer, to check the composition of the residual gases and their contribution to the total pressure of the system. The samples treated by the various methods represent a small chamber of diameter 100 mm and length 400 mm. They possess welds with edges fused under argon and are equipped with "Conflat"-type flanges. They are machined

Table 4.1 Outgassing rates of 1/8 in.-thick stainless steel and aluminum after different processing (Young, 1969) [4-16]

Sample	Material and surface treatment	Outgassing rate after 24 h at room temperature $\text{Torr} \cdot \text{L}/(\text{s} \cdot \text{cm}^2)$
A	304 stainless steel (S. S.) cleaned by glass-bead shot blasting (32 μin. surface, baked 30 h at 250 °C).	2×10^{-12}
B	304 S. S., 4–6 μin. electropolished surface (baked 30 h at 250 °C).	3×10^{-12}
C	304 S. S., 20–25 μin. electropolished surface (baked 30 h at 250 °C).	2×10^{-12}
D	Sample B (above) baked with inner and outer surface exposed to air 16 h at 250 °C (plus an additional 15 h at 250 °C under vacuum).	5×10^{-13}
E	304 stainless steel (S. S.) cleaned by glass-bead shot blasting, then baked with inner and outer surface exposed to air 61 h at 450 °C (additional 15 h at 250 °C under vacuum).	3×10^{-13}
F	Sample E (above) after removal of inner and outer oxide by glass-bead shot blasting (additional 15 h at 250 °C under vacuum).	3×10^{-13}
G	304 S. S., 20–25 μin. electropolished surface baked 30 h at 250 °C plus 17 h at 450 °C (continuation of sample C above).	4×10^{-13}
H	Aluminum, type 1100, cleaned with detergent, rinsed in acetone, baked 15 h at 250 °C under vacuum.	4×10^{-13}

in 316 L stainless steel (AFNOR Z 2 CND 17-13). The surface state is rough drawn for the tubes and standard ultrahigh vacuum type for the flanges. Altogether the geometrical surface area under vacuum is $1300\,cm^2$.

Before undergoing any heat treatment each sample was subjected to the following cleaning sequence: (a) 2 h degreasing with perchlorethylene vapor at 125 °C. (b) washing for 1 h in Diversey 708 cleaner at 55 °C (in an ultrasonic cell), and (c) rinsing with clean water and drying.

Outgassing rates of stainless steel after different heat-treatment processing are presented in **Table 4.2** [4-17].

Yoshimura et al. (1990) [4-18] investigated the feature of electropolished and mechanically polished stainless-steel surfaces using a scanning electron microscope (SEM) and an Auger electron spectrometer (AES).

Microstructure [4-18]

Three kinds of sample surfaces [belt-polished, buff-polished and electropolished (EP)] were observed as secondary-electron images (SEI) and backscattered-electron images (BEI) using a scanning electron microscope (JSM-T330, JEOL Ltd.).

Table 4.2 Outgassing rates of 316L stainless steel after different processing as a function of the various gases observed (Nuvolone, 1977) [4-17]

Sample	Surface treatment	Outgassing rates 10^{-13} Torr · L/(s · cm^2)				
		H_2	H_2O	CO	A	CO_2
A	Pumped under vacuum for 75 h	670	430	65	···	10
	50 h vacuum bakeout at 150 °C	290	13	4.5	···	0.3
B	40 h vacuum bakeout at 300 °C	62	0.5	1.7	···	0.01
C	Degassed at 400 °C for 20 h in a vacuum furnace (5×10^{-9} Torr)	14	0.2	0.33	0.12	0.08
	Degassed at 800 °C for 2 h in a vacuum furnace (5×10^{-9} Torr)	2.7	···	0.05	···	0.04
D	Exposed to atmosphere for 5 months pumped under vacuum for 24 h.	···	55	50	···	10
	20 h vacuum bakeout at 150 °C	2.5	···	0.06	···	0.03
	2 h. in air at atmospheric pressure at 400 °C	13	···	0.84	···	0.3
E	Exposed to atmosphere for 5 months, pumped under vacuum for 24 h.	···	60	52	···	25
	20 h vacuum bakeout at 150 °C	13	0.56	0.28	···	0.13
F	20 h in oxygen at 200 Torr at 400 °C	450	190	···	92	···
	20 h vacuum bakeout at 150 °C	3.9	0.07	0.3	0.38	···
G	2 h in oxygen at 20 Torr at 400 °C	···	15	10	6.5	···
	20 h vacuum bakeout at 150 °C	···	0.67	0.48	0.34	···
H	2 h in oxygen at 2 Torr at 400 °C	···	12	39	14	···
	20 h vacuum bakeout at 150 °C	4.3	2.4	0.27	1.5	···

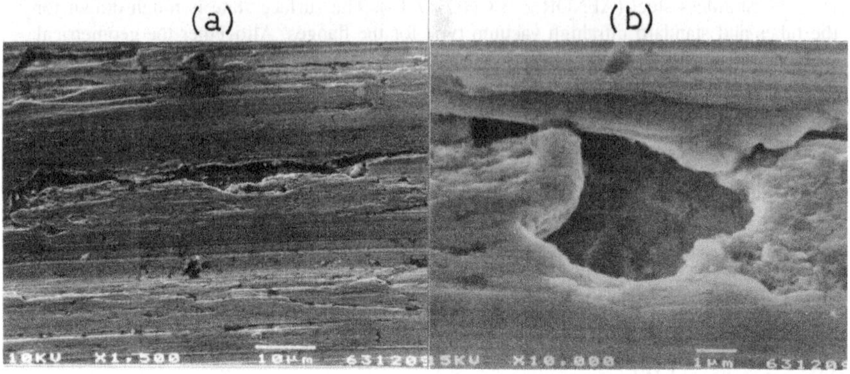

Fig. 4.13 SEM micrographs of the selected area of the belt-polished surface. **(a)** Area with wide fissures, and **(b)** fissure containing dust (Yoshimura et al., 1990) [4-18]

The belt-polished surface had many fissures whose sizes and structures varied depending on the area selected. SEM micrographs of high magnification revealed that some of the fissures contained dust, in which a large number of gas molecules (may be liquid or paste state) must be trapped. SEM micrographs of selected areas of the belt-polished surface are presented in **Fig. 4.13 (a)** and **(b)**.

The buff-polished surface also has many fissures along the polishing direction, whose sizes are smaller than those of the belt-polished surface. SEM micrographs of high magnification also revealed some fissures containing dust. Such micrographs are presented in **Fig. 4.14 (a)** and **(b)**.

The microstructure of the EP surface was almost independent of the area selected. SEM and BEM micrographs of the EP surface are presented in **Fig. 4.15 (a)** and **(b)**, respectively. There were no fissures on the surface. The BEM micrograph clearly shows the difference among backscattering electron yields of individual crystal faces, which means the EP surface was extremely flat, smooth, and clean.

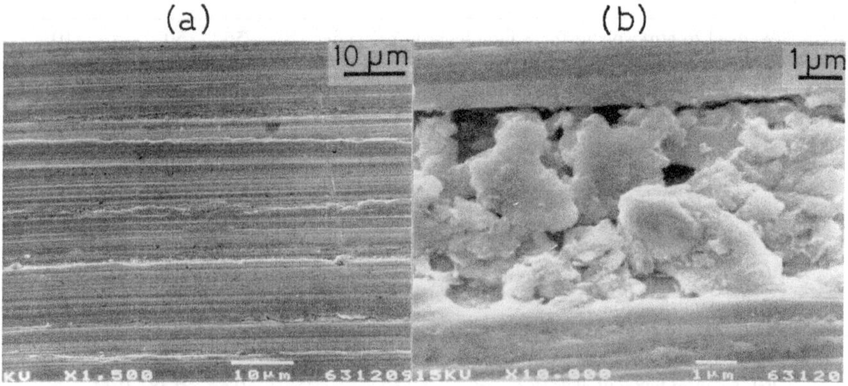

Fig. 4.14 SEM micrographs of the buff-polished surface. **(a)** Typical area, and **(b)** fissure containing dust (Yoshimura et al., 1990) [4-18]

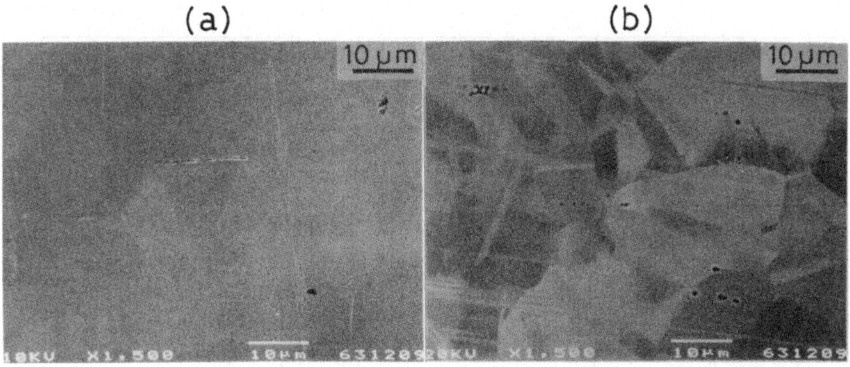

Fig. 4.15 **(a)** SEM micrograph and **(b)** BEM micrograph of the EP surface (Yoshimura et al., 1990) [4-18]

Elemental Features [4-18]

AES analysis was conducted for the as-received, buff-polished, and EP surfaces to clarify their elemental features, by concentrating the interest on the O and C intensities.

AES spectra on the outer surfaces and AES depth profiles for sample surfaces were analyzed using an Auger electron spectrometer (JAMP-30, JEOL Ltd.). An AES spectrum and AES depth profile of the as-received surface are presented in **Fig. 4.16 (a)** and **(b)**, respectively. Those of the buff-polished and EP surfaces are presented in **Fig 4.17** and **4.18**, respectively.

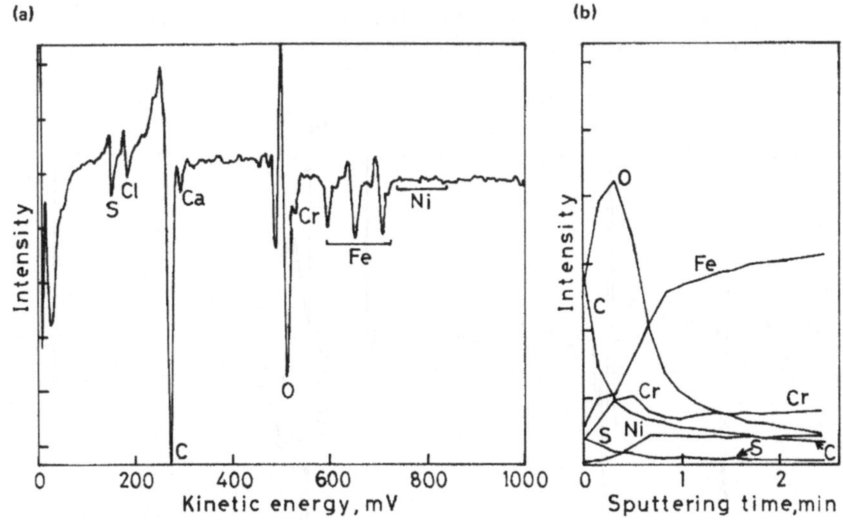

Fig. 4.16 **(a)** AES spectrum and **(b)** AES depth profile of the as-received surface. Sputtering conditions were as follows: Sputter ion; Ar^+, energy; 3 keV, normal incidence. The sputtering rate was unknown for the sample surfaces, though it was known as 10 nm/min for SiO_2 (Yoshimura et al., 1990) [4-18]

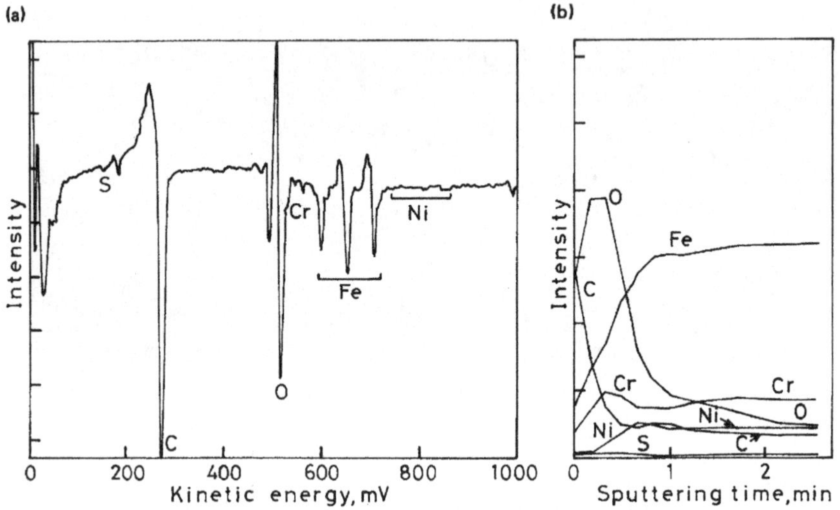

Fig. 4.17 **(a)** AES spectrum and **(b)** AES depth profile of the buff-polished surface. Sputtering conditions: Sputter ion; Ar^+, energy; 3 keV, normal incidence (Yoshimura et al., 1990) [4-18]

By comparing the spectra of **Fig 4.16** and **4.17**, one can recognize that the as-received and buff-polished surfaces were almost the same in elemental features. Elemental features of the EP surface, in comparison with the buff-polished surface, are as follows.

(1) The O intensity on the outer surface was much higher than that on the buff-polished surface, which means that the oxide layer of the EP surface was finer in microstructure than that of the buff-polished surface. The O intensity was reduced much more rapidly

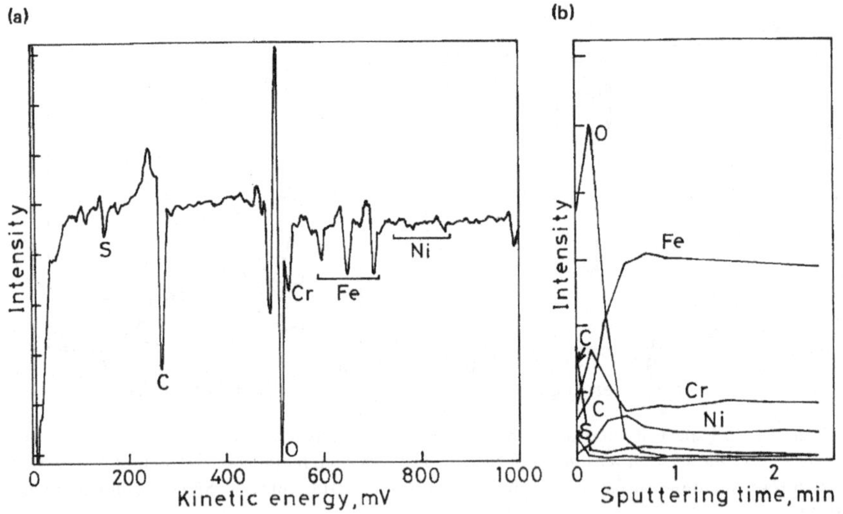

Fig. 4.18 **(a)** AES spectrum and **(b)** AES depth profile of the EP surface. Sputtering conditions: Sputter ion; Ar^+, energy; 3 keV, normal incidence (Yoshimura et al., 1990) [4-18]

to a negligible level with sputtering, in comparison with that of the buff-polished surface, which means that the oxide layer of the EP surface was much thinner than that of the buff-polished surface.

(2) The C intensity on the outer surface was much lower than that on the buff-polished surface; it rapidly reduced to zero with sputtering. The EP surface was cleaner than the buff-polished surface, in respect of the C-based contamination.

(3) The Cr and S intensities on the outer surface were both higher than those on the buff-polished surface, respectively.

Conclusions [4-18]
Microstructures and elemental features of the EP surface were investigated with SEM and AES, respectively, in comparison with those of as-received, belt-polished, and buff-polished surfaces. No fissures were observed on the EP surface by SEM. The oxide layer of the EP surface was cleaner and thinner, and finer in microstructure than those of mechanically polished surfaces. This is the reason why the EP surface showed an extremely low outgassing rate.

Tohyama et al. (1990) [4-19] presented an article, "Outgassing characteristics of electropolished stainless steel".

Abstract [4-19]: SUS316L stainless steels, which have various contents of gas and the amounts of nonmetallic inclusions, have been melted in various conditions and rolled to plates. The sample of 8 mm × 8 mm × 1 mm were prepared from these plates, and all surfaces were finally electropolished. The amounts of outgassing gas by thermal desorption and the outgassing rate of H_2, H_2O, CO, CO_2 after baked at 1123 K in a vacuum of 10^{-7} Pa were measured by a quadrupole mass spectrometer. The molecular amount of H_2 of all the samples was larger by 10–10^2 times, than those of H_2O, CO, CO_2 and there was no correlation between the H_2 molecular amounts of outgassing by heating and the H_2 content in steel. But the outgassing rate of H_2 was reduced with decreasing both the H_2 content and the amounts of nonmetallic inclusions. Especially, the outgassing rate of H_2 was attributed to the inclusions because the hydrogen was found to be entrapped by them on the electropolished surface. Clean steel, which has been produced in special clean melt, showed two orders of magnitude lower in rate than normal steel. The outgassing rates of CO and CO_2 were reduced with decreasing C content in steel.

Figure 4.19 shows an example of nonmetallic inclusion on the electropolished surface.

Yoshimura et al. (1991) [4-20] studied the surface of a vacuum-fired (1050 °C) stainless-steel plate using a SEM and an AES. "The grain boundaries of the surface were vague and shallow, which occurred due to elemental diffusion at high temperature in vacuum (**Fig. 4.20**). AES spectra and AES depth profiles of the vacuum-fired surface and the as-received surface were also measured, showing that the newly formed oxide layer of the vacuum-fired surface is thinner, and can be said to be higher in density than the native layer of an "as-received" surface (**Fig. 4.21**) [4-20]."

Tsukui et al. (1993) [4-21] investigated the two processes, electrochemically polishing and high temperature baking, and their combination in order to reduce the outgassing rate of wall material. "The inner surfaces of the stainless-steel chamber processed by two optimum ways to minimize its outgassing rate have been characterized by means of surface analysis and measurements of gas-emission

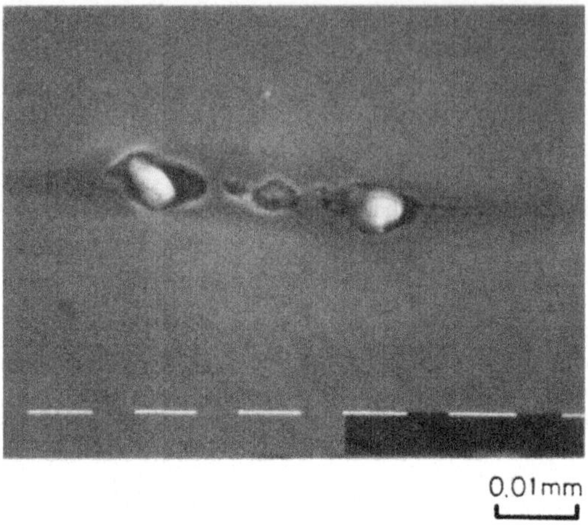

Fig. 4.19 Example of nonmetallic inclusion on the electropolished surface (Tohyama et al., 1990) [4-19]

rate. It has been concluded that in order to accomplish an extremely high vacuum (XHV; $< 10^{-12}$ Torr), the inner surface should be electrochemically polished after the complete baking of the material at 1300 K [4-21]."

Moore (1998) [4-22] discussed the atmospheric permeation of an austenitic stainless-steel wall. He claimed that the wall thickness of austenitic stainless steel, now in common use in vacuum chambers, is based on an overestimate of permeation of atmospheric hydrogen through the walls. "Actual room-temperature permeation is estimated to be less than the accepted estimate by several orders of magnitude. Thus, vacuum systems may be designed with substantially thinner walls, as thin as they can be made to withstand atmospheric pressure. These in turn will permit much simpler degassing procedures and the achievement of lower outgassing rates in practice [4-22]."

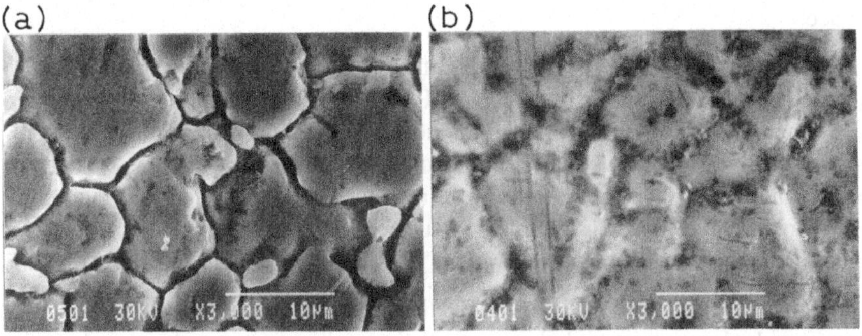

Fig. 4.20 SEM micrographs. **(a)** As-received surface, and **(b)** vacuum fired surface (Yoshimura et al., 1991) [4-20]

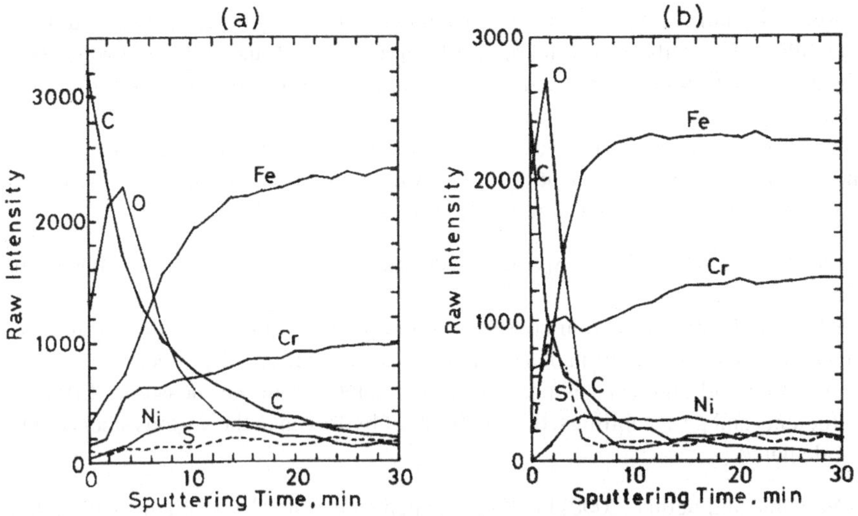

Fig. 4.21 AES depth profiles. **(a)** As-received surface, and **(b)** vacuum fired surface. Sputtering conditions were as follows: Sputter ion; Ar^+, energy; 3 keV, normal incidence. The sputtering rate was unknown for the both sample surfaces, though it was known as 10 nm/min for SiO_2 (Yoshimura et al., 1991) [4-20]

Moore (2001) [4-23] proposed and recommended thin-walled vacuum chambers of austenitic stainless steel for reducing the time, effort, and cost needed to outgas the chambers to reach the desired and specified vacuum level. Room-temperature, single-sided outgassing rates for various times and various wall thicknesses are presented in **Table 4.3** [4-23].

Table 4.3 Room temperature, single sided outgassing rates for various times and wall thickness. Outgassing rates are for single sided outgassing. The equations correspond to those of Calder and Lewin for two sided outgassing, and for twice the wall thickness. For $t > 0.1d^2D^{-1}$, where the residual atomic hydrogen concentration is a quarter sine wave with peak at the air surface and zero at the vacuum surface, then $Q = 2c_0Dd^{-1} \times \exp\left[-\pi^2(2d)^{-2}Dt\right]$ where Q =outgassing rate, $\mathrm{Torr \cdot L \cdot s^{-1} \cdot cm^{-2}}$; c_0 =initial uniform hydrogen concentration ($= 0.3\,\mathrm{Torr \cdot L/cm^3}$ as given by Calder and Lewin); D =room temperature diffusion coefficient= 5×10^{-14} $\mathrm{cm^2/s}$; d =wall thickness, cm; t =time under vacuum, seconds. The concentration may be reduced orders of magnitude by a high temperature bake. Following the bake, the concentration has returned to a uniform distribution. The subsequent outgassing rate is NOT constant, as commonly presumed, but follows the same transient shape as the original, with zero time reset to the end of bake (Moore, 2001) [4-23]

Wall thickness	Time to begin	Time to reach outgassing rate of:			
cm	Exponential	10^{-13}	10^{-15}	10^{-17}	$\mathrm{T \cdot L/(s \cdot cm^2)}$
0.0001	5.556 h	7.51	11.84	16.15	days
0.001	23.15 days	1.47	2.65	3.83	years
0.01	2314.5 days	0.874	2.06	3.24	centuries
0.1	634.2 yr	2.83	14.7	26.5	millenia

Nemanič and Šetina (1999) [4-24] studied outgassing in thin-wall stainless-steel cells. "The cells (wall thickness 0.15 mm, AISI304 and AISI316) were used to test the relation between thermal treatment and the resulting room-temperature outgassing rate q_{out}. The cells were cleaned and baked in situ at about 205 °C for about 16 h. The pressure rise was recorded by a spinning rotor gauge after sealoff. The initial pressure-rise slope at a H_2 equivalent pressure of about 10^{-5} mbar was in the order of $q_{out} = 1 \times 10^{-14}\,\text{mbar} \cdot \text{L}(\text{H}_2) \cdot \text{s}^{-1} \cdot \text{cm}^{-2}$ [4-24]."

Eschbach et al. (1963) [4-25] conducted permeability measurements with gaseous hydrogen for various steels.

Abstract [4-25]: Diffusion coefficients and permeabilities of hydrogen in low-carbon steel and chromium-nickel steels were determined and the effect of structure, composition and ambient gas pressure investigated. From the measured diffusion coefficients and permeabilities, the solubility of hydrogen in the various steels could be calculated.

Nemanič and Šetina (2000) [4-26] presented an article, "Experiments with a thin walled stainless-steel vacuum chamber."

Abstract [4-26]: Two identical stainless-steel chambers (AISI type 304, volume 12 L) with uniform wall thickness (0.6 mm) were constructed. Each was equipped with a miniature ion getter pump and spinning rotor gauge (SRG) thimble. They were pumped down and initially degassed by a moderate bakeout (2.5 h, 150 °C) using a turbomolecular pump system. The gas accumulation method was applied for the determination of q_{out} and the total amount of released gases. A capacitance manometer was used during bakeout but a **SRG** was applied after the chamber was sealed off. The q_{out} at room temperature after initial bakeout was in the order of $\cong 3 \times 10^{-12}\,\text{mbar} \cdot \text{L}/(\text{cm}^2 \cdot \text{s})$. SRG measurements of q_{out} at 50 °C over several days showed a tendency of slight decrease. The bakeout of the chamber was being repeated at a duration of 72 h at 200 °C. The amount of released hydrogen was much smaller than expected, whereas the decrease of q_{out} was noticeable. In fact, it was one tenth of the value before the bakeout. The results show the benefit of using a moderately thin wall and agree with the model of recombination-limited hydrogen outgassing from stainless steel.

Bernardini et al. (1998) [4-27] presented an article, "Air bake-out to reduce hydrogen outgassing from stainless steel."

Abstract [4-27]: Heating the raw material at 400 °C in air was suggested as a money saving alternative to the classical vacuum heating at 950 ° C. We report the results of hydrogen content analysis performed on stainless steel samples submitted to different treatments, and also the measurement performed on the prototype tube (1.2 m-diameter, 48 m-long). We concluded that air bake-out drives out most of the hydrogen absorbed in the bulk stainless steel, while the presence of the oxide layer does not reduce the hydrogen outgassing.

Ishikawa and Nemanič (2003) [4-28] presented an article, "An overview of methods to suppress hydrogen outgassing rate from austenitic stainless steels with reference to UHV and extremely high vacuum (EXV)."

They concluded as follows:

"Several papers in this field show constant interest to prepare material with the lowest q_{out}, which in turn would simplify the path to UHV and EXV. The impression is that the driving force for investigations is not just the achievement of the lowest ultimate pressure, but simply cost effective solutions in relation to several applications of UHV. Such solutions always require some compromise between possible issues [4-28]."

"The achievement of moderate values of $q_{out} = 1 \times 10^{-13}$ mbar · L · cm^{-2} · s^{-1} can be obtained today quite routinely, neglecting the true rate-limiting mechanism. To go beyond this limit, much longer processing time is required to outgas the material than has been usually stated. Among the engineering solutions to overcome the problem of a high q_{out}, a different constructional concept of reducing the wall thickness to the limit of structural integrity seems to be the simplest one. In conventional chamber design, a well-controlled adlayer formation, like dense Cr oxide or TiN, seems to be the most promising [4-28]."

Zajec and Nemanič (2005) [4-29] presented an article, "Hydrogen pumping by austenitic stainless steel."

Abstract [4-29]: In the present study, hydrogen sorption and desorption kinetics close to equilibrium were investigated in a pinched-off AISI 316 steel cell by a sensitive pressure-rise method. The pressure was monitored with a spinning rotor gauge (SRG) just before the pinch-off and after it for 6 months at two stabilized temperatures: 25 °C and 50 °C. The preprocessing of the cell (of uniform wall thickness 0.15 mm volume 125 cm^3, and inner surface 460 cm^2) consisted of baking at 200 °C for 109 h with several evacuation cycles to ultrahigh vacuum (UHV). The quantity of released hydrogen during the bake-out procedure equaled the average concentration change $\Delta C = 2.8 \times 10^{17}$ at. H/cm^3. After the pinch-off intentionally done in the high vacuum range where hydrogen represented the residual atmosphere, surprisingly the hydrogen pressure slowly declined from the initial $p(328\,K) = 3.7 \times 10^{-4}$ mbar, with an initial rate $dp/dt = -5.5 \times 10^{-11}$mbar/s and later attained a stable value, which could be termed the equilibrium. In similar reported experiments, where valving-off began in the UHV, the dp/dt was always positive and constant over several orders of magnitude in pressure. During 6 months of measurements, a sudden temperature jump from 25 °C to 55 °C or back was applied a few times to investigate the stability of the equilibrium or the impact on the pressure course. The most plausible explanation of the results is given along with discussion whether hydrogen permeated through the cell wall or if it was absorbed in the cell wall.

Aluminum Alloy, Copper and Titanium

Tsukahara (2000) [4-30] reviewed outgassing rates of various metals (stainless steel, aluminum alloy, copper and copper alloy, and titanium) pretreated with various technologies. Yoshimura (2003) [4-31] reviewed some articles and listed many articles on surface treatments for various kinds of metals (stainless steel, aluminum alloy, copper and copper alloy, and titanium) for ultrahigh-vacuum system construction.

Recently, metals other than stainless steel, such as aluminum alloy and titanium, were investigated as construction materials for ultrahigh-vacuum systems.

Aluminum alloys clearly have advantages that they are light in weight and can be machined easily. So, many large chambers for film formation systems are recently made of aluminum alloys.

Sato et al. (2002) [4-32] presented an article, "Anodized film for vacuum equipment. Their research intends to create the suitable protective layer on aluminum alloy for semiconductor processing systems.

Abstract [4-32]: We developed new methods for anodizing aluminum surfaces of plasma process chambers, one in sulfuric acid, and another in oxalic acid, in order to reduce particulate generation. The anodized films show a lower density of cracks than conventionally anodized films even after exposing to high temperature. Therefore, they have a better performance of small number in particulate. Analysis by X-ray diffraction, scanning electron microscopy, infrared spectroscopy and electron probe microanalysis reveals that both films are amorphous and that a cracking mechanism depends on the electrolyte and anodizing process. It is also found that number of cracks and total amount of outgassing increase after sealing process.

Figure 4.22 shows the cross-sectional scanning electron micrographs of newly developed anodized films.

Aluminum alloys are preferred materials for the vacuum system of electron storage rings because of their low residual radioactivity. Many reports on glow-discharge cleaning for aluminum alloys have been presented.

The gaskets of the International ConFlat® flange (ICF), which are widely used in ultrahigh vacuum systems, are made of oxygen-free copper. It has been reported that ESD and PSD of oxygen-free copper are very low.

Titanium plates are exclusively used as the cathodes of sputter ion pumps. Outgassing of titanium (thermally-stimulated, electron-stimulated and photon-stimulated desorption) is considered to be very low.

Permeation Through Elastomer Seals

Viton O-rings (fluoroelastomer) and Kalrez O-rings (perfluoroelastomer) are selectively used as seals in the high-vacuum system of an electron microscope. Water-vapor molecules in the atmosphere permeate elastomer seals such as Viton O-rings, and as a result, H_2O molecules become the main residual gas in high-vacuum systems using elastomer seals [4-38].

Csernatony (1966) [4-33, 4-34, 4-35, 4-36] and Csernatony and Crawley (1967) [4-37] presented various properties of Viton-A elastomers based on his experimental results.

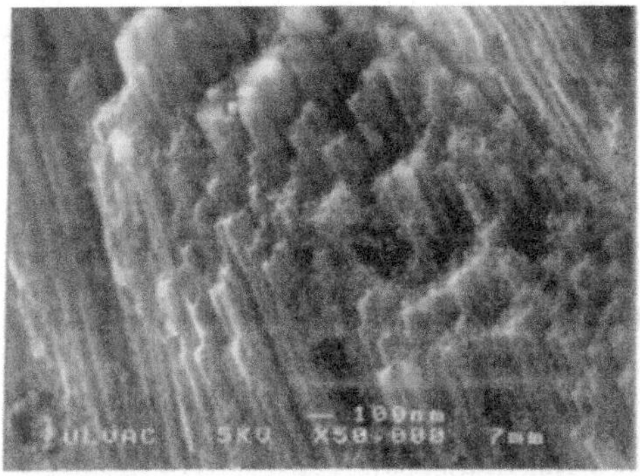

(a)

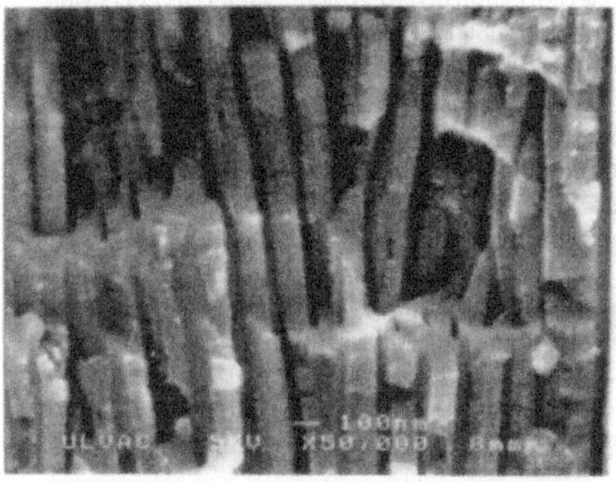

(b)

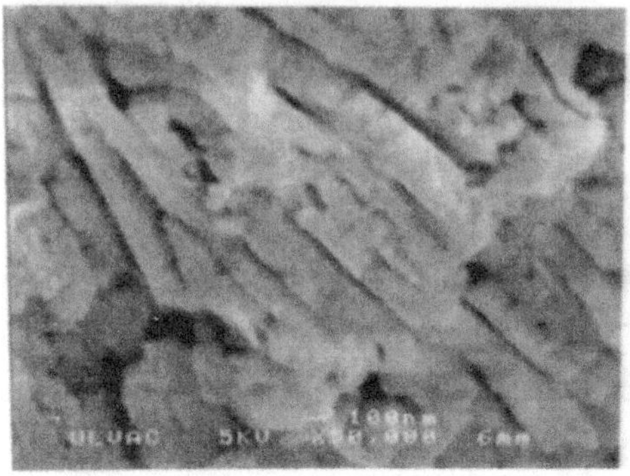

(c)

Fig. 4.22 Cross-sectional scanning electron micrographs. **(a)** Newly developed anodized film (A6061) in sulfuric acid. **(b)** Newly developed anodized film (A5052) in oxalic acid. (c) Newly developed anodized film (A6061) in oxalic acid (Sato et al., 2002) [4-32]

Csernatony and Crawley (1967) [4-37] reported the practical application of Viton O-ring seals in high vacuum. "The importance of preventing atmospheric permeation leads to the adoption of double O ring seals, with an evacuated interspace, as an alternative to all-metal gaskets. When using double O ring seals, pressure of about 10^{-10} Torr can readily be reached, with very low organic contamination levels [4-37]."

Yoshimura (1989) [4-38] presented the data on water vapor permeation through Viton O-ring seals.

Water Vapor Permeation Through Viton O-ring Seals[4-38]
The experimental setup is shown in **Fig. 4.23**. A stainless-steel test dome (TD) with a quadrupole mass spectrometer (MS, Anelva Ltd.) and Bayard-Alpert gauge (BA1) is evacuated with the first pump (DP1, Diffstak 100/300, 280 L/s, Santovac-5, Edwards Ltd.) followed by the second one (DP2, 2.5 in., Santovac-5). A thick Viton O ring (Co-seal, 4 in., Edwards Ltd.) is fitted between TD and DP1. BA1 is sealed with a small Viton O ring, P-14 (Mitsubishi Cable Industries Ltd.). MS is sealed with a copper gasket. The following experiment (experiment A) was performed in summer for about one month. The room temperature and humidity during the period were about 25 °C and 75% on an average, respectively. The pumping system had been already running for several months before the experiment was started.

The TD including MS was first baked (about 100 °C, 3 days) in situ under high vacuum. The bakeout temperature of the Co-seal was guessed to be nearly 50 °C. Both BA1 and MS were switched on (emission current, 1 mA) when baking was stopped, and never switched off during the long experiment period to avoid the adsorption and desorption effect which occurred if they were switched on and off. Another gauge BA2 was intermittently switched

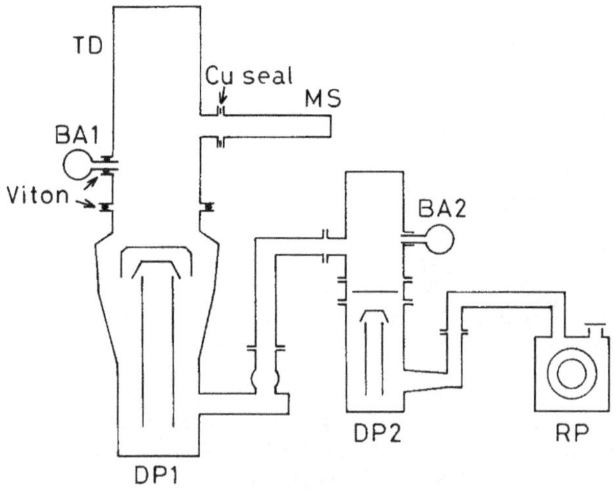

Fig. 4.23 Experimental setup. A thick Viton O ring (Co-seal, 4 in.) is used between the test dome (TD) and the first diffusion pump (DP1, Diffstak 100/300). In experiment B, Diffstak of DP1 was replaced by another diffusion pump (4 in.) with a cold cap and a water-cooled chevron baffle (Yoshimura, 1989) [4-38]

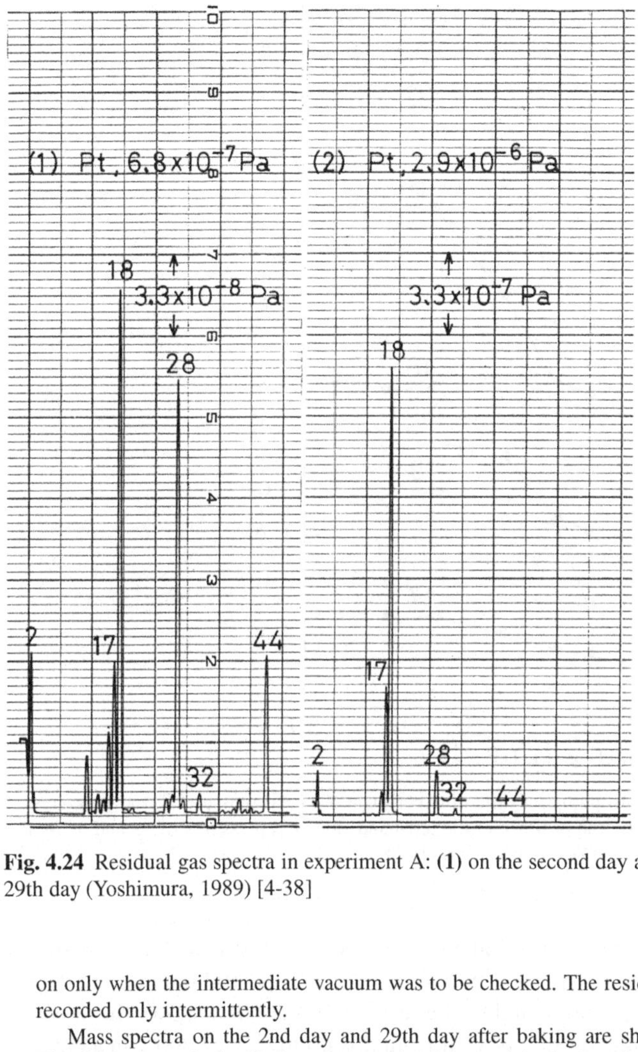

Fig. 4.24 Residual gas spectra in experiment A: **(1)** on the second day after baking and **(2)** on the 29th day (Yoshimura, 1989) [4-38]

on only when the intermediate vacuum was to be checked. The residual gas spectrum was recorded only intermittently.

Mass spectra on the 2nd day and 29th day after baking are shown in **(1)** and **(2)** of **Fig. 4.24**, respectively. Hydrocarbon peaks decreased in the first few days, and were kept in minimum levels. On the other hand, the H_2O peaks [masses 17 (OH^+), 18 (H_2O^+), and 19 (H_3O^+)] were kept almost constant in the first several days, and then increased gradually until saturated. The increasing H_2O pressure was reflected on the total pressure. The O_2 peak (32) also increased with pumping time until saturated. The H_2 peak (2) increased slightly with time. Other atmospheric gases, CO and N_2 (28), Ar (40), and CO_2 (44), on the other hand, did not increase in peak amplitude with time. The rise of the total pressure (P_t), H_2O pressure (17, 18, 19), and O_2 pressure (32), as a function of pumping time, are presented in **Fig. 4.25**.

Chernatony (1977) [4-39] reviewed the recent advances in elastomer technology for UHV applications.

Abstract [4-39]: The advances made in vacuum-oriented elastomer performance during the development of the latest UHV compatible elastomer (Kalrez) are briefly

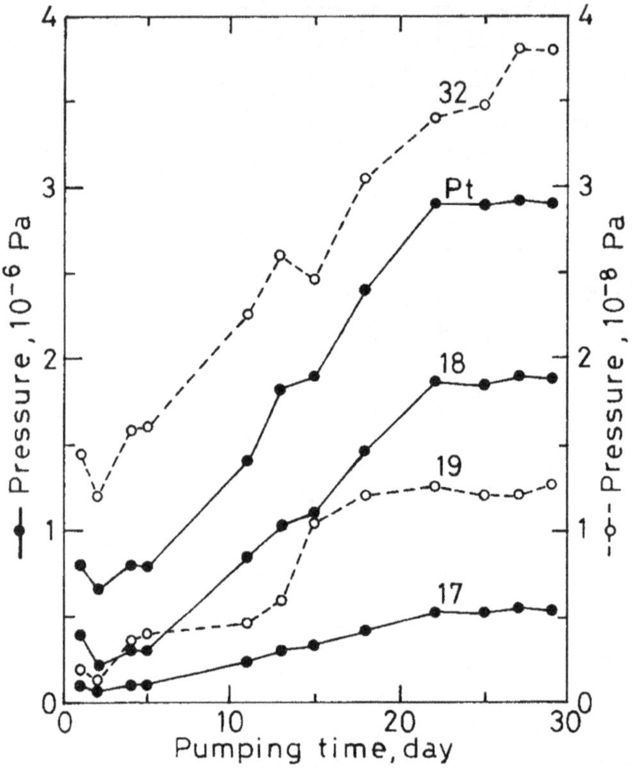

Fig. 4.25 Pressure-rises with pumping time in experiment A. Total pressure P_t and pressures of masses 17 and 18 are read by the ordinate of the left-hand side, and pressures of 19 and 32 by the ordinate of the right-hand side (Yoshimura, 1989) [4-38]

reviewed by correlating their chemical character with the physical characteristics of relevance to UHV technology. Total pressure and mass-spectral data on the degradation characteristics of Kalrez and of the latest Viton (E60C) shows the superior performance of the first and establishes it from this point of view to be in a class of its own for UHV-oriented applications.

Peacock (1980) [4-40] described the practical selection of elastomer materials for vacuum seals, intending to introduce some of the considerations involved in choosing vacuum seals to the vaccum worker. “Early reports on the new perfluoroelastomer Kalrez have shown problems with compression set and adhesion, but these difficulties may have resulted from improper groove design. Outgassing rates for unbaked and baked polymers are summarized from many references. Permeation data for various polymers and gases are also reported [4-40].”

Table 4.4 Outgassing rates for unbaked and baked polymers in $\text{Torr} \cdot \text{L} \cdot \text{s}^{-1} \cdot \text{cm}^{-2}$ (Peacock, 1980) [4-40]

Polymer	Unbaked, 1 h pumping	Baked, ultimate	References
Fluoroelastomer	4×10^{-7}–2×10^{-5}	3×10^{-11}–2×10^{-9}	3, 5, 6, 7, 8, 9, 10, 15
Buna-N	2×10^{-7}–3×10^{-6}	-	1, 11
Neoprene	5×10^{-5}–3×10^{-4}	-	1, 10, 12
Butyl	2×10^{-6}–1×10^{-5}	-	1, 10
Polyurethane	5×10^{-7}	-	10, 11
Silicone	3×10^{-6}–2×10^{-5}	-	10, 12, 13
Perfluoroelastomer	3×10^{-9}	3×10^{-11}–3×10^{-10}	4, 14, 15
Teflon	2×10^{-8}–4×10^{-6}	-	1, 5, 9, 11, 12, 16
KEL-F	4×10^{-8}	3.5×10^{-10}	1
Polyimide	8×10^{-7}	3×10^{-11}	8

Outgassing [4-40]

Since the elastomer seals in many vacuum systems are a major contributor to the gas load, consideration of outgassing and its reduction is important. **Table 4.4** shows the range of outgassing rates in the literature for unbaked polymer seal materials after one hour of pumping, and for baked elastomers where values could be found. The data for baked elastomers demonstrate that it is possible to reach very low outgassing rates under good conditions.

Permeation [4-40]
It is often difficult in practice to separate the effects of diffusion, permeation, and outgassing. With polymer materials most observed outgassing is actually gas diffusion from the interior of the sample. Numerous authors [2, 17, 18, 19, 20, 21] have discussed permeation and diffusion in polymers, and interested readers are referred to the literature. If the discussion is restricted to the equilibrium case, the permeation rates can be compared for the various gases and polymers of interest. In this simple case, the rate of gas permeation, Q, in sccm/s through a diffusion barrier is:

$$Q = KA(P_1 - P_2)/d,$$

where K is the permeation constant in $\text{sccm} \cdot \text{s}^{-1} \cdot \text{cm}^{-2} \cdot \text{cm} \cdot \text{atm}^{-1}$, A is the area of the barrier in cm^2, P_1 is the high side pressure, and P_2 is the low side pressure in atmospheres, and d is the barrier thickness in cm. Of the various units used in the literature, these are as common as any, and aid in comparing results of various authors.

Table 4.5 contains permeation data for some polymers. The references contain information for many other polymers. Two examples can illustrate the practical importance of permeation. Probably anyone who has leak tested a vacuum system using O-rings has observed the leak signals due to permeation. Permeation can be a serious annoyance at times, for example when the test object is "bagged." It is difficult to tell a leak from permeation unless the helium can be masked away from the elastomer seals. If a signal is observed in less than a minute after exposure to probe gas, it is usually a leak. A signal which starts after a minute and slowly increases may be considered to be permeation.

Table 4.5 Permeation data for various polymers and gases [2, 3, 22, 23, 24, 25]. The temperature range is 20 °C–30 ° C. The units are sccm · s^{-1} · cm^{-2} · cm · atm^{-1}. $1 sccms^{-1} (atm \cdot cm^3 \cdot s^{-1}) =$ 100 Pa · L/s(Peacock, 1980) [4-40]

Polymer	Helium ($K \times 10^8$)	Nitrogen ($K \times 10^8$)	Oxygen ($K \times 10^8$)	Carbon dioxide ($K \times 10^8$)	Water ($K \times 10^8$)
Fluoroelastomer	9–16	0.05–0.3	1.0–1.1	5.8–6.0	40
Buna-N	5.2–6	0.2–2.0	0.7–6.0	5.7–48	760
Buna-S	18	4.8–5	13	94	1800
Neoprene	10–11	0.8–1.2	3–4	19–20	1400
Butyl	5.2–8	0.24–0.35	1.0–1.3	4–5.2	30–150
Polyurethane	-	0.4–1.1	1.1–3.6	10–30	260–9500
Propyl		7	20	90	-
Silicone	-	-	76–460	460–2300	8000
TEFLON	-	0.14	0.04	0.12	27
KEL-F	-	0.004–0.3	0.02–0.7	0.04–1	-
Polyimide	1.9	0.03	0.1	0.2	-

References

1. B. B. Dayton, *1959 6th National Symposium on Vacuum Technology Transactions* (Pargamon, New York, 1960), p. 101.
2. W. G. Perkins, *J. Vac. Sci. Technol.* **10**, 543 (1973).
3. A. Lebovits, *Mod. Plast.* **43**, 139 (1966).
4. L. Firth and D. G. Stringer, "*KALREZ Perfluoroelastomer for Use in UHF Systems*", Culham Laboratory Report RFX/VA/N15. Unpublished.
5. R. Jaeckel, *1961 Transactions of the 8th National Vacuum Symposium combined with the 2nd International Congress on Vacuum Science and Technology*, (Pargamon, New York, 1962), p. 17.
6. A. Schram, *Le Vide* **103**, 55 (1963).
7. L. de Csernatony, *Vacuum* **16**, 129 (1966).
8. P. W. Hait, *Vacuum* **17**, 547 (1967).
9. R. S. Barton and R. P. Govier, *J. Vac. Sci. Technol.* **2**, 113 (1965).
10. I. Farkass and E. J. Barry, *1960 7th National Symposium on Vacuum Technology Transactions* (Pargamon, New York, 1961), p. 35.
11. F. Markley, R. Roman, and R. Voseck, *1961 Transactions of the 8th National Vacuum Symposium combined with the Second International Congress on Vacuum Science and Technology*, (Pargamon, New York, 1962), p. 78.
12. D. J. Santeler, *1958 5th National Symposium on Vacuum Technology Transactions*, (Pargamon, New York, 1959), p. 1.
13. R. J. Elsey, *Vacuum* **25**, 347 (1975).
14. M. F. Zabielski and P. R. Blaszuk, *J. Vac. Sci. Technol.* **13**, 644 (1976).
15. L. de Chernatony, *Vacuum* **27**, 605 (1977).
16. G. Thieme, *Vacuum* **13**, 137 (1963).
17. R. M. Barrier, *Diffusion in and through solids*. (Cambridge U. P., Cambridge, 1951).
18. N. Buchner, *Kunststoffe* **49**, 401 (1959).
19. C. J. van Amerongen, *J. Polym. Sci.* **5**, 307 (1950).
20. C. E. Rogers in "*Engineering Design for Plastics*", edited by E. Baez (Reinhold, New York, 1964).

21. "*Diffusion in Polymers*", edited by J. Crank and G. S. Park (Academic, New York, 1968).
22. J. E. Ayer, D. R. Schmitt, and R. M. Mayfield, *J. Appl. Polym. Sci.* **3**, 1 (1960).
23. A. W. Myers, V. Tammela, V. Stannett, and M. Szwarc, *Mod Plast.* **37**, 139 (1960).
24. R. A. Pasternak, M. V. Christensen, and J. Heller, *Macromolecules* **3**, 366 (1970).
25. "*The Oil, Solvent, Chemical Resistance and Permeability of Commercial Elastomers*", 2-8426 Adiprene-L Permanent File (E. I. du Pont de Nemours and Co., Wilmington, DE).

Yoshimura (1985) [4-41] measured net outgassing rates $K(t)$ per unit surface area of typical nonmetallic materials after t−hour evacuation, measured by the differential pressure rise method. The data are summarized in **Table 4.6**. (For the principle of the differential pressure rise method, see the next Section on "Methods for Measuring Outgassing Rates.)

Table 4.6 Outgassing rates $K(t)$ (Torr · L · s^{-1} · cm^{-2}) of typical nonmetallic materials after t−hour evacuation (Yoshimura, 1985) [4-41]

Materials, conditions, pretreatments	Outgassing rates K(Torr · L · s^{-1} · cm^{-2}) $10^{10}K(3)$	$10^{10}K(7)$	$10^{10}K(10)$
Elastomers:			
Natural rubber, thick	32 000	23 000	20 000
Nitril, O-ring	4 500	2 300	2 000
Nitril, O-ring, vacuum bakeout (100 °C, 3 h)	3.0	1.2	0.9
Viton, O-ring	4 000	2 000	1 600
Viton, O-ring, vacuum bakeout (100 °C, 3 h)	0.6	0.3	0.18
Viton, O-ring, exposed to dried air (0.5 h) following vacuum bakeout (100 °C, 3 h)	80	20	15
Plastics:			
Polyvinylchloride, thick	18 000	7 500	5 000
Electron microscope film	15 000	2 500	800
Teflon, O-ring	25	4	1.5
Teflon, O-ring, vacuum bakeout (100 °C, 3 h)	0.07	0.005	0.003
Tefron, exposed to dried air (0.5 h) following vacuum bakeout (100 °C, 3 h)	0.4	0.02	0.01
Ceramics:			
Alumina, stick, porous	300	60	40
Alumina, stick, porous, vacuum bakeout (100 °C, 3 h)	0.3	0.05	0.04
Alumina, stick, porous, exposed to dried air (0.5 h) following vacuum bakeout (100 °C, 3 h)	50	1	0.8
Alumina, stick, fine	0.12	0.02	0.01
Steatite, stick	90	7	6
Steatite, stick, vacuum bakeout (100 °C, 3 h)	0.4	0.015	0.01
Steatite, stick, exposed to dried air (0.5 h) following vacuum bakeout (100 °C, 3 h)	5	0.04	0.03

Evaporation

Vapor pressure data for typical elements is important for obtaining an ultrahigh vacuum with the bakeout treatment. Some elements of construction material might evaporate with a considerable rate at high temperatures. When a screw made of brass containing Zn is degassed at high temperature around 300 °C (573 K), zinc (Zn) shows about 10^{-3} Torr of vapor pressure, causing a trouble. If a screw of brass is bombarded with high-energy electrons, a trouble may occur due to evaporation of zinc.

Honig (1957) [4-42] presented an article, "Vapor pressure data for the more common elements."

Summary [4-42]: The most recent data available concerning vapor pressures, melting and boiling points, and heats of sublimation have been selected, tabulated, and plotted for 57 elements, many of which are of special interest to workers in the fields of electronics and high-vacuum technique. It has been found convenient to present vapor pressure data graphically as plots of log$p$ (mm Hg) versus log T (K), and also to tabulate the absolute temperatures for fixed pressures. This collection contains data published or available before March 1, 1957.

Figures 4.26 and **4.27** present the vapor pressure data in graphical form. The curves were placed on two separate sheets in such a way as to minimize interference. The circle point shown on most curves is the melting point. Where the melting point falls outside the pressure range of the graph, the letters "s" (solid) or "l" (liquid) have been appended to the chemical symbol [4-42].

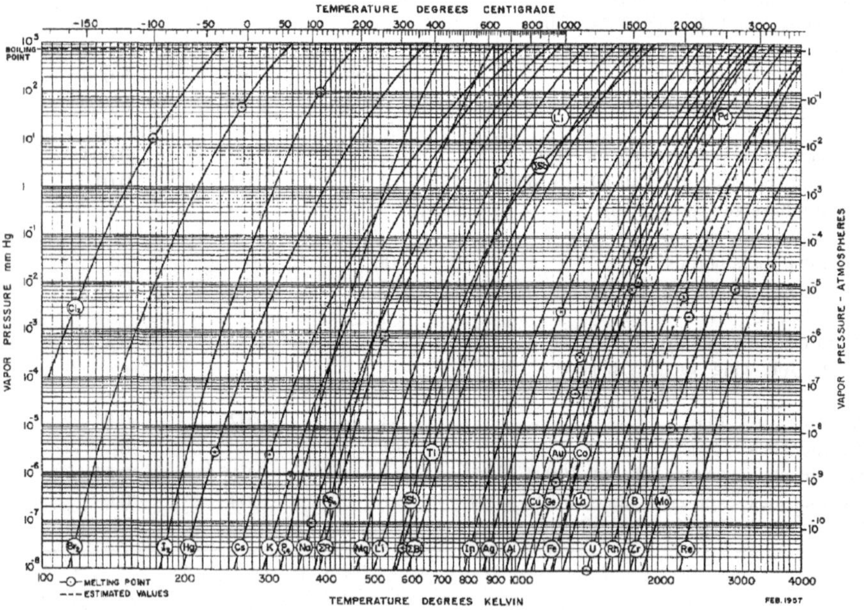

Fig. 4.26 Vapor pressure curves for the more common elements (1) (Honig, 1957) [4-42]

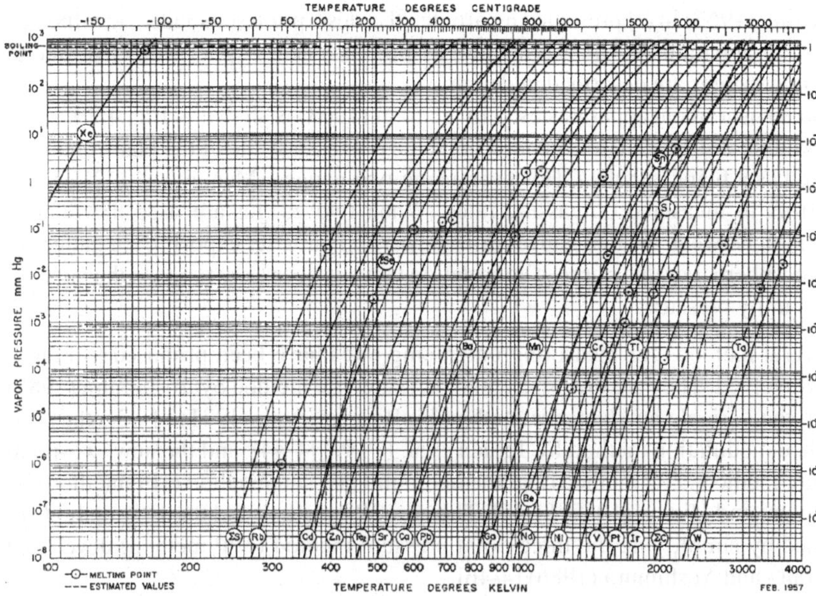

Fig. 4.27 Vapor pressure curves for the more common elements (2) (Honig, 1957) [4-42]

Methods for Measuring Outgassing Rates

Engineers who engage in designing high-vacuum systems of instruments must often estimate the outgassing rates of some new materials or newly pretreated materials.

Dayton (1959) [4-1] measured outgassing rates of various materials using an apparatus in which two pumping speeds can be selected (**Fig. 4.28**).

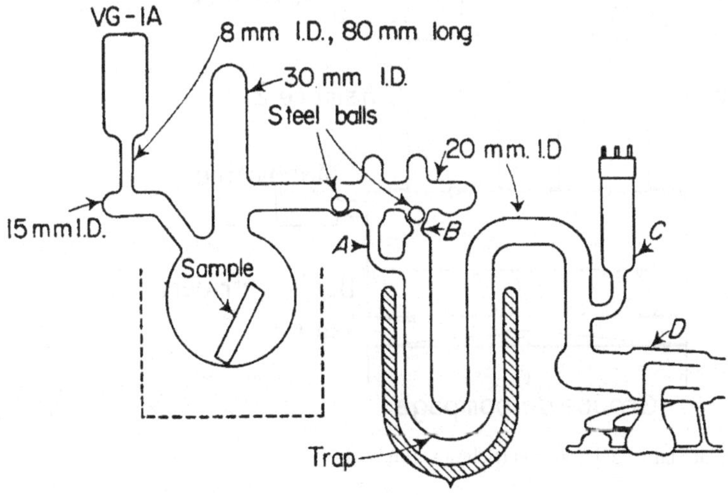

Fig. 4.28 Modified Zabel apparatus (Dayton, 1959) [4-1]

Schram (1963) [4-7] presented an article, "La désorption sous vide", in which an interesting apparatus for measuring outgassing rates and a systematic error due to using two gauges are described in details.

The measurement apparatus used by Schram (1963) is shown in **Fig. 4.29**.

A sample was loaded in Chambre 1 and the outgassing rate of the blank chamber are used as a blank-run data. However, using two gauges (Manomètre 1, 2) caused big errors in measured outgassing rates. When two gauges are used the relative calibration between the two gauges must be carried out after exposing the gauges to the atmosphere. For this purpose an isolation valve is necessary to make the vacuum field of unity pressure.

Several methods for measuring outgassing rates have been proposed to reduce errors involved. They are the differential pressure-rise method by Yoshimura (1985) [4-41], the variable conductance method by Berman et al. (1971) [4-43], the conductance modulation method by Terada et al. (1989) [4-44], and the new two-point pressure method (2PP method) and the one-point pressure method (1PP method) by Yoshimura and Hirano (1989) [4-45]. The new 2PP method and the 1PP method are obtained by simplifying the three-point pressure method (3PP method) developed by Hirano and Yoshimura (1986) [4-46].

When one measures the outgassing rates of sample material inserted inside a vacuum chamber, one must eliminate the error due to the outgassing rate of chamber walls and the pumping function of the ionization vacuum gauge used. The differential pressure-rise method presented by Yoshimura (1985) [4-41] has an outstanding advantage that the error due to chamber walls and the vacuum gauge is much reduced.

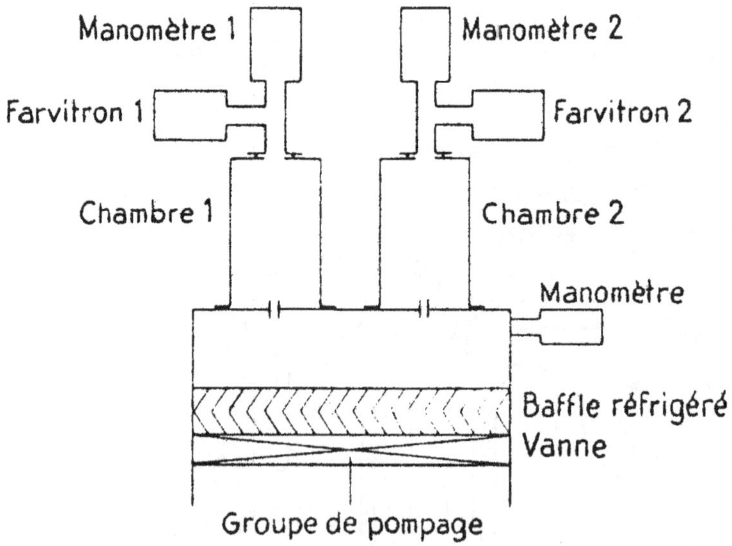

Fig. 4.29 Mesure des taux de désorption (Schram, 1963) [4-7]

Differential Pressure-rise Method [4-41]

Principle [4-41]
Figure 4.30 (a) show the test-dome system schematically, where both test domes, TD1 and TD2, are made of homogeneous glass, including greaseless cocks C1 and C2. TD2 with a sample with surface area A_S has its volume V_2 and area A_2, both of which are much smaller than volume V_1 and area A_1 of the empty dome TD1, respectively. Typical pressure-rise curves in measuring procedures are presented in **Fig. 4.30 (b)**.

The test-dome system with a sample is evacuated to a high vacuum. After an evacuation time t_1, the cocks C1 and C2 are sequentially handled to measure a pressure-rise rate $(dP/dt)_1$ at a reference pressure P_r for a combined domes of TD1 and TD2 including the sample and another rate $(dP/dt)_2$ at the same pressure P_r for the empty dome TD1, as shown in **Fig. 4.30 (b)**. The total time required to measure the successive two pressure-rise rate, $(dP/dt)_1$ and $(dP/dt)_2$, is only a few minutes, which is negligibly short comparing with the evacuation time t_1. This process is repeated after scheduled evacuation times t_2, t_3 and so on.

Denote net outgassing rates per unit surface area of the domes TD1 and TD2 at a reference pressure P_r for the measurement of $(dP/dt)_1$ as q_{11} and q_{21}, respectively, and a net rates per unit surface area of the domes TD1 at P_r for the measurement of $(dP/dt)_2$ as q_{12}. And denote pumping speeds of the vacuum gauge (VG) at P_r for the measurements

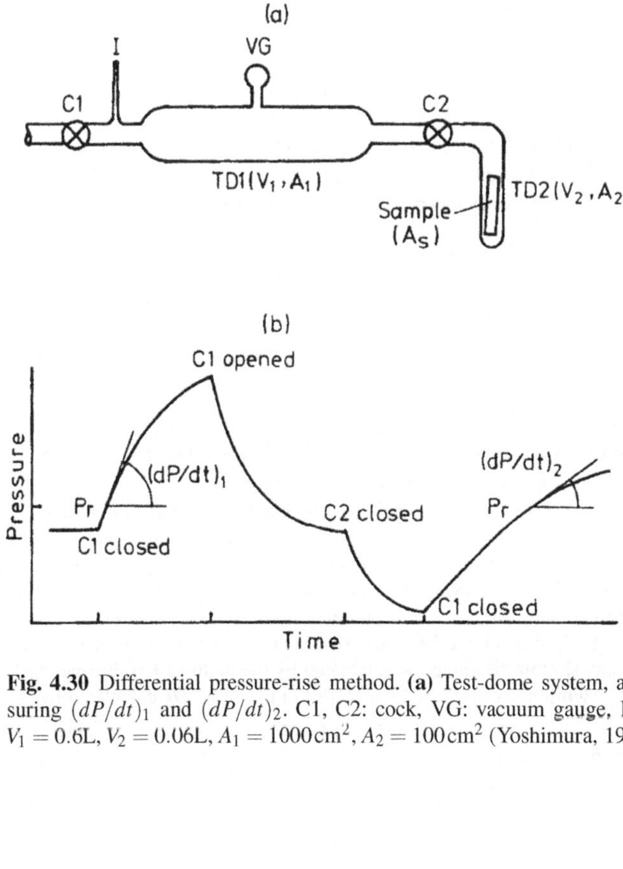

Fig. 4.30 Differential pressure-rise method. **(a)** Test-dome system, and **(b)** procedures for measuring $(dP/dt)_1$ and $(dP/dt)_2$. C1, C2: cock, VG: vacuum gauge, I: gas inlet. Parameters are: $V_1 = 0.6$L, $V_2 = 0.06$L, $A_1 = 1000\,\text{cm}^2$, $A_2 = 100\,\text{cm}^2$ (Yoshimura, 1985) [4-41]

of $(dP/dt)_1$ and $(dP/dt)_2$ as S_{g1} and S_{g2}, respectively. Then, the following equations are obtained:

$$(V_1+V_2)\left(\frac{dP}{dt}\right)_1 = Q_S + q_{11}A_1 + q_{21}A_2 - P_r S_{g1}, \tag{4.14}$$

$$V_1\left(\frac{dP}{dt}\right)_2 = q_{12}A_1 - P_r S_{g2}, \tag{4.15}$$

where Q_S is the net outgassing rate of the sample at P_r.

Assume that $q_{11} = q_{21} = q_{12} = q$ and $S_{g1} = S_{g2} = S_g$, then Eq. (4.14) and Eq. (4.15) reduce to

$$(V_1+V_2)\left(\frac{dP}{dt}\right)_1 = Q_S + (A_1+A_2)q - P_r S_g, \tag{4.16}$$

$$V_1\left(\frac{dP}{dt}\right)_{02} = A_1 q - P_r S_g, \tag{4.17}$$

respectively. The assumption introduced above would be acceptable because the histories of both domes under vacuum would almost the same with each other and the same gauge VG is used for measuring the pressure-rise rates $(dP/dt)_1$ and $(dP/dt)_2$.

When the left-hand side of Eq. (4.14), $(V_1+V_2)\left(\frac{dP}{dt}\right)_1$, is calculated as Q_S, as usually done in conventional pressure-rise methods, a large error is involved due to the outgassing of both domes and the pumping function of the gauge VG.

Now, the outgassing rate Q_S can be calculated more accurately by two differential methods, as described below.

Method 1:
The following equation is derived from Eqs. (4.16) and (4.17):

$$(V_1+V_2)\left(\frac{dP}{dt}\right)_1 - V_1\left(\frac{dP}{dt}\right)_2 = Q_S + A_2 q. \tag{4.18}$$

That is, Q_S is calculated from the left-hand side of Eq. (4.18), though an error A_2q, the outgassing rate of TD2, is involved in the calculated value. The error A_2q is much smaller than the error involved in the value conventionally calculated from the left-hand side of Eq. (4.14).

Method 2:
The following equation is derived from Eqs. (4.16) and (4.17):

$$(V_1+V_2)\left(\frac{dP}{dt}\right)_1 - V_1\frac{A_1+A_2}{A_1}\left(\frac{dP}{dt}\right)_2 = Q_S + P_r S_g \frac{A_2}{A_1}. \tag{4.19}$$

That is, Q_S is calculated from the left-hand side of Eq. (4.19), though an error $P_r S_g A_2/A_1$ due to the pumping speed of VG is involved in the calculated value. The error $P_r S_g \frac{A_2}{A_1}$ becomes small because the ratio A_2/A_1 is selected as small as 0.1.

Now, compare the errors involved in the values Q_S calculated by the two differential methods described above.

The parameters related to the errors are: A_1, 1000 cm^2, A_2, 100cm^2; P_r, 10^{-8}–10^{-5} Torr; q, 10^{-11}–10^{-8} Torr $\cdot$ L $\cdot$ s^{-1} $\cdot$ cm^{-2}. A Bayard-Alpert (B-A) gauge with an emission of 0.5 mA was used to measure pressure-rise rates. The pumping speed S_g of the B-A gauge has been estimated as about 0.01 L/s as described later (see Yoshimura, 1970 [6-12] in Chap. 6). Therefore, the error A_2q involved in the value Q_S calculated by the method 1 is estimated as 10^{-9}–10^{-6} Torr $\cdot$ L $\cdot$ s^{-1}, while the error $P_r S_g \frac{A_2}{A_1}$ involved in the value Q_S calculated by the method 2 is estimated as 10^{-11}–10^{-8} Torr $\cdot$ L $\cdot$ s^{-1}. As a result, the method 2 is superior to method 1 from the view point of error.

Estimation of the Characteristic Values P_X and K_0 of a Solid Material [4-41]
The net outgassing K per unit surface area of a solid material could be expressed as

$$K = K_0\left(1 - P/P_X\right). \tag{4.44}$$

The characteristic values, internal pressure P_X and free outgassing rate K_0 per unit surface area of a solid material depend on the history of the solid under vacuum. The values P_X and K_0 could be estimated by two different K values, K_1 and K_2, measured at two different pressures P_1 and P_2 as

$$P_X = (K_1 P_2 - K_2 P_1)/(K_1 - K_2), \tag{4.45}$$

$$K_0 = (P_1 K_2 - P_2 K_1)/(P_1 - P_2). \tag{4.46}$$

The characteristic values P_X and K_0 of Viton O-rings at a certain condition were estimated using two different K values measure at two different pressures.

Variable Conductance Method [4-43]

Berman et al. (1971) [4-43] presented an article, "Corrections in outgassing rate measurements by the variable conductance method."

Theory [4-43]
The general disposition of the system to measure the outgassing rate by the variable conductance method is shown in **Fig. 4.31**.

If the permeation through the walls of the chamber II and the leak in the same chamber are negligible and no temperature gradient exists between the two chambers, the equations:

$$Q_1 - S_1 p_1 = C(p_1 - p_2) \tag{4.20}$$

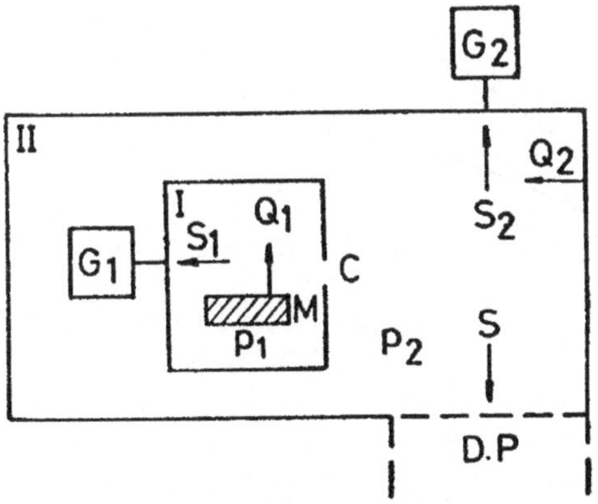

Fig. 4.31 Sketch of the measuring system. M, material; Q_1, outgassing rate of M (The outgassing rate of the walls of chamber I is negligible); Q_2, outgassing rate of the walls of chamber II; G_1, G_2, ionization gauges; S_1, S_2, pumping speed of the gauges G_1 and G_2; C, conductance of aperture adjustable to achieve different conductances; p_1, p_2, pressures in chamber I and II, respectively; S, pumping speed of the system; D.P, duffusion pump (Berman et al., 1971) [4-43]

$$Q_1 - S_1 p_1 + Q_2 = p_2 S_2 + p_2 S \tag{4.21}$$

describe the system when the equilibrium pressure is reached. Further, if Q_1 (Torr · L/s) and C (L/s), the pressures in chamber I and II may be written:

$$p_1 = \frac{1}{1+\frac{S_1}{C}} \left(\frac{Q_1}{C} + p_2 \right) \text{ (Torr)} \tag{4.22}$$

$$p_2 = \frac{Q_1 - S_1 p_1 + Q_2}{S + S_2}. \text{ (Torr)} \tag{4.23}$$

From Eqs. (4.20), (4.21) and (4.23) the normalized equation

$$p_1 = \frac{1}{1+\frac{S_1}{C}+\frac{S_1}{S+S_2}} \left(\frac{Q_1}{C} + \frac{Q_1+Q_2}{S+S_2} \right) \text{ (Torr)} \tag{4.24}$$

is obtained.

Let us assume that:

(a) the outgassing rate (Q_2) of the walls if the chamber II is so chosen as not to obscure the outgassing rate (Q_1) of the material, and both (Q_1) and (Q_2) remain constant from the beginning of the evacuation of the system until the end of the measures.
(b) the value of the conductances is so chosen as to lead always to a ratio $\frac{S_1}{C} \leq 0.1$.
(c) the pumping speed of the gauges (S_1, S_2) is at least two orders of magnitude less than the pumping speed (S) of the system, and S_1, S_2 and S are constant in time.

In the light of these assumptions, Eq. (4.24) becomes:

$$p_1 = Q_1 \left(\frac{1}{C} + \frac{1}{S} \right). \text{ (Torr)} \tag{4.25}$$

Employed by Oatley [1] to determine the speed of a vacuum pump, Eq. (4.25) represents a straight line in a p_1 and $1/C$ system of axes (1, **Fig. 4.32**).

The line intersects the axes at:

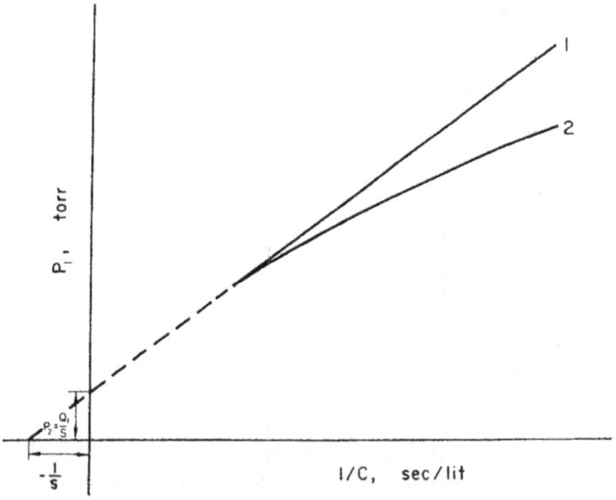

Fig. 4.32 General (2) and particular (1) aspect of Eq. (4.24). Pressure p_1 is supposed to be obtained at the same moment for different values of $1/C$ (Berman et al., 1971) [4-43]

$$p_1 = \frac{Q_1}{S} = p_2 \text{ when } \frac{1}{C} = 0$$

$$\frac{1}{C} = -\frac{1}{S} \text{ when } p_1 = 0.$$

Its slope reads the value of the outgassing rate Q_1.

If all the assumptions made before remain unchanged, besides the values of the conductances, chosen this time so as $\frac{S_1}{C} > 0.1$, Eq. (4.24) will represent a curve (2, **Fig. 4.32**). The tangent at any point to this curve will show a lower slope, than the slope of the straight line, hence at lower outgassing rate.

Comparing Eqs. (4.24) and (4.25) it is obvious that the expression

$$\varphi\left(\frac{S_1}{C}\right) = \frac{1}{1 + \frac{S_1}{C} + \frac{S_1}{S+S_2}}$$

which differentiates them from each other is responsible for bending the line (1) to the curve (2) (**Fig. 4.32**).

The importance of $\varphi\left(\frac{S_1}{C}\right)$ may be clearly pointed out if its variation as a function of $\left(\frac{S_1}{C}\right)$ is plotted (**Fig. 4.33**).

From the graph in **Fig. 4.33** it results:

—for the same values of the ratio $\frac{S_1}{C}$, $\varphi\left(\frac{S_1}{C}\right)$ depicts more curvature the higher the pumping speed of the gauge (S_1, S_2).

—variations of the pumping speed (S) over a large range of orders of magnitude ($10 \cdots 10^3$ L/s) does not influence the aspect of the curve.

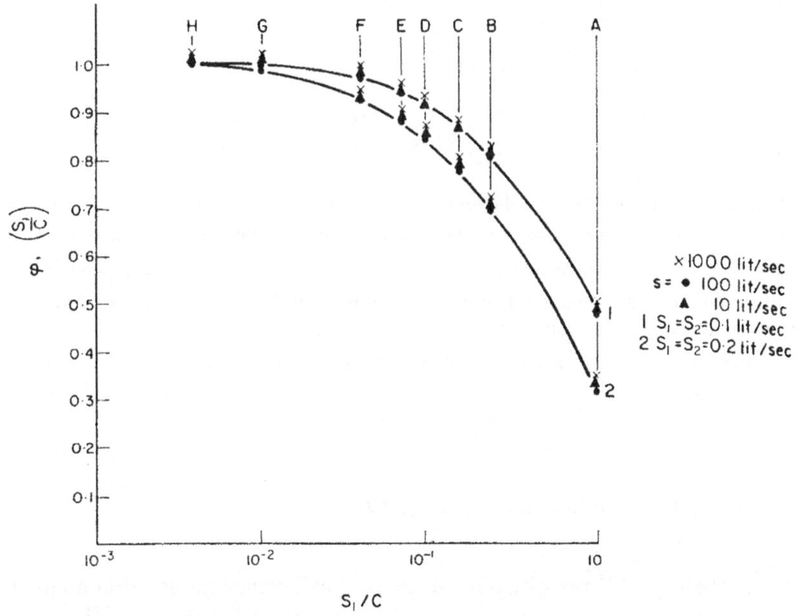

Fig. 4.33 Variation of $\varphi\left(\frac{S_1}{C}\right)$ as a function of $\frac{S_1}{C}$. Calculated values (Berman et al., 1971) [4-43]

Therefore, the larger the values the ratio $\frac{S_1}{C}$ takes, the more bent-down line (2) (**Fig. 4.32**) would be, and the smaller the slope at any point of the curve would be in comparison with the straight line (1).

If the value of $\varphi\left(\frac{S_1}{C}\right)$ is known for different values of C, the curve Eq. (4.24) may be easily corrected to the straight line Eq. (4.25) and consequently the outgassing rate deduced.

The value of $\varphi\left(\frac{S_1}{C}\right)$ can be obtained without knowing any characteristics of the vacuum system employed. The differential of Eq. (4.24) gives the slope for the values of C for which:

$$\frac{dp_1}{d(1/C)} = \frac{Q_1 - Q_2\frac{S_1}{S+S_2}}{\left(1+\frac{S_1}{C}+\frac{S_1}{S+S_2}\right)^2}. \tag{4.26}$$

Taking $\frac{dp_1}{d(1/C)} = Q_1''$ and $Q_1 - S_1 p_1 = C(p_1 - p_2) = Q_1'$ and substituting in the first part of Eq. (4.24) p_1 with the value given by (4.24)

$$Q_1'' = \frac{Q_1 - Q_2\frac{S_1}{S+S_2}}{1+\frac{S_1}{C}+\frac{S_1}{S+S_2}}. \text{ (Torr}\cdot\text{L/s)} \tag{4.27}$$

Both in Eqs. (4.26) and (4.27) $Q_2\frac{S_1}{S+S_2}$ may be neglected on account of the assumptions accepted before. Then:

$$Q_1'' = \frac{Q_1}{\left(1+\frac{S_1}{C}+\frac{S_1}{S+S_2}\right)^2} \text{ (Torr}\cdot\text{L/s)} \tag{4.28}$$

$$Q_1' = \frac{Q_1}{\left(1+\frac{S_1}{C}+\frac{S_1}{S+S_2}\right)}. \text{(Torr}\cdot\text{L/s))} \tag{4.29}$$

Equalizing Q_1 (4.28) and (4.29) it results:

$$1+\frac{S_1}{C}+\frac{S_1}{S+S_2} = \frac{Q_1'}{Q_1''}$$

Q_1' may be calculated for a certain value of C from Eq. (4.20) when p_1 and p_2 are known. Q_1' may be measured from the slope of the tangent to the curve at the selected value of C.

Calculating $\varphi\left(\frac{S_1}{C}\right)$ for the conductances employed and multiplying the corresponding values of p_1 with the values found this way, a straight line will be found. The slope of this line shows the real outgassing rate.

If the characteristics of the vacuum system are known, namely, S, S_1 and $S_2 \varphi\left(\frac{S_1}{C}\right)$ may be easily calculated.

Conductance Modulation Method [4-44]

Terada et al. (1989) [4-44] presented an article, "Conductance modulation method for the measurement of the pumping speed and outgassing rate of pumps in ultrahigh vaccum."

Principle [4-44]

Figure 4.34 shows the outline of the vacuum system considered in this analysis. The system consists of a chamber, a pump, and an orifice system inserted between them. The conductance of the orifice system is changed from C_A to C_B by the operation from outside of the vacuum chamber. The pumping speed S_A at the entrance of the orifice system corresponding to S_P is given as a combination of S_P and C_A,

$$\frac{1}{S_A} \cong \frac{1}{C_A} + \frac{1}{S_P}; \tag{4.30}$$

for S_B,

$$\frac{1}{S_B} \cong \frac{1}{C_B} + \frac{1}{S_P}. \tag{4.31}$$

These equations are approximate expressions because we have neglected a small correction for the difference in the diameter at the joint between the orifice assembly and the pump. The accurate equations are given in the previous publication [1]. The pressure P_A in the chamber pumped through C_A is

$$P_A = \frac{Q_i}{S_A} + \frac{Q_W}{S_A} + \frac{Q_P}{S_P} \tag{4.32}$$

$$= \frac{Q_i}{S_A} + P_{A0}. \tag{4.33}$$

Here P_{A0} is the ultimate pressure in the test dome evacuated through the orifice system A. The similar relation also holds for P_B. If Q_i, Q_W, and Q_P are constant during the process of the CM method, the following relation is obtained:

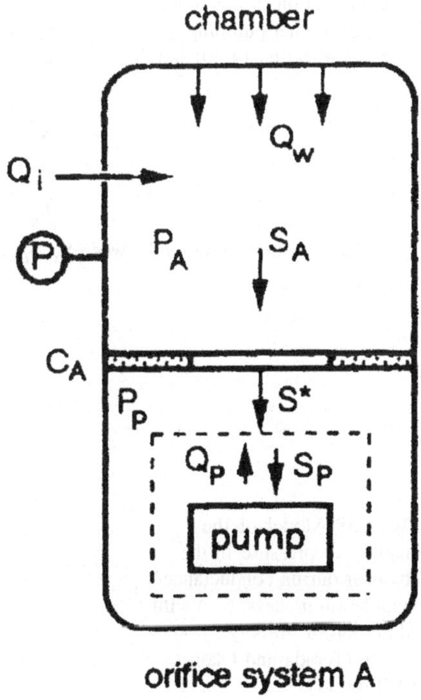

Fig. 4.34 Gas balance in UHV system. S_P: Intrinsic pumping speed. S^*: Net pumping speed. Q_W: Outgassing rate from the wall. Q_P: Outgassing rate from the pump. Q_i: Gas admission rate from outside of the system. C_A: Conductance of the orifice system (Terada and Okano, 1989) [4-44]

$$\frac{(P_A - P_{A0})}{(P_B - P_{B0})} = \frac{S_B}{S_A}; \tag{4.34}$$

here P_{A0} and P_{B0} are the pressures measured at $Q_i = 0$. The values of S_F, S_A, and S_B are obtained by using Eqs. (4.30), (4.31), and (4.34). **Figure 4.35** shows schematically the pressure changes during the conductance modulation at finite Q_i and $Q_i = 0$. Q_i: Gas admission rate from outside of the system.

The difference of the pressures P_{A0} and P_{B0} at $Q_i = 0$ relates to Q_W as follows:

$$P_{A0} - P_{B0} = \left(\frac{1}{S_A} - \frac{1}{S_B}\right) Q_W. \tag{4.35}$$

The value of Q_P is calculated from the known value of Q_W as

$$Q_P = \left(P_{A0} - \frac{Q_W}{S_A}\right) S_P. \tag{4.36}$$

The calculations described above assume that S_P is constant for the pressures ranging from P_{A0} to P_A. In case this assumption is not valid, the value of S_P obtained from Eq. (4.30), (4.31), and (4.34) means an averaged value in this pressure range. If we are interested in the variations of S_P at its operating pressures, S_P at specific pressure can be determined from the pressure differences, ΔP_A and ΔP_B, corresponding to the small increment of Q_i, ΔQ_i. This procedure is equivalent to the substitution of ΔP_A and ΔP_B for $(P_A - P_{A0})$ and $(P_B - P_{B0})$ in Eq. (4.34), respectively. For small enough ΔQ_i, S_P at specific pressure is obtained as

$$S_P = \frac{C_A C_B (\Delta P_B - \Delta P_A)}{C_A \Delta P_A - C_B \Delta P_B}. \tag{4.37}$$

The net pumping speed S^* is obtained from the relation about the combination of the combination of the conductance with S^* and the steady flow condition. We define S_A^* and S_B^* as the net pumping speeds at the entrances of the orifice system A and B, respectively. Similar to Eqs. (4.30) and (4.31), S_A^* and S_B^* are expressed as

$$\frac{1}{S_A^*} \cong \frac{1}{C_A} + \frac{1}{S^*}, \tag{4.38}$$

$$\frac{1}{S_B^*} \cong \frac{1}{C_B} + \frac{1}{S^*}. \tag{4.39}$$

We obtain the relation between S_A^*, S_B^*, P_A, and P_B from the steady flow condition as

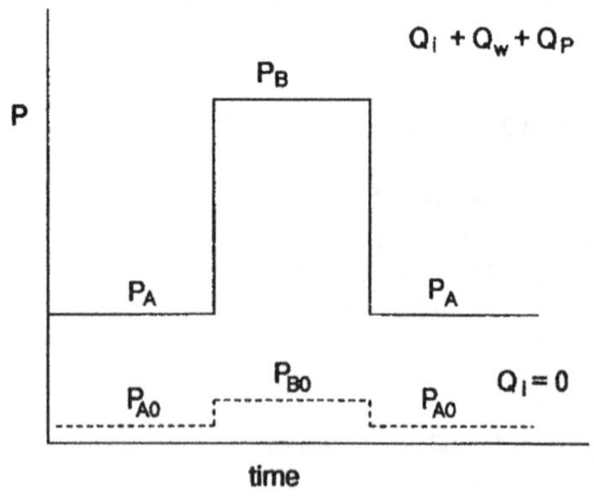

Fig. 4.35 Model of the changes of pressure in the chamber during conductance modulation process: (—) with admission of finits Q_i; (---) $Q_i = 0$ (Terada and Okano, 1989) [4-44]

$$\frac{P_A}{P_B} = \frac{S_B^*}{S_A^*}. \tag{4.40}$$

The value of S^* is calculated from Eqs. (4.38), (4.39), and (4.40) in the same way as S_P. At $Q_i = 0$, we can obtain S^* at ultimate pressure of the system S_0^* by substituting P_{A0} and P_{B0} into Eq. (4.40). This is an important feature of the CM method.

The scheme of the CM method is summarized as follows:

(1) S^* is determined from the ratio of P_A and P_B.
(2) S_P is determined from the ratio $(P_A - P_{A0})/(P_B - P_{B0})$. The obtained S_P is the averaged value for the pressures at conductance modulations.
(3) S_P at specific pressure can be obtained for small enough ΔQ_i by using (4.37).
(4) With the measured values of S_P and S_0^*, we can calculate the outgassing rates Q_W and Q_P at specific pressures.

Two-Point Pressure Method and One-Point Pressure Method [4-45]

The one-point pressure method (the 1PP method) is the simplest method for measuring the outgassing rates of metal pipes, which is derived from the new two-point pressure method (the 2PP method) for measuring the outgassing rate [4-45].

Principle [4-45]

The 2PP method is derived from the 3PP method [1]. So, let us first review the principle of the three-point pressure method (the 3PP method) [1]. (The 3PP method is described in detail in Chap. 2)

Consider an outgassing pipe of conductance C with an unknown outgassing rate Q_W, accompanied with an unknown introduced leak rate Q_L. The pipe is evacuated by an unknown pumping speed, as shown in **Fig. 4.36 (a)**, and we assume the same outgassing rate per unit length of the pipe regardless of the position. Then, the pressure P_k at a position indicated by the fraction k $(0 < k < 1)$ is given by Eq. (4.41).

Next, consider a simplified system of an outgassing pipe without a leak rate Q_L, as shown in **Fig. 4.36 (b)**. When the outgassing rate of the end plate is negligibly low, which is applicable in general, the pressure P_k at k is given by

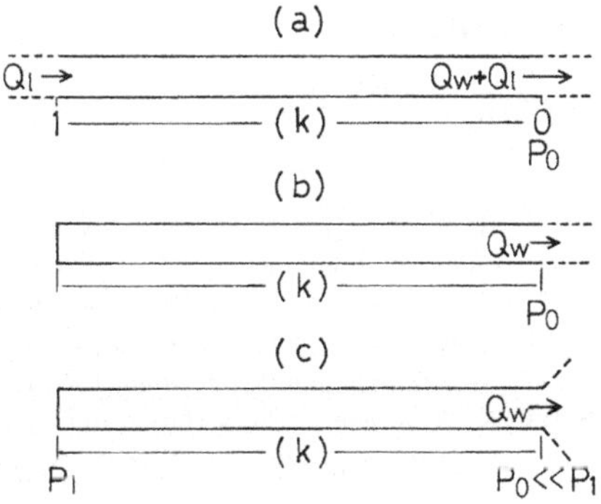

Fig. 4.36 Outgassing pipe systems. **(a)** Pipe of a conductance C with a leak rate Q_L, **(b)** pipe with an end plate whose outgassing rate is negligibly small, and **(c)** pipe evacuated by a high pumping speed (Yoshimura and Hirano, 1989) [4-45]

$$P_k = P_0 + \frac{k(1-k/2)Q_W}{C}. \tag{4.41}$$

Equation (4.41) contains two unknown factors P_0 and Q_W. And so, the value Q_W can be calculated from Eq. (4.41) using two pressures P_{k1} and P_{k2} measured at two different positions k_1 and k_2, respectively, as

$$Q_W = \frac{C(P_{k2} - P_{k1})}{(k_2 - k_1)\left[1 - (k_1 + k_2)/2\right]}. \tag{4.42}$$

We call this measuring method based on Eq. (4.42), the new two-point pressure method (or simply the new 2PP method).

Finally, let us consider the case that the pipe is evacuated by an effective pumping speed much higher than the pipe conductance C, as shown in **Fig. 4.36 (c)**. In this case, one could treat the pressure P_0 at the pipe port as zero. And consequently, one-point pressure P_k at k can derive the outgassing rate Q_W simply as

$$Q_W = \frac{CP_k}{k(1-k/2)}. \tag{4.43}$$

We call this method based on Eq. (4.43), the one-point pressure method (or simply the 1PP method) [4.45].

Comment:

When one sets k to 0.8, (4.43) reduces to

$$Q_W = \frac{CP_{0.8}}{0.8(1-0.4)} = 2.08CP_{0.8} \cong 2CP_{0.8}.$$

Reference

1. H. Hirano and N. Yoshimura, *J. Vac. Sci. Technol.* A **4**, 2526 (1986).

Yoshimura (1990) [4-47] presented an article, "Discussion on methods for measuring the outgassing rate," where the differential pressure-rise method and the 1PP method are reviewed in details.

References

Mechanism

4-1. B. B. Dayton, "Relations between size of vacuum chamber, outgassing rate, and required pumping speed", 1959 *6th National Symposium on Vacuum Technology Transactions* (Pergamon Press, 1960), pp. 101–119.
4-2. B. B. Dayton, "Outgassing rate of contaminated metal surfaces", *Transactions of the 8th National Vacuum Symposium*, 1961 (Pergamon Press, 1962), pp. 42–57.
4-3. K. W. Rogers, "The variation in outgassing rate with the time of exposure and pumping", *Transactions of the 10th National Vacuum Symposium*, 1963 (Macmillan, New York, 1964) pp. 84–87.
4-4. B. B. Dayton, "The effect of bake-out on the degassing of metals", *Transactions of the 9th National Vacuum Symposium*, 1962 (Macmillan, New York, 1963), pp. 293–300.

4-5. N. Yoshimura, H. Hirano, K. Ohara, and I. Ando, "Outgassing characteristics of an electropolished stainless-steel pipe with an operating extractor ionization gauge", *J. Vac. Sci. Technol. A* **9** (4), pp. 2315–2318 (1991).
4-6. P. A. Redhead, "Effects of readsorption on outgassing rate measurements", *J. Vac. Sci. Technol. A* **14** (4), pp. 2599–2609 (1996). Erratum: Effects of readsorption on outgassing rate measurements [*J. Vac. Sci. Technol. A* **14**, p. 2599 (1996)], *J. Vac. Sci. Technol. A* **15** (4), p. 2455 (1997).
4-7. A. Schram, "La désorption sous vide", *Le Vide* **103**, pp. 55–68 (1963) (in French).
4-8. R. Calder and G. Lewin, "Reduction of stainless-steel outgassing in ultra-high vacuum", *Brit. J. Appl. Phys.* **18**, pp. 1459–1472 (1967).
4-9. D. J. Santeler, "Estimating the gas partial pressure due to diffusive outgassing", *J. Vac. Sci. Technol. A* **10** (4), pp. 1879–1883 (1992).
4-10. M. Li and H. F. Dylla, "Model for the outgassing of water from metal surfaces", *J. Vac. Sci. Technol. A* **11** (4), pp. 1702–1707 (1993).
4-11. B. C. Moore, "Recombination limited outgassing of stainless steel", *J. Vac. Sci. Technol. A* **13** (3), pp. 545–548 (1995).
4-12. K. Akaishi, M. Nagasuga, and Y. Funato, "True and measured outgassing rates of a vacuum chamber with a reversibly adsorbed phase", *J. Vac. Sci. Technol. A* **19** (1), pp. 365–371 (2001).

Data

4-13. R. O. Adams, "A review of the stainless steel surface", *J. Vac. Sci. Technol. A* **1** (1), pp. 12–18 (1983).
4-14. T. Fujita, "Stainless steel as a vacuum industrial material", *Shinku (J. Vac. Soc. Japan)* **19** (9), pp. 293–303 (1976).
4-15. Y. Ishimori, N. Yoshimura, S. Hasegawa, and H. Oikawa, "Outgassing rates of stainless steel and mild steel after different pretreatments", *Shinku (J. Vac. Soc. Japan)* **14** (8), pp. 295–301 (1971) (in Japanese).
4-16. J. R. Young, "Outgassing characteristics of stainless steel and aluminum with different surface treatments", *J. Vac. Sci. Technol.* **6** (3), pp. 398–400 (1969).
4-17. R. Nuvolone, "Technology of low-pressure systems–establishment of optimum conditions to obtain low degassing rates on 316L stainless steel by heat treatments", *J. Vac. Sci. Technol.* **14** (5), pp. 1210–1212 (1977).
4-18. N. Yoshimura, T. Sato, S. Adachi, and T. Kanazawa, "Outgassing characteristics and microstructure of an electropolished stainless steel surface", *J. Vac. Sci. Technol. A* **8** (2), pp. 924–929 (1990).
4-19. A. Tohyama, T. Yamada, Y. Hirohata, and T. Yamashina, "Outgassing characteristics of electropolished stainless steel", *J. Japan Inst. Metals*, **54** (3), pp. 247–254 (1990) (in Japanese).
4-20. N. Yoshimura, H. Hirano, T. Sato, I. Ando, and S. Adachi, "Outgassing characteristics and microstructure of a "vacuum fired" (1050 °C) stainless steel surface", *J. Vac. Sci. Technol. A* **9** (4), pp. 2326–2330 (1991).
4-21. K. Tsukui, R. Hasunuma, K. Endo, T. Osaka, and I. Ohdomari, "Treatment of the wall materials of extremely high vacuum chamber for dynamical surface analysis", *J. Vac. Sci. Technol. A* **11** (2), pp. 417–421 (1993).
4-22. B. C. Moore, "Atmospheric permeation of austenitic stainless steel", *J. Vac. Sci. Technol. A* **16** (5), pp. 3114–3118 (1998).
4-23. B. C. Moore, "Thin-walled vacuum chambers of austenitic stainless steel", *J. Vac. Sci. Technol. A* **19** (1) pp. 228–231 (2001).
4-24. V. Nemanič and J. Šetina, "Outgassing in thin wall stainless steel cells", *J. Vac. Sci. Technol. A* **17** (3), pp. 1040–1046 (1999).
4-25. H. L. Eschbach, F. Gross, and S. Schulien, "Permeability measurements with gaseous hydrogen for various steels", *Vacuum* **13**, pp. 543–547 (1963).

4-26. V. Nemanič and J. Šetina, "Experiments with a thin walled stainless-steel vacuum chamber", *J. Vac. Sci. Technol. A* **18** (4), pp. 1789–1793 (2000).
4-27. M. Bernardini, S. Braccini, R. De Salvo, A. Di Virgilio, A. Gaddi, A. Gennai, G. Genuini, A. Giazotto, G. Losurdo, H. B. Pan, A. Pasqualetti, D. Passuello, P. Popolizio, F. Raffaelli, G. Torelli, Z. Zhang, C. Bradaschia, R. Del Fabbro, I. Ferrante, F. Fidecaro, P. La Penna, S. Mancini, R. Poggiani, P. Narducci, A. Solina, and R. Valentini, "Air bake-out to reduce hydrogen outgassing from stainless steel", *J. Vac. Sci. Technol. A* **16** (1), pp. 188–193 (1998).
4-28. Y. Ishikawa and V. Nemanič, "An overview of methods to suppress outgassing rate from austenitic stainless steel with reference to UHV and EXV", *Vacuum* **69**, pp. 501–512 (2003).
4-29. B. Zajec and V. Nemanič, "Hydrogen pumping by austenitic stainless steel", *J. Vac. Sci. Technol. A* **23** (2) pp. 322–329 (2005).
4-30. S. Tsukahara, "Hydrogen and metals for ultrahigh vacuum construction materials", *Appl.Phys.* (The Japan Society of Applied Physics) **69** (1), pp. 22–28 (2000) (in Japanese).
4-31. N. Yoshimura, "Ultrahigh vacuum technology in electron microscopes: Chapter 3 Outgassing of constituent materials", *Shinku (J. Vac. Soc. Japan)* **46** (6), pp. 529–535 (2003) (in Japanese).
4-32. H. Sato, H. Nakamura, S. Tsukahara, Y. Ishikawa, S. Misawa, Y. Takahashi, and S. Inayoshi, "Anodized film for vacuum equipment", *Shinku (J. Vac. Soc. Japan)* **45** (5), pp. 438–442 (2002) (in Japanese).
4-33. L. de Csernatony, "The properties of Viton 'A' elastomers I. Determination of the sorption and air solubility characteristics of Viton 'A' and their effect in vacuum applications of the material", *Vacuum* **16** (1), pp. 13–15 (1966).
4-34. L. de Csernatony, "The properties of Viton 'A' elastomers II. The influence of permeation, diffusion and solubility of gases on the gas emission rate from an O-ring used as an atmospheric seal or high vacuum immersed", *Vacuum* **16** (3), pp. 129–134 (1966).
4-35. L. de Csernatony, "The properties of Viton 'A' elastomers III. Steady state and transient activated gas emission processes from Viton 'A'", *Vacuum* **16** (5), pp. 247–251 (1966).
4-36. L. de Csernatony, "The properties of Viton 'A' elastomers IV. The influence of solid-gas interaction at Viton 'A' and stainless steel surfaces on gas evolution rates in high vacuum", *Vacuum* **16** (8), pp. 427–431 (1966).
4-37. L. de Csernatony and D. J. Crawley, "The properties of Viton 'A' elastomers. Part V. The practical application of Viton 'A' seals in high vacuum", *Vacuum* **17** (10), pp. 551–554 (1967).
4-38. N. Yoshimura, "Water vapor permeation through Viton O ring seals", *J. Vac. Sci. Technol. A* **7** (1), pp. 110–112 (1989).
4-39. L. de Chernatony, "Recent advances in elastomer technology for UHV applications", *Vacuum* **27** (10/11), pp. 605–609 (1977).
4-40. R. N. Peacock, "Practical selection of elastomer materials for vacuum seals", *J. Vac. Sci. Technol.* **17** (1), pp. 330–336 (1980).
4-41. N. Yoshimura, "A differential pressure-rise method for measuring the net outgassing rates of a solid material and for estimating its characteristic values as a gas source", *J. Vac. Sci. Technol. A* **3** (6), pp. 2177–2183 (1985).

Evaporation

4-42. R. E. Honig, "Vapor pressure data for the more common elements", *RCA Review*, June, pp. 195–204 (1957).

Measurement Methods

4-43. A. Berman, I. Hausman, and A. Roth, "Corrections in outgassing rate measurements by the variable conductance method", *Vacuum* **21** (9), pp. 373–377 (1971).

4-44. K. Terada, T. Okano, and Y. Tuzi, "Conductance modulation method for the measurement of the pumping speed and outgassing rate of pumps in ultrahigh vacuum", *J. Vac. Sci. Technol. A* **7** (3), pp. 2397–2402 (1989).
4-45. N. Yoshimura and H. Hirano, "Two-point pressure method for measuring the outgassing rate", *J. Vac. Sci. Technol. A* **7** (6), pp. 3351–3355 (1989).
4-46. H. Hirano and N. Yoshimura, "A three-point-pressure method for measuring the gas-flow rate through a conducting pipe", *J. Vac. Sci. Technol. A* **4** (6), pp. 2526–2530 (1986).
4-47. N. Yoshimura, "Discussion on methods for measuring the outgassing rate", *Shinku (J. Vac. Soc. Japan)* **33** (5), pp. 475–481 (1990) (in Japanese).

Other Articles

Mechanism

*4-1. I. Ando and N. Yoshimura, "Estimation of hydrogen outgassing rate of stainless steel", *Shinku (J. Vac. Soc. Japan)* **33** (3), pp. 185–187 (1990) (in Japanese).
*4-2. G. Horikoshi, "Physical understanding of gas desorption mechanisms", *J. Vac. Sci. Technol. A* **5** (4), pp.2501–2506 (1987).
*4-3. B. B. Dayton, "Outgassing rate of preconditioned vacuum systems after short exposure to the atmosphere: Outgassing rate measurements on Viton-A and copper", *J. Vac. Sci. Technol. A* **13** (2), pp.451–461 (1995).
*4-4. K. Akaishi, Y. Kubota, O. Motojima, M. Nakasuga, Y. Funato, and M. Mushiaki, "Experimental study on the scaling law of the outgassing rate with a pumping parameter", *J. Vac. Sci. Technol. A* **15** (2), pp. 258–264 (1997).
*4-5. K. Akaishi "Solution of the outgassing equation for the pump down of an unbaked vacuum system", *J. Vac. Sci. Technol. A* **17** (1), pp. 229–234 (1999).
*4-6. R. J. Elsey, "Outgassing of vacuum materials-I", *Vacuum* **25** (7), pp. 299–306 (1975).
*4-7. R. J. Elsey, "Outgassing of vacuum materials-II", *Vacuum* **25** (8), pp. 347–361 (1975).
*4-8. M. D. Malev, "Gas absorption and outgassing of metals", *Vacuum* **23** (2), pp. 43–50 (1973).
*4-9. N. Yoshimura, I. Ando, T. Sato, and S. Adachi, "Microstructure and elemental features of vacuum-fired (1050 °C) stainless steel surface", *Shinku (J. Vac. Soc. Japan)* **33** (5), pp. 525–529 (1990) (in Japanese).
*4-10. N. Yoshimura, T. Sato, S. Adachi, and T. Kanazawa, "Microstructure and elemental features of various stainless steel surfaces", *Shinku (J. Vac. Soc. Japan)* **33** (3), pp. 182–184 (1990) (in Japanese).

Measurement Method

*4-11. N. Yoshimura, H. Oikawa, and O. Mikami, "Measurement of outgassing rates from materials by differential pressure rise method", *Shinku (J. Vac. Soc. Japan)* **13** (1), pp. 23–28 (1970) (in Japanese).
*4-12. H. Hirano and N. Yoshimura, "A three-point-pressure method for measuring the gas-flow rate through a conducting pipe", *Shinku (J. Vac. Soc. Japan)* **30** (6), pp. 531–537 (1987) (in Japanese).
*4-13. M. Minato and Y. Itoh, "Measurement of outgassing rate by conductance modulation method", *Shinku (J. Vac. Soc. Japan)* **36** (3), pp. 175–177 (1993) (in Japanese).
*4-14. K. Saito, Y. Sato, S. Inayoshi, Y. Yang, and S. Tsukahara, "Outgassing measurement by switching between two pumping paths", *Shinku (J. Vac. Soc. Japan)* **38** (4), pp. 449–454 (1995) (in Japanese).
*4-15. S. Komiya, Y. Sugiyama, M. Kobayashi, and Y. Tuzi, "Direct-molecular-beam-method for mass selective outgassing rate measurement", *J. Vac. Sci. Technol.* **16** (2), pp. 689–691 (1979).

*4-16. K. Saito, "Outgassing from vacuum materials", *Shinku (J. Vac. Soc. Japan)* **40** (11), pp. 835–840 (1997) (in Japanese).
*4-17. P. A. Redhead, "Recommended practices for measuring and reporting outgassing data", *J. Vac. Sci. Technol. A* **20** (5), pp. 1667–1675 (2002).

Data

*4-18. R. R. Addiss, Jr., L. Pensak, and N. J. Scott, "Evaluation of a new fluoroelastomer as a gasketing material for high vacuum systems", *1960 7th National Symposium on Vacuun Technology Transactions* (Pergamon Press, 1961), pp. 39–44.
*4-19. F. J. Schittko, "Measurement of gas emission from solid surfaces", *Vacuum* **13**, pp. 525–537 (1963).
*4-20. W. Beckmann, "Gas desorption of some rubber-type materials", *Vacuum* **13**, pp. 349–357 (1963).
*4-21. B. H. Colwell, "Outgassing rates of refractory and electrical insulating materials used in high vacuum furnaces", *Vacuum* **20** (11), pp. 481–490 (1970).
*4-22. W. G. Perkins, "Permeation and outgassing of vacuum materials", *J. Vac. Sci. Technol.* **10** (4), pp. 543–556 (1973).
*4-23. G. F. Weston, "Materials for ultrahigh vacuum", *Vacuum* **25** (11/12), pp. 469–484 (1975).
*4-24. E. D. Erikson, T. G. Beat, D. D. Berger, and B. A. Frazier, "Vacuum outgassing of various materials", *J. Vac. Sci. Technol. A* **2** (2), pp. 206–210 (1984).
*4-25. S. S. Rosenblum, "Vacuum outgassing rates of plastics and composites for electrical insulators", *J. Vac. Sci. Technol. A* **4** (1), pp. 107–110 (1986).
*4-26. T. Kubo, Y. Satoh, and Y. Saito, "Outgassing rate measurement of the electrical cables and the elastomer/plastomer materials", *Shinku (J. Vac. Soc. Japan)* **41** (3), pp. 217–221 (1989) (in Japanese)
*4-27. Y. Ishikawa, Y. Koguchi, and K. Odaka, "Outgassing rate of some austenitic stainless steels", *J. Vac. Sci. Technol. A* **9** (2), pp. 250–253 (1991).
*4-28. J.-P. Bacher, C. Benvenuti, P. Chiggiato, M.-P. Reinert, S. Sgobba, and A.-M. Brass, "Thermal desorption study of selected austenitic stainless steels", *J. Vac. Sci. Technol. A* **21** (1), pp. 167–174 (2003).
*4-29. H. C. Hseuh and Xiuhua Cui, "Outgassing and desorption of the stainless-steel beam tubes after different degassing treatments", *J. Vac. Sci. Technol. A* **7** (3), pp. 2418–2422 (1989).
*4-30. S. Kato, M. Aono, K. Sato, and Y. Baba, "Achievement of extreme high vacuum in the order of 10^{-10} Pa without baking of test chamber", *J. Vac. Sci. Technol. A* **8** (3), pp. 2860–2864 (1990).
*4-31. S. Watanabe, S. Kurokouchi, and M. Aono, "Pumping properties using an electrolytic polished stainless steel vacuum chamber", *J. Vac. Sci. Technol. A* **16** (5), pp. 3084–3087 (1998).
*4-32. K. Odaka, Y. Ishikawa, and M. Furuse, "Effect of baking temperature and air exposure on the outgassing rate of type 316L stainless steel", *J. Vac. Sci. Technol. A* **5** (5), pp. 2902–2906 (1987).
*4-33. K. Odaka and S. Ueda, "Outgassing reduction of type 304 stainless steel by surface oxidation in air", *J. Vac. Sci. Technol. A* **13** (3), pp. 520–523 (1995).
*4-34. Y. Ishikawa and T. Yoshimura, "Importance of the surface oxide layer in the reduction of outgassing from stainless steels", *J. Vac. Sci. Technol. A* **13** (4), pp. 1847–1852 (1995).
*4-35. K. Sugiyama, T. Ohmi, M. Morita, Y. Nakahara, and N. Miki, "Low outgassing and anticorrosive metal surface treatment for ultrahigh vacuum equipment", *J. Vac. Sci. Technol. A* **8** (4), pp. 3337–3340 (1996).
*4-36. T. Ohmi, Y. Nakagawa, M. Nakamura, A. Ohki, and T. Koyama, "Formation of chromium oxide on 316L austenitic stainless steel", *J. Vac. Sci. Technol. A* **14** (4), pp. 2505–2510 (1996).

*4-37. I. Chun, B. Cho, and S. Chung, "Outgassing rate characteristic of a stainless-steel extreme high vacuum system", *J. Vac. Sci. Technol. A* **14** (4), pp. 2636–2640 (1996).
*4-38. I. Chun, B. Cho, and S. Chung, "Effect of the Cr-rich oxide surface on fast pumpdown to ultrahigh vacuum", *J. Vac. Sci. Technol. A* **15** (5), pp. 2518–2520 (1997).
*4-39. S. Watanabe, S. Kurokouchi, S. Kato, and M. Aono, "Achievement of extremely high vacuum in an electrolytically polished stainless steel vacuum chamber", *J. Vac. Sci. Technol. A* **16** (4), pp. 2711–2717 (1998).
*4-40. D. Fujita, "Surface oxide film on stainless steel and its effects on ultrahigh vacuum", *Shinku (J. Vac. Soc. Japan)* **45** (5), pp. 402–408 (2002) (in Japanese).
*4-41. H. F. Dylla, D. M. Manos, and P. H. LaMarche, "Correlation of outgassing of stainless steel and aluminum with various surface treatments", *J. Vac. Sci. Technol. A* **11** (5), pp. 2623–2636 (1993).
*4-42. K. Yoshihara, M. Tosa, and K. Nii, "Surface precipitation of boron nitride on the surface of type 304 stainless steels doped with nitrogen, boron, and cerium", *J. Vac. Sci. Technol. A* **3** (4), pp. 1804–1808 (1985).
*4-43. A. Itakura, M. Tosa, S. Ikeda, and K. Yoshihara, "Hydrogen permeation properties and surface structure of BN-coated stainless steel membrane", *Vacuum* **47** (6–8), pp. 697–700 (1996).
*4-44. D. Fujita and T. Homma, "Characterization and thermal desorption spectroscopy study on a new, low outgassing material surface for improved ultrahigh vacuum uses", *J. Vac. Sci. Technol. A* **6** (2), pp. 230–234 (1988).
*4-45. K. Saito, S. Inayoshi, Y. Ikeda, Y. Yang, and S. Tsukahara, "TiN thin film on stainless steel for extremely high vacuum material", *J. Vac. Sci. Technol. A* **13** (3), pp. 556–561 (1995).
*4-46. M. Sato, M. Nishiura, M. Oishi, M. Minato, Y. Sakuma, Y. Ikeda, K. Saito, S. Misawa, and S. Tukahara, "A new system of TiN coating on interior surface of cylindrical vacuum chamber by hollow cathode discharge method", *Vacuum* **47** (6–8), pp. 753–756 (1996).
*4-47. Y. Saito, Y. Ogawa, G. Horikoshi, N. Matuda, R. Takahashi, and M. Fukushima, "Vacuum system of the 300 m gravitational wave laser interferometer in Japan (TAMA300)", *Vacuum* **53**, pp. 353–356 (1999).
*4-48. S. Seal, R. Nardelli, A. Kale, V. Desai, and E. Armacanqui, "Role of surface chemistry on the nature of passive oxide film growth on Fe–Cr (low and high) steels at high temperatures", *J. Vac. Sci. Technol. A* **17** (4), pp. 1109–1115 (1999).

Aluminum Alloy

*4-49. H. Ishimaru, "All-aluminum-alloy ultrahigh vacuum system for a large-scale electron-positron collider", *J. Vac. Sci. Technol. A* **2** (2), pp. 1170–1175 (1984).
*4-50. J. R. Chen, K. Narushima, and H. Ishimaru, "Thermal outgassing from aluminum alloy vacuum chambers", *J. Vac. Sci. Technol. A* **3** (6), pp. 2188–2191 (1985).
*4-51. M. Suemitsu, T. Kaneko, and N. Miyamoto, "Aluminum alloy ultrahigh vacuum chamber for molecular beam epitaxy", *J. Vac. Sci. Technol. A* **5** (1), pp. 37–43 (1987).
*4-52. M. Suemitsu, H. Shimoyamada, N. Miyamoto, T. Tokai, Y. Moriya, H. Ikeda, and H. Yokoyama, "Ultrahigh-vacuum compatible mirror-polished aluminum-alloy surface: Observation of surface-roughness-correlated outgassing rates", *J. Vac. Sci. Technol. A* **10** (3), pp. 570–572 (1992).
*4-53. T. Momose, H. Yoshida, Z. Sherverni, T. Ebina, K. Tatenuma, and Y. Ikushima, "Surface cleaning on aluminum for ultrahigh vacuum using supercritical fluid CO_2 with H_2O and NaCl as additives", *J. Vac. Sci. Technol. A* **17** (4), pp. 1391–1393 (1999).
*4-54. K. Tatenuma, T. Momose, and H. Ishimaru, "Quick acquisition of clean ultrahigh vacuum by chemical process technology", *J. Vac. Sci. Technol. A* **11** (4), pp. 1719–1724 (1993).
*4-55. N. Schindler, T. Riemann, and Chr. Edelmann, "Some investigations on the effective short time outgassing depth of metals", *J. Vac. Sci. Technol. A* **16** (6), pp. 3569–3577 (1998).

*4-56. M. Suemitsu, "Ultrathin oxide of aluminum—formation of ultrahigh-vacuum-compatible surfaces by alcoholic-lathing method—", *Shinku (J. Vac. Soc. Japan)* **45** (5), pp. 415–421 (2002) (in Japanese).

Copper and Copper Alloy

*4-57. C. L. Foerster, H. Halama, G. Korn, M. Calderon, and W. Barletta, "Desorption measurements of copper and copper alloys for PEP-II", *Vacuum* **44** (5–7), pp. 489–491 (1993).
*4-58. Y. Hori, M. Kobayashi, and Y. Takiyama, "Vacuum characteristics of an oxygen-free high-conductivity copper duct at the KEK Photon Factory ring", *J. Vac. Sci. Technol. A* **12** (4), pp. 1644–1647 (1994).
*4-59. R. A. Rosenberg, M. W. McDowell, and J. R. Noonan, "X-ray photoelectron spectroscopy analysis of aluminum and copper cleaning procedures for the Advanced Photon Source", *J. Vac. Sci. Technol. A* **12** (4), pp. 1755–1759 (1994).
*4-60. F. Watanabe, M. Suemitsu, and N. Miyamoto, "In situ deoxidization/oxidization of a copper surface: A new concept for attaining ultralow outgassing rates from a vacuum wall", *J. Vac. Sci. Technol. A* **13** (1), pp. 147–150 (1995).
*4-61. F. Watanabe, Y. Koyatsu, and H. Miki, "Attaining an ultralow outgassing rate of 10^{-12} $P{\cdot}am^3{\cdot}s^{-1}{\cdot}m^{-2}$ from an oxide-free high conductivity copper chamber with beryllium-copper-alloy flanges", *J. Vac. Sci. Technol. A* **13** (5), pp. 2587–2591 (1995).
*4-62. F. Watanabe, "Mechanism of ultralow outgassing rates in pure copper and chromium-copper alloy vacuum chambers: Reexamination by the pressure-rise method", *J. Vac. Sci. Technol. A* **19** (2), pp. 640–645 (2001).

Titanium

*4-63. M. Minato and Y. Itoh, "Vacuum characteristics of titanium", *J. Vac. Sci. Technol. A* **13** (3), pp. 540–544 (1995).
*4-64. Y. Itoh and M. Minato, " Effect of different surface treatments on the outgassing characteristics of titanium", *Shinku (J. Vac. Soc. Japan)* **40** (3), pp. 248–250 (1997) (in Japanese).
*4-65. T. Homma, S. Akiya, T. Suzuki, and M. Minato, "Polishing of titanium surface II— A modified structure of the surface by polishing and H_2, H_2O-desorption", *Shinku (J. Vac. Soc. Japan)* **40** (6), pp. 513–517 (1997) (in Japanese).
*4-66. T. Homma, "A new vacuum function of metal surfaces with extreme thin oxide films", *Shinku (J. Vac. Soc. Japan)* **45** (5), pp. 391–394 (2002) (in Japanese).
*4-67. Y. Takahashi, K. Higuchi, M. Kaneko, K. Sugisaki, and Y. Saito, "Mechanical properties and outgassing rate measurement of titanium hydro-formed bellows", *Shinku (J. Vac. Soc. Japan)* **45** (6), pp. 533–536 (2002) (in Japanese).
*4-68. T. Nishiba, Y. Harada, M. Kawahara, M. Nose, and Y. Saito, "Mechanical and vacuum properties of hydroformed titanium bellows", *Shinku (J. Vac. Soc. Japan)* **45** (7), pp. 590–594 (2002) (in Japanese).
*4-69. Y. Morimoto, A. Takemura, Y. Muroo, M. Uota, Y. Sato, and Y. Saito, "Outgassing characteristics of titanium with surface treatment of oxidation", *Shinku (J. Vac. Soc. Japan)* **45** (8), pp. 665–669 (2002) (in Japanese).

Adsorption/Desorption

*4-70. P. A. Redhead, "Modeling the pump-down of a reversible absorbed phase. I. Monolayer and submonolayer initial coverage", *J. Vac. Sci. Technol. A* **13** (2), pp. 467–475 (1995).

Chapter 5
Phenomena Induced by Electron Irradiation

When energetic particles such as electrons bombard solid surfaces many phenomena occur. When the energetic ions bombard the surfaces of solid materials, the phenomena called "sputtering" occurs. Ion-beam sputtering is used for thinning the specimen or sample in the transmission electron microscope (TEM) and Auger electron spectrometer (AES).

"Sputtering", as well as "vacuum evaporation", is the basic process in depositing thin-films. This book, however, omits "sputtering" and "vacuum evaporation" from the phenomen in this chapter. See other book [5-1], for studying "sputtering".

The phenomena, such as electron stimulated desorption (ESD), photon stimulated desorption (PSD), polymerization of hydrocarbon molecules, darkening in secondary electron images, and etching of carbonaceous specimen, often occur in scientific instruments like electron microscopes.

5A Electron/Photon Stimulated Desorption (ESD/PSD)

Many kinds of scientific instruments such as an Auger electron spectrometer (AES) and scanning electron microscope (SEM) use a field-emission gun (FEG) to make a fine electron probe. In a field-emission gun (FEG), an extraction electrode just below an FE emitter is bombarded with electrons from the emitter, showing electron-stimulated desorption (ESD) and causing pressure-rise in the vicinity of an FE emitter. Also, in a multi-stage accelerating tube for creating a high-voltage electron beam many electrodes will be bombarded with primary, secondary, and reflected electrons and positive and negative ions, which might induce microdischarge (see Chap. 7). In electron-beam systems such as an electron microscope, ESD is one of the most troublesome phenomena. For high-energy electron accelerators photon-stimulated desorption (PSD) induces outgassing.

Achard et al. (1979) [5A-1] measured the electron and ion induced neutral gas desorption coefficients at 1.4 keV for stainless steel, Inconel 600, Inconel 718, titanium alloy, OFHC copper and an aluminum alloy, after bakeout at temperatures ranging from 150 °C to 600 °C.

Abstract [5A-1]: The surface composition was also determined by Auger spectroscopy after heating at various temperatures up to 1000 °C. The similarity of the results from all the materials investigated was explained in terms of a porous surface oxide layer which provided a reservoir for desorbable gas.

Electron induced desorption coefficients for 316 L+N stainless steel, titanium alloy, and Inconel 600 are presented in **Fig. 5A.1, 5A.2, 5A.3**. It is noted that electron induced desorption coefficients for Inconel 600 is smallest among the investigated materials. Inconel 600 could be one of the candidate materials for electron-extracting electrode in a field emission gun.

Gómez-Goñi and Mathewson (1997) [5A-2] measured the temperature dependence of electron induced gas desorption yields from stainless steel, copper, and aluminum surfaces for electrons of 300 eV energy.

Abstract [5A-2]: The main gases desorbed were H_2, CH_4, H_2O, CO, and CO_2. The effect of increasing the temperature on the gas desorption yields was measured and the only yields affected were those of H_2O and H_2, while the others remained essentially constant. The dependence on the dose and the temperature using two identical samples was also measured, and yields were found to be proportional to the dose to the power α, with α positive. A diffusion model for H_2 was applied to interpret the experimental results, showing that the model explains the results, if we consider that the H_2 concentration across the bulk of the material decreases exponentially.

In this paper [5A-2] many experimental data and rearranged data, listed below, are presented.

*Electron induced gas desorption yields for the samples (SS, Cu, Al) as a function of the dose for an electron beam of 300 eV energy at set temperatures:

For SS 316 L+N (at 36 °C and 200 °C).

Fig. 5A.1 Electron induced desorption coefficients for 316L+N stainless steel (Achard et al., 1979) [5A-1]

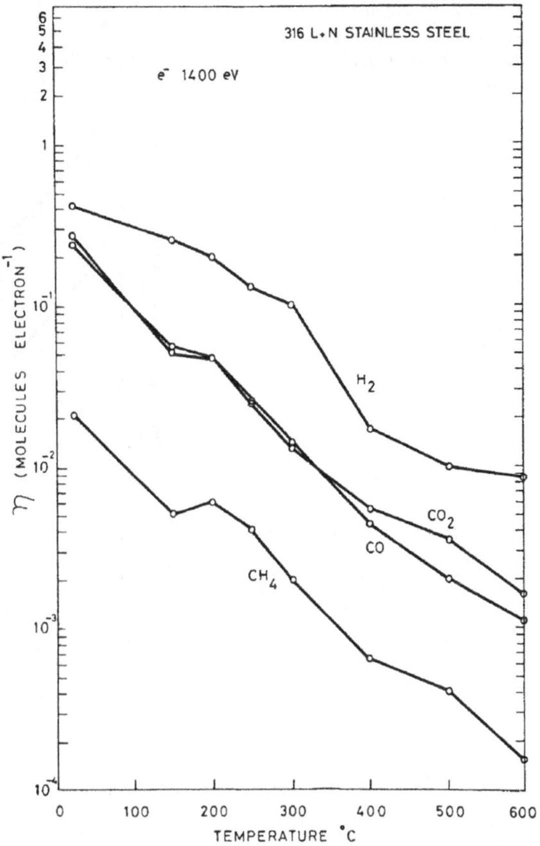

For OFHC Cu (at 40 °C and 200 °C).
For anticorodal Al (at 38 °C and 200 °C).

*Electron induced gas desorption yields for the samples(SS, Cu, Al) as a function of the desorbed molecules expressed in monolayers (ML) (1 ML = 2×10^{15} molecules/cm^2):

For SS 316 L+N (at 36 °C and 200 °C). (presented in **Fig. 5A.4**).
For OFHC Cu (at 40 °C and 200 °C).
For anticorodal Al (at 38 °C and 200 °C).

*Electron induced gas desorption yields with a diffusion model with constant H_2 concentration and exponential decay H_2 concentration for the samples (SS, Cu, Al):

For SS 316 L+N (at 36 °C and 200 °C).
For OFHC Cu (at 40 °C and 200 °C).
For anticorodal Al (at 38 °C and 200 °C).

Ueda et al. (1990) [5A-3] measured photodesorption yields due to synchrotron radiation (SR) from type 316 stainless steel, type A6063 aluminum alloy and high purity oxygen free copper (ASTM Class-1 OFC) test chambers.

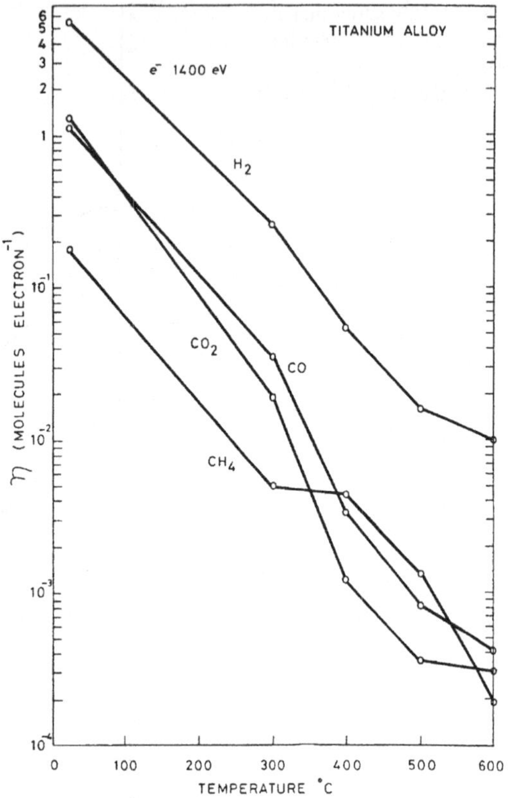

Fig. 5A.2 Electron induced desorption coefficients for titanium alloy (Achard et al., 1979) [5A-1]

Abstract [5A-3]: The yield of OFC finished with machining is reduced to 9×10^{-6} molecules/photon (N_2 equivalent) which is the lowest value throughout the experiments. Stainless steel treated with electrolytic polishing and 48 h prebaking at 450 °C shows the second lowest yield. The yield of aluminum alloy is somewhat higher than the others.

Hsiung et al. (2001) [5A-4] studied the exposure-dose-dependent photon-stimulated desorption (PSD) phenomena by using synchrotron light at a critical energy of 2.14 keV from the 1.5 GeV Taiwan Light Source.

Abstract [5A-4]: The result shows a decrease of the PSD yield for each gas and a decreased yield of photoemission at higher beam doses on the samples. The transient curves of pressure rise for each PSD molecule illustrate a longer delay time τ of the peak after exposing the surface with higher beam dose D. The delay time of H_2O and O_2 are longer than that of CO, CO_2, CH_4, etc., in the case of both Al and Cu samples. In this work, it is found that $\tau \sim D^{\alpha}$, with $0.5 < \alpha < 1$, in cases where $D < 2000$ mA·h. The discrepancy between the Al and Cu samples is not much different at higher beam doses. The processes of the chemical reactions by

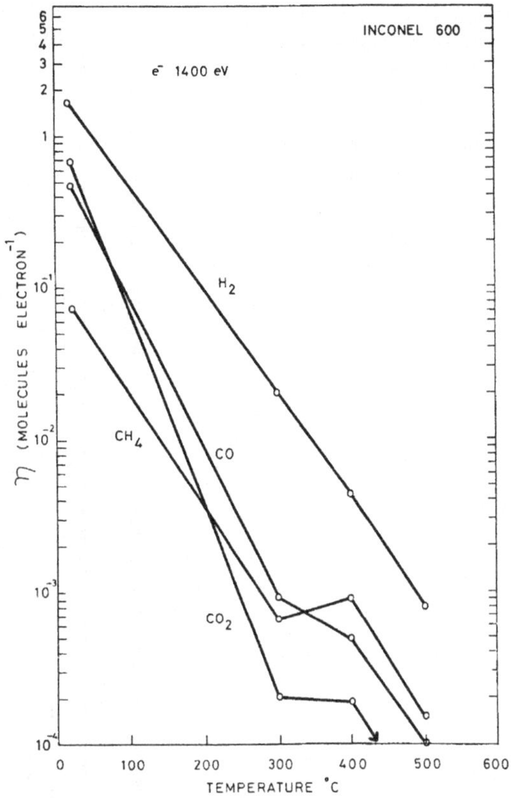

Fig. 5A.3 Electron induced desorption coefficients for Inconel 600 (Achard et al., 1979) [5A-1]

the photoelectrons and diffusion in the surface oxide layer are suspected to conduct the dose-dependent PSD phenomena.

Chen et al. (2002) [5A-5] studied the photon-stimulated desorption from an aluminum surface after water vapor exposure.

Abstract [5A-5]: Aluminum samples, following exposure to water vapor, were irradiated by synchrotron light with critical photon energy of 2.1 keV. The signal corresponding to H_2O gas exponentially decreased during irradiation. Experiments were performed under various conditions of target bias, sample temperature, photon flux, and photon dose to clarify the H_2O PSD behavior.

When chamber walls or electrodes are irradiated with electrons or photons, electron-stimulated desorption (ESD) or photon-stimulated desorption (PSD) occurs. Glow discharge cleaning (GDC) is effective in reducing ESD and PSD.

The following three methods of GDC have been separately investigated: (1) Using inert gases (Ar or He) for sputtering. (2) Using H_2 gas for reduction. (3) Using O_2 gas for burning. GDC as a surface treatment is widely applied to particle accelerators and partially to scientific instruments such as electron microscopes.

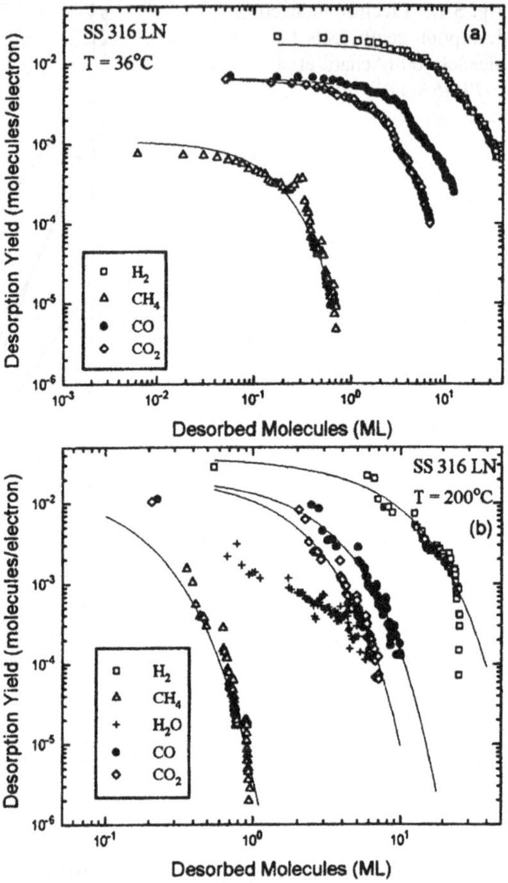

Fig. 5A.4 Electron induced gas desorption yields for a 316 L+N stainless steel sample as a function of the desorbed molecules expressed in monolayers (ML) (1 ML = 2×10^{15} molecules/cm^2). In *solid lines*, empirical fits to $\eta = \eta_0 \exp(-\theta/\theta_0)$ are shown. (**a**) T = 36 °C, (**b**) T = 200 °C (Gómez-Goñi and Mathewson, 1997) [5A-2]

Korzec et al. (1994) [5A-6] examined the removal of different lubricants (oils and greases) from metal surfaces by use of a capacitively coupled 13.56-MHz low-pressure oxygen discharge.

Abstract [5A-6]: The influence of rf power (up to 1 kW), pressure (from 0.01 to 1 mbar), and magnetic field (up to 20 mT) on the process duration and surface cleanliness has been investigated. Results indicate a strong influence of the lubricant temperature on the process duration. The removal rate of N62 oil at 140 °C is twice as high as that at 50 °C. For temperatures higher than 140 °C the creation of a hardly removal solid residue occurs. The influence of the initial lubricant area and lubricant thickness on the process duration has been investigated. The increase of the lubricant area results in an increase of the process time (loading effect for cleaning process). The increase of the lubricant film thickness results in a proportional increase of the process time.

The schematic of the experimental setup is shown in **Fig. 5A.5**.

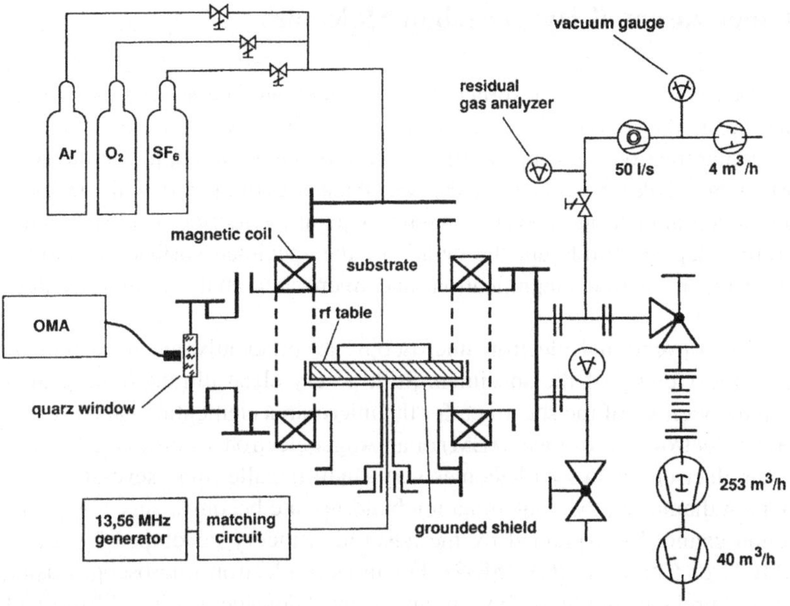

Fig. 5A.5 Experimental setup (Korzec et al., 1994) [5A-6]

5B Polymerization of Hydrocarbon Molecules

When hydrocarbon molecules adsorbed on a specimen surface are irradiated with an electron beam, they polymerize, making a contamination deposit, which is called specimen contamination build-up. If silicone-hydrocarbon molecules or fluorine-hydrocarbon molecules adsorbed on the specimen are irradiated with an electron beam, a contamination deposit containing silicon or fluorine results. When a contamination deposit builds up, the details of the specimen surface are hidden when observing it at high magnification, and micro-area analysis also becomes impossible.

Today, for conventional electron microscopy, a sufficiently clean vacuum is achieved in the vicinity of the specimen. Still, a very clean ultrahigh vacuum is required in the vicinity of the specimen for the micro-area elemental analysis using an analytical electron microscope (AEM) or an Auger electron spectrometer (AES).

When an electron-beam spot becomes very small (smaller than several tens of nanometers in diameter), the contamination build-up rate becomes very high. This phenomenon cannot be explained by the adsorption theory developed by Ennos (1953) [5B-1] or Christy (1960) [5B-2]. For modern electron microscopes using a fine electron probe in an ultrahigh vacuum, one must consider surface diffusion of molecules, as well as adsorption of molecules.

Key words: Anti-contamination device (ACD): Fins, cooled with liquid nitrogen, surrounding the specimen.

Electron-beam shower (EBS): Wide electron beam irradiating the specimen to desorb the gas molecules adsorbed on the specimen surface.

Transport of Hydrocarbon Molecules in High Vacuum

Ennos (1953 [5B-1] presented an article, "The origin of specimen contamination in the electron microscope." He concluded that the contamination is formed by the interaction of the electrons with organic molecules adsorbed on the bombarded surfaces, the molecules being replenished from the vapor phase.

Experiment Using a Cold Trap Surrounding a Target [5B-1]

A stainless-steel target was cleaned and mounted inside a long, narrow glass liquid-air trap of the design shown in **Fig. 5B.1**. It was supported by a 1/16 in. diameter steel rod, attached at the other end to the electron gun, so as not to touch the walls of the trap. Electrons from the 2 kV gun were directed onto the target and, by means of the magnetically operated shutter, two equal electron bombardments were given to adjacent areas of the target, the first with the trap not cooled, the second with it containing liquid air. While the normal brown deposit was obtained from the first bombardment, only very slight contamination (detected by breathing on the target) was evident for the second run.

Now the liquid-air trap will prevent all condensable vapor molecules from reaching the surface of the target except those molecules which pass straight through the open end of the trap and strike the target without first colliding with the trap wall. The very small

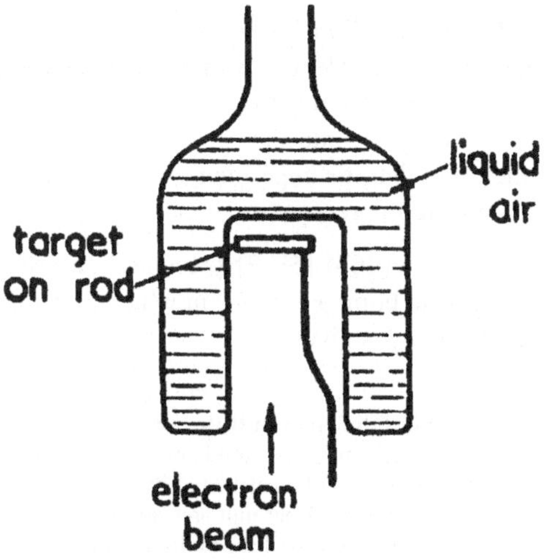

Fig. 5B.1 Design of liquid-air trap (Ennos, 1953) [5B-1]

contamination observed when the liquid-air trap was in operation could be accounted for by the finite aperture. From the dimensions of the trap, a reduction of contamination of between ten and twenty times would be expected. Migration of organic molecules would not be prevented but might be slowed down by a drop in temperature of the target and supports caused by radiation exchange. A thermocouple attached to the target showed this temperature drop to be negligible, however. It appears then that very little contamination by migration of molecules takes place [5B-1].

Experiment Heating the Wires Supporting a Target [5B-1]
The target in the second instance consisted of a clean glass microscope cover-slip, suspended by fine tungsten wires from two thicker tungsten wires which could be heated electrically (see **Fig. 5B.2**). Hydrocarbon molecules which might otherwise migrate over the hot wires would be either carbonized or evaporated off. Contamination formed on the target to the same extent whether or not the tungsten wires were heated, provided the target did not become heated by radiation (the support wires were made long to prevent this). On depositing gold on the glass by evaporation (and thus creating a fresh surface) further contamination still appeared on electron bombardment. This experiment again indicates that surface migration, if occurring at all, plays a very small part in the contamination process [5B-1].

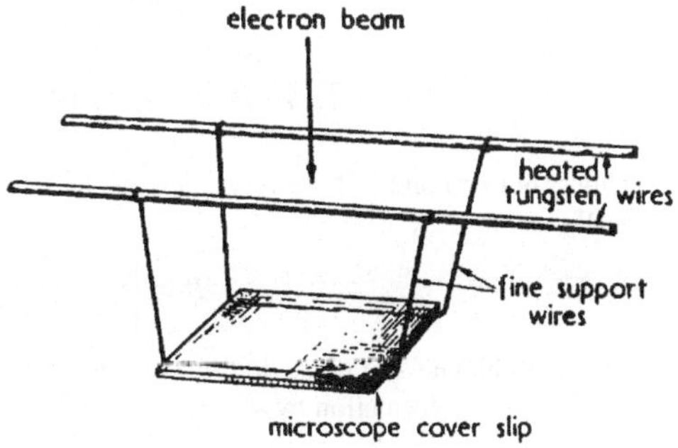

Fig. 5B.2 Target consisting of glass microscope cover-slip (Ennos, 1953) [5B-1]

Comment: Ennos (1953) [5B-1] is the pioneer who investigated contamination build-up induced by electron-beam irradiation. He showed by understandable experiments that contamination can be accounted for by the arrival of organic molecules from the vapor onto the irradiated area directly. However, we must take into consideration that his experiments were conducted under a high vacuum with organic vapor of high pressure in 1953.

Christy (1960) [5B-2] presented an article, "Formation of thin polymer films by electron bombardment," in which a theoretical expression for the contamination build-up rate is given.

Theory of Film Growth Rate [5B-2]

The rate of formation of solid polymer film on the substrate under the influence of electron bombardment can be calculated by a simple phenomenological theory. We assume there are N oil molecules per unit area adsorbed on the surface, with a mean time of stay τ. An electron hitting one of these molecules can tie it down permanently by forming a cross link, incorporating the molecule into the solid polymer film. Let P be the number of molecules per unit area which have been cross-linked, so that dP/dt is the rate of growth of the film. Then, assuming a first-order rate process,

$$dP/dt = \sigma f N\,, \tag{5B.1}$$

where σ is the cross-section for the cross-linking collision and f is the number of electrons per unit area per unit time. Since every such cross-linking event removes an oil molecule from the number N on the surface,

$$dN/dt = F - N/\tau - dP/dt\,, \tag{5B.2}$$

where F is the number of oil molecules which impinge on the surface from the vapor, per unit area per unit time.

On substituting Eq. (5B.1) into (5B.2) and integrating, we get

$$N = \left\{\frac{F}{\sigma f + 1/\tau}\right\}\{1 - K\exp - (\sigma f + 1/\tau)t\}\,, \tag{5B.3}$$

where K depends on the number of oil molecules initially existed on the surface. We now assume the surface is not completely covered with oil, so that $N < 1/a$, where α is the area of an oil molecule. Then this assumption requires $F < (\sigma/a)f + 1/a\tau$. The opposite case will be examined later.

On inserting Eq. (5B.3) into Eq. (5B.1) and integrating, we get

$$P = \frac{F}{1 + (1/\sigma\tau f)}\left[t + \frac{K}{\sigma f + 1/\tau}\{\exp - (\sigma f + 1/\tau)t - 1\}\right].$$

Unless both σf and $1/\tau$ are very small compared to unity (in which case P depends quadratically on t),

$$P \cong \left[\frac{F}{1 + (1/\sigma\tau f)}\right]t\,. \tag{5B.4}$$

The total thickness of the solid film is vP, where v is the volume of one molecule. Defining the rate of film formation by $R \equiv vP/t$, we have

$$R = \frac{vF}{1+(1/\sigma\tau f)}, \text{ when } F < (\sigma/\alpha)f + (1/\alpha\tau) . \quad (5B.5)$$

According to this result, at high current density or low temperature, when $\sigma\tau f >> 1$, the rate saturates at a value vF determined by the effective vapor pressure of the oil. At very low current density or high temperature, on the other hand, the rate is approximately $\sigma\tau f vF$, depending also on current density and temperature.

In the case not yet considered, when the flux of oil molecules is so large that there is always a monolayer of oil, $N = 1/a$ should be put into Eq. (5B.1) so that

$$R = (\sigma/a)vf, \text{ when } F \geq (\sigma/a)f + (1/a\tau) . \quad (5B.6)$$

This result is independent of the oil pressure, since there is an oil molecule waiting to be hit by every electron. We have seen that in the opposite extreme when the electron flux is so high that every oil molecule is struck before it escapes from the surface, the rate is independent of electron current density.

Transport of Hydrocarbon Molecules in Ultrahigh Vacuum

Some researchers {[Müller (1971) [5B-3], Knox (1976) [5B-4], Fourie [5B-6, 5B-7, 5B-8, 5B-9], Wall (1980) [5B-10], Yoshimura et al. (1983) [5B-11]} threw light on surface diffusion as the main mechanism of replenishing hydrocarbon molecules.

Müller (1971) [5B-3] investigated the contamination-rate in dependence of the probe diameter and electron density theoretically and experimentally using a Siemens Elimiskop 101 (electron microscope). "Experimental results showed that the smaller the spot size of the electron probe is, the higher is the contamination build-up rate. It was shown that when the probe size is very small, the contamination build-up rate is proportional to $1/r^2$, where r is the radius of the electron probe used."

Knox (1976) [5B-4] studied the specimen contamination in an electron microscope in an attempt to clarify the processes involved in contamination deposition.

Abstract [5B-4]: A very narrow electron beam (3.5 nm diameter in SEM mode) has been used to make small peaks of contamination on the surface of a thin carbon film. These peaks have been imaged and the rates of growth of the height, diameter and volume determined. It is concluded that a portion of the contamination can be related to the interaction of the electron beam with the adsorbed surface layer. Surface migration of adsorbed material is indicated as the supply mechanism. Significant deposition of material in regions not exposed to the beam is reported for the first time. A method for determining the thickness of thin amorphous carbon film in the range of a hundred to a few thousand Å is described.

Reimer and Wächter (1978) [5B-5] measured the contamination of different supporting films by irradiating a circular area a few μm diameter and by using an electron probe of a few tens of nm diameter.

Abstract [5B-5]: Contamination rings thus generated can be explained by diffusion of the hydrocarbon molecules on the specimen surface and by adsorption from the vacuum. The influence of specimen heating and charging are discussed. Some specimen cleaning methods have been examined. The best results were obtained by immersion of the specimen and the cartridge in methanol.

Fourie (1976) [5B-6] presented an article, "Contamination phenomena in cryopumped TEM and ultra-high vacuum field-emission STEM systems."

Abstract [5B-6]: For a stationary circular beam a ring-like contamination deposit is formed on the specimen. The profile of the ring is a function of the current density (J) in the beam. For very low J the contamination deposit is uniform, whereas for high J a narrow peripheral ring is observed. With a conductor film as specimen, the thickness has a negligible influence on the profile of the ring. However, for dielectric films the profile is very sensitive to thickness. For a given J, the peripheral ring will decrease in width with increasing thickness of specimen until a condition of zero contamination is attained. A grid bar in the vicinity of the illuminated region will cause partial contamination to occur under these conditions but only on the side nearest the grid bar. These results can be explained by a model where an electric field is generated by the escape of secondary electrons from the illuminated region. The incoming molecules experience an electric dipole interaction with the electric field and are subjected to a deflecting force which, when large enough, will totally deflect the molecules.

Fourie (1978/79) [5B-7] discussed the use of a low-energy anti-contamination electron (LEACE) gun for eliminating surface-originating contamination in electron microscopes. "Examples are shown of an orders-of-magnitude reduction in contamination rates in both the stationary and scanning electron probe modes by means of the LEACE-technique."

Fourie (1979) [5B-8] described the experiments which support the theory of the generation of an electric field by the interaction of primary electrons with thin foils.

Abstract [5B-8]: It is shown that the electric field is due to a positive charge on the foil resulting from the ejection of secondary electrons by the transmitted primary electrons. It is further shown that the electric field strength increases with the primary current, I_P. Experiments which indicate that secondary electron generation occurs up to about 1.5 μm from the primary-beam perimeter, are discussed.

A low-energy anti-contamination electron (LEACE) gun is described. It is shown that contamination can be eliminated totally with the use of this gun and an efficient cryo anti-contamination device.

Fourie (1981) [5B-9] described the experiments which produce phenomena that can be understood only in terms of the electric-field theory of contamination.

Abstract [5B-9]: The experiments show; that the rate of contamination around a stationary electron beam of 10 nm diameter increases with time within the same grid square but reverts to the original rate when the beam is moved to a fresh grid square; that the volume of contamination per given electron dosage is reduced about one hundred fold when a stationary beam is expanded from 3 to 30 nm diameter; that the rate of contamination in a stationary electron beam exceeds, by far, that in a scanning electron beam of identical dimensions and beam current; that contamination cones on insulator foils have a vastly different and etched like appearance as compared with those on conductor foils; that contamination produced by means of ultraviolet light on a quartz window is greatly enhanced in the presence of a high and inhomogeneous electric field; and finally, that ultraviolet radiation etching can be produced on a quartz window at elevated temperatures in the presence of high and inhomogeneous electric fields.

Wall (1980) [5B-10] presented an article, "Contamination in the STEM at ultrahigh vacuum."

Abstract [5B-10]: Contamination observed in the STEM on carbon specimen supports can be described quantitatively by a model wherein surface diffusing contaminants are cross-linked to the substrate under the influence of the electron beam. This model predicts the observed $1/r^2$ dependence of contamination thickness on beam radius r.

Surface Diffusion [5B-10]

Two cases are of particular interest: the steady-state rate of transport of molecules diffusing from a source of radius r_e (the grid) to a sink of radius r_o (the beam), and the time course of establishing this equilibrium state. Assuming n_0 molecules per unit area have a characteristic diffusion coefficient D_S, Jacobs [1] gives the rate of transport:

$$\frac{dn}{dt} = \frac{2\pi D_S n_0}{\ln(r_e/r_o)}, \tag{5B.7}$$

where dn/dt is the number of molecules reaching r_0 per unit time. Here we are assuming that all molecules reaching r_0 become cross-linked before they diffuse away again. If the beam current is not sufficiently high to cross-link all the arriving molecules, the contamination rate will be reduced. Hart et al. [2] introduced a correction factor $[1+1/(D\sigma)]$ to account for this:

$$\frac{dn}{dt} = \frac{2\pi D_S n_0}{[1+1/(D\sigma)]\ln(r_e/r_0)}, \tag{5B.8}$$

where D is the dose in electrons per unit area, and σ is the cross-section for formation of a cross-link. If we multiply both sides of this equation by $\bar{v}$, the volume per molecule, and denote that $\bar{v}n_0 = h_0$, the effective thickness of the diffusing layer, we have:

$$\frac{dV}{dt} = \bar{v}\frac{dn}{dt} = \frac{2\pi D_S h_0}{[1+1/(D\sigma)]\ln(r_e/r_0)}, \tag{5B.9}$$

where dV/dt is the volume of contaminant deposited per unit time.

Several points are worth noting about Eq. (5B.8). At high dose ($D >> 1/\sigma$, that is, $D\sigma >> 1$) the number of molecules deposited per unit time is nearly independent of beam size. The volume (mass) deposited per unit time being constant implies that the apparent thickness deposited per unit time is proportional to $1/r_0^2$. At low dose ($D\sigma << 1$) the amount of contamination deposited [integral of Eq. (5B.8)] is proportional to the dose. When $D\sigma = 1$, contamination is 50% of the maximal rate.

The interpretation of r_0 in Eq. (5B.8) as the beam size is not exact since the radius of interaction for inelastic scattering (cross-linking) is ~20 Å [3, 4]. For small beams we should replace r_0 by $\sqrt{(r_0^2+20^2)}$ (in Å). For larger beams the contamination may not be deposited uniformly over the beam cross-section. The magnitude of this effect can be estimated using the Einstein equation:

$$\bar{x}^2 = 4D_S t, \tag{5B.10}$$

where $\bar{x}$ is the mean distance traveled by a molecule in time t. If we consider an annulus of width $\bar{x}$ around the perimeter of a beam of radius r_0, we can calculate the distance $\bar{x}$ traversed by a molecule before it receives a dose $1/\sigma$. Noting that $dD/dt = I/\pi r_0^2$, where I is the beam current, we see that:

$$\frac{dD}{dt} \times t = \frac{I\bar{x}^2}{4\pi r_0^2 D_S} = \frac{1}{\sigma}$$

$$\text{or}: \frac{\bar{x}}{r_0} = \sqrt{\frac{4\pi D_S}{I\sigma}} \quad \left(\frac{\bar{x}}{r_0} << 1\right). \tag{5B.11}$$

If r_0 is much greater than 20 Å and $\bar{x}/r_0 < 1$, the contamination will have a "crater" appearance. Increasing beam current or decreasing D_S will lead to "craters" with sharper edges and deeper central depressions. Muller [5] has calculated the shape of these craters. The same effect may occur around the edge of a raster scan under similar circumstances if the repetition rate of the scan raster is fast compared to the time required for a molecule to diffuse from the edge to the center of the raster.

References

1. M. H. Jacobs, *Diffusion Process*, Springer-Verlag, Berlin, pp.114–120, 1967.
2. R. K. Hart, T. F. Kassner, and J. K. Maurin, "The contamination of surfaces during high-energy irradiation", *Phil. Mag.* **21**, pp. 453–467, 1970.
3. M. Isaacson and J. P. Langmore, "Determination of the non-localization of the inelastic scattering of electrons by electron microscopy", *Optik* **41**, pp. 92–96, 1974.
4. M. Isaacson and J. P. Langmore, and J. Wall, "The preparation and observation of biological specimens for the high resolution STEM", In: *SEM*/1974, IIT Research Institute, Chicago, IL 60616, pp. 19–26.
5. Karl-Heinz Müller, "Elektronen-Mikroschreiber mit gesehwindigkeitsgesteuerter Strahlfürung. I", *Optik* **33**, pp. 296–311, 1971.

Yoshimura et al. (1983) [5B-11] presented an article, "Mechanism of contamination build-up induced by fine electron-probe irradiation."

Experiments [5B-11]
To clarify the transfer mechanism of contaminating molecules and the dominant source of them, carbon thin films were irradiated with a fine electron probe of 12 nm diameter, a current density of about 7.6×10^2 A/cm^2 and an accelerating voltage of 200 kV, using three types of anti-contamination devices, individually.

Pre-treatment of Specimen and Specimen Cartridge [5B-11]
The specimen and the specimen cartridge were first immersed in Freon 113 ($C_2F_3Cl_3$) liquid for about 1 h, to chemically eliminate the hydrocarbon molecules adsorbed on their surfaces. The pre-treated specimen was then placed in the specimen chamber of the microscope column. Furthermore, the specimen was pretreated with an electron-beam shower (EBS) for 5 min to obtain a clean surface before the irradiation experiment. In the EBS treatment, the beam diameter is about 2 mm, current density about 2×10^{-5} A/cm^2 and accelerating voltage 200 kV [5B-11].

Three Types of Anti-contamination Devices (ACDs) [5B-11]
ACDs were used to control the hydrocarbon partial pressures around the specimen. The ACDs consisted of fins with different shapes, as illustrated in **Fig. 5B.3**. The fins can be cooled down to about –120 °C by liquid nitrogen.

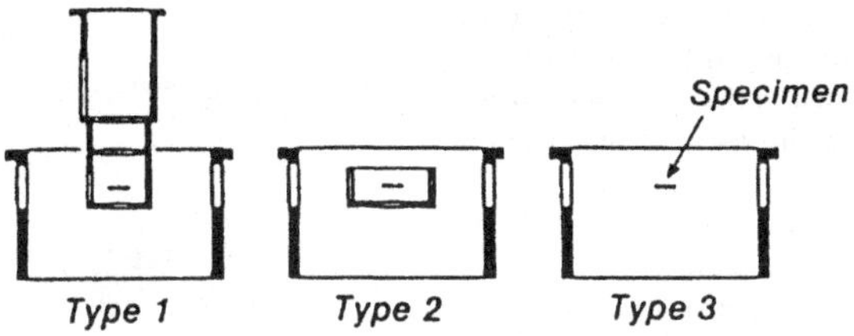

Fig. 5B.3 Three types of ACDs with different fins (Yoshimura et al., 1983) [5B-11]

Procedures for Irradiation and Observation [5B-11]
Typical procedures of irradiation and observation are presented in **Fig. 5B.4**. At a time T_1 after the EBS treatment, each of five adjacent spots on the specimen was irradiated successively for a certain time Δt. Then, this process was performed with different elapsed time, T_2, T_3 and so forth. After each process, the irradiated areas were observed as SEIs (scanning electron images) in each irradiation series.

Experimental Results [5B-11]
Two minutes after the EBS treatment, each of 25 spots on the specimen was sequentially irradiated with a fine electron probe for a time of 20 s without using an ACD. The same irradiation process was performed with different elapsed time, 20 and 80 min. Following each series of irradiation processes, SEIs of the irradiated areas were observed with the specimen surface inclined 45° to the electron beam, as presented in **Fig. 5B.5 (a–c)**.

In our experiments, the contamination rate (CR) was measured in mL/min by calculating the volume of the conical deposit on one side as proposed by Knox (1976) [1] and Harada et al. (1979) [2].

Next, experiments were done using the type-1 ACD. Each of 25 spots on the specimen was irradiated sequentially for 50 s, a few minutes after the EBS treatment. The SEIs of the irradiated area showed that contamination deposits did not form on the irradiated areas, but darkening of SEIs on the irradiated areas was observed. A similar process of irradiation was performed with the type-1 ACD for 200 s, at the elapsed times of 0.5, 1.0, 1.5, 2.0, 2.5 and 3 h after the EBS treatment. Contamination deposit build-up induced by a fine electron probe was prevented for 2.0 h after the EBS treatment. Small contamination deposits built up 2.5 and 3.0 h after the EBS treatment. The contamination deposit was larger for longer elapsed times. Other two experiments were done in a similar manner with the types 2 and 3 ACD, individually.

The growth rate of a conical deposit was estimated in mL/min for some typical deposits observed without and with types 1, 2 and 3 ACD, individually. The CRs as a function of the elapsed time T after the EBS treatment are presented in **Fig. 5B.6**. In each curve, the growth

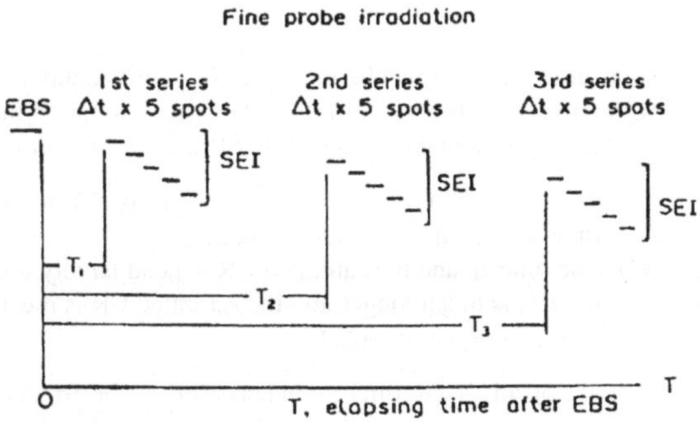

Fig. 5B.4 Typical procedures for fine-probe irradiation and SEI observation (Yoshimura et al., 1983) [5B-11]

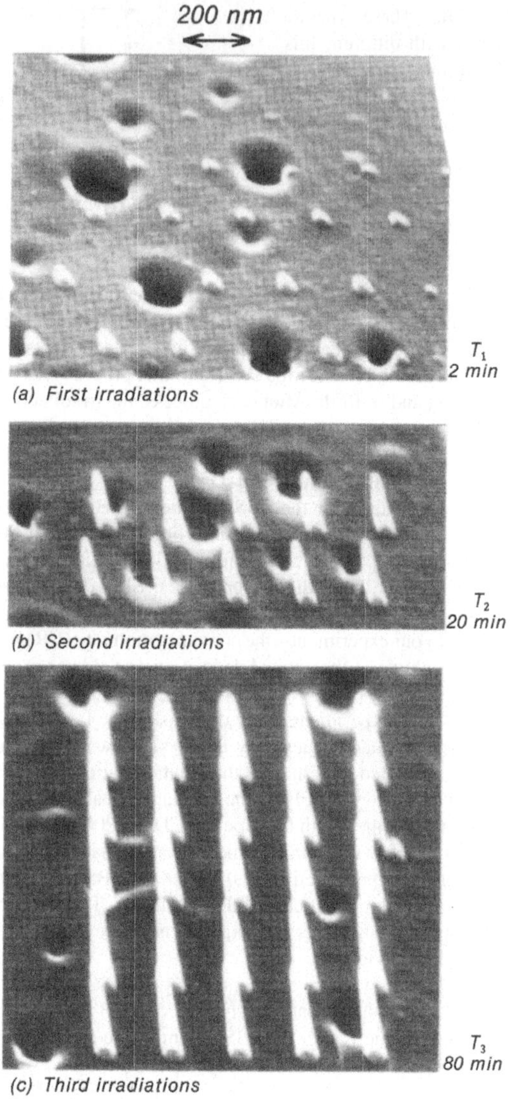

Fig. 5B.5 Secondary-electron images (SEIs) of the irradiated area in the specimen surface inclined 45° to the electron beam, in three series. In each series, the irradiation started from the extreme right spot in the first row. Small conical images show the contamination deposits which are arranged along horizontal, parallel lines. Large black images show the holes arranged irregularly. The times of 2 min, 20 min and 80 min at the right side of each figure show the elapsed time after the EBS treatment for each series. The fine-probe irradiation was conducted without an ACD (Yoshimura et al., 1983) [5B-11]

starts after a certain incubation time (T_0), zero contamination period, and its CR increases with the elapsed time T until it reaches each saturation rate.

The experimental results of **Fig. 5B.6** show the following characteristics:

(1) The incubation time T_0 exists when an ACD is used. Each CR increases with elapsed time and reaches its saturated rate.
(2) The time T_0 and the saturated CR depend largely upon the type of the ACD used. The time T_0 is much longer and the saturated CR is much lower with the type-1 ACD than with other types of ACD.

First of all, if the direct condensation to the irradiated area is the dominant transfer mechanism, the CR should be independent of the elapsed time after the EBS treatment,

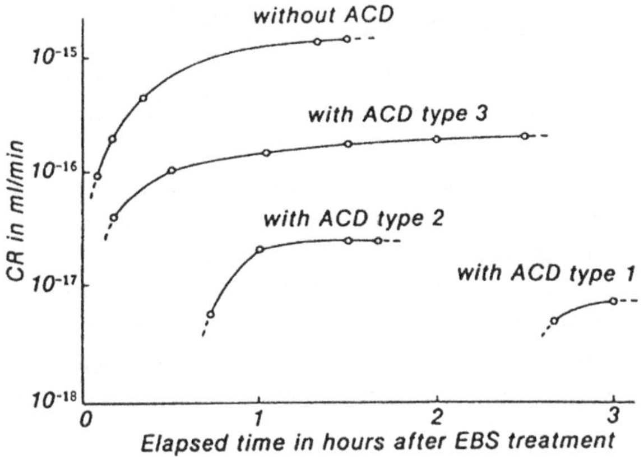

Fig. 5B.6 Contamination rates as a function of the elapsed time T after the EBS treatment, obtained without and with types 1, 2 and 3 ACD, individually (Yoshimura et al., 1983) [5B-11]

and so the time T_0 should not be observed. Next, if contaminating molecules come from the relatively dirty parts in contact with the specimen by surface diffusion, as Wall [5B-10] considered, the time T_0 and the saturated CR should be independent of the type of ACD. The surface diffusion following the adsorption of gases upon the whole specimen surface can explain our experimental results. A clean specimen surface is hit by residual hydrocarbon gases even under an ultrahigh vacuum. Some of the impinging molecules remain on the surface as adsorbates. The amount of adsorbed molecules is known to increase to the saturated amount with the holding time in vacuum. Contaminating molecules will be supplied to the sink by the surface diffusion of adsorbed molecules on the area surrounding the sink. However, in an elapsed time shorter than T_0 after the EBS treatment the supply rate will be too low to build up a detectable deposit. After the time T_0, the CR will increase with the increase of the amount of adsorbed molecules. Moreover, the time T_0 will increase with the decrease of residual gas pressure in the vicinity of the specimen, depending on the amount of adsorbed molecules, which is smaller under lower pressure.

When the saturated amount of adsorbed molecules is predominantly governed by the equilibrium between adsorption and desorption of molecules, the growth rate or the CR (height/time) $\propto 2\pi r_0/\pi r^2 \propto 1/r_0$, because the rate of diffusing molecules into the sink is proportional to the circumference of the sink, $2\pi r_0$. On the other hand, when a certain fraction of the hydrocarbon molecules impinging on the whole specimen surface is polymerized in a sink of radius r_0 regardless of any mechanism accepted, CR (height/time) $\propto K/\pi r_0^2 \propto 1/r_0^2$ (K: constant). In this case the saturated amount of adsorbed molecules and the concentration gradient of them on the surface are predominantly governed by the sink radius r_0. Both cases discussed above are considered to be the extreme cases. In reality, the CR is considered to be proportional to $1/r_0$ to $1/r_0^2$.

It can be concluded that in clean ultrahigh vacuum, the model of surface diffusion of adsorbed molecules on the whole specimen surface is most acceptable for the presented experimental results [5B-11].

References

1. W. A. Knox, *Ultramicroscopy* **1**, 175 (1976).
2. Y. Harada, K. Tomita, T. Watabe, H. Watanabe, and T. Etoh, *Scanning Electron Microscopy*/1979/*II*, 103.

Materials to be Polymerized by Electron Beam Irradiation

Ennos (1954) [5B-12] evaluated various vacuum-system materials which probably cause an electron-induced contamination. "Various materials which are used in the construction of kinetic vacuum systems were introduced into a clean system, and the thickness of the contamination deposit formed on a surface bombarded by electrons was measured after removing the target and depositing silver over its surface by vacuum evaporation. The height of the resultant steps in the silver film caused by contamination deposition was evaluated using the multi-beam interference method. In order of descending contaminating power are diffusion pump oil (hydrocarbon), vacuum grease (hydrocarbon), various rubber gaskets, silicone pump oil, and vacuum wax."

Yoshimura and Oikawa (1970) [5B-13] evaluated typical DP oils and vacuum greases using a method similar to that adopted by Ennos [5B-12]. "Silicone oil (DC-705/704) and silicone grease cause heavy contamination, and their deposits show charging when irradiated with an electron beam."

Ambrose et al. (1972) [5B-14] demonstrated the reduction of polymer growth in electron microscopes by use of a fluorocarbon-oxide pump fluid. "The rotary pump was replaced by one charged with a fluorocarbon-oxide lubricant and the fluid charge in the diffusion pump replaced by a fluorocarbon-oxide evaporant. Both specimen and electrode contamination were greatly reduced by the use of the new fluid."

Holland et al. (1973) [5B-15] studied the behavior of perfluoropolyether [FOMB LIN (tradename of Montedison SpA) and other vacuum fluids under ion and electron bombardment."

Abstract [5B-15]: A dc glow discharge was sustained in hydrogen, oxygen, helium, air, argon and Freon 14 between plane electrodes previously smeared with perfluoropolyether. After the discharge both the cathode and the anode surfaces remained visually clean except when hydrogen was used as a discharge gas when a polymer-like film was formed on the cathode and a brown film deposited on the anode. For comparison Apiezon C, Silicone 704, Santovac 5 and Dow Corning FS.1265, were also exposed to argon and hydrogen dc glow discharges, when solid deposits were produced on both anode and cathode. Perfluoropolyether fluid was then smeared in the quartz bottle of an rf ion source and discharges were run in hydrogen and argon; considerable etching was observed on the inside of the bottle, as seen in **Fig. 5B.7**. By comparison a smear of Santovac 5 in the bottle did not give rise to etching but left a tarry deposit.

Bauer and Speidel (1977) [5B-16] presented an article, "Reduction of the contamination rate in the STEM at a pressure of 10^{-5} Torr."

Abstract [5B-16]: In contrast with wide beam irradiation as used in the TEM, the contamination in the STEM, where a narrow beam is used, is mainly supplied by surface migration. Heating and successive cooling of the specimen chamber does reduce the concentration of residual gas molecules, but has no essential influence on

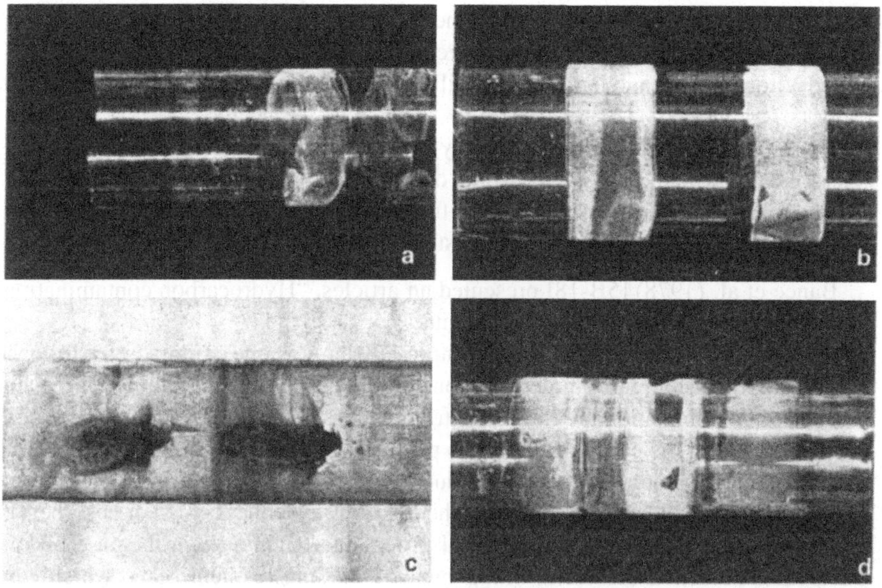

Fig. 5B.7 (**a**) Hydrogen discharge, perfluoropolyether fluid. (**b**) Argon discharge, perfluoropolyether fluid (silver film coated on outside of bottle beneath coupling electrodes and subsequently removed before photography). (**c**) Argon discharge, polyphenylether fluid. (**d**) Argon discharge, perfluoropolyether fluid (showing white P.T.F.E.-like deposit which was removed before photographing in (**b**) (Holland et al., 1973) [5B-15]

their area concentration on the surface, so that the rate of contamination on a scanning area of $260 \times 260\,\text{mm}^2$ on a thin carbon film could not be reduced by more than a factor of 0.8. Replacing the Viton gaskets in the vicinity of the specimen by Ag-In-gaskets, the contamination rate could be reduced to 1/3. After maintaining a glow discharge in N_2-gas and after a few minutes in O_2-gas at a pressure of 10^{-3} Torr in the specimen chamber the contamination rate was reduced to 1/30 of the original value. Micrographs are shown to demonstrate the resultant improvement of resolution and the harmlessness of the procedure for biological specimens.

Miller (1978) [5B-17] presented an article, "SEM vacuum techniques and contamination management."

Abstract [5B-17]: Contamination sources are: adsorbed hydrocarbons on chamber surfaces, including the specimen itself, elastomers and plastics, vacuum grease and pump oils.

The principal approaches to contamination management are:

- Ultrahigh vacuum system construction
- SEM modifications and techniques
- Use of PFPE fluids and lubricants

UHV system expense and relative inconvenience are not justified for SEM contamination control unless surface chemical analysis capability is needed. Without UHV, effective SEM vacuum system features include liquid nitrogen diffusion-pump traps, cold specimen stages and shrouds, molecular sieve traps on oil sealed pumps, pre-baked Viton O-rings and inert gas backfilling or purges and use of clean gloves and tools.

Hydrocarbons can be desorbed by oxygen discharges, ultrasonics and ultraviolet plus solvents, which often involve some disassembly for thorough cleaning.

Initial installation of or conversion to fluorocarbon (PFPE) fluids offers near zero contamination for conventional SEM systems.

Bance et al. (1978) [5B-18] presented an articles, "Hydrocarbon contamination in vacuum dependent scientific instruments."

Abstract [5B-18]: Pumping systems normally employ one or more of the following: rotary, diffusion, turbomolecular, ion, sublimation and cryo pumps. The main sources of contamination are directly attributable to these pumps, the construction materials used in the equipment and the processing of these materials. In an ultra-clean system the sample itself may introduce hydrocarbon contamination. Reduction of hydrocarbon contamination involves the use of traps (either cooled or absorbent), selection of special pump fluids, special processing techniques in the case of ion pumps and selection of system components, including resilient seals, which can be suitably processed to reduce contamination. Special techniques may be used to clean samples in situ. Experiments are described which show the sources of system contamination and remedies are suggested based on practical experience gained in design and operation of vacuum systems for electron microscopes, electron spectrometers, scanning transmission electron microscopes and mass spectrometers.

5C Darkening in Secondary Electron Images in SEM

In scanning electron microscopy, a secondary electron image (SEI) of the area covered with a polymerized contamination layer of specimen sometimes shows darkening because the secondary-electron emission coefficient of the polymerized contamination layer is smaller than that of clean specimen surface. This phenomenon is usually called specimen contamination in SEIs. Also, the SEI of a clean specimen surface, which has been irradiated with a fine electron probe, sometimes shows darkening. This darkening is caused due to the electron stimulated desorption (ESD) of the irradiated area, as demonstrated by Yoshimura et al. [5B-11].

Le Gressus et al. (1979) [5C-1] presented an interesting phenomena concerning darkening in SEI in a conventional high-vacuum SEM. Two examples of darkening in SEI were presented. One example (**Fig. 5C.1**) shows that darkening occurs on the zone of the specimen that was pre-etched with ions, and the other (**Fig. 5C.2**) shows that on a pre-sputtered specimen, darkening would not occur even in the area pre-scanned with the electron beam [5C-1].

When a specimen surface was sputtered beforehand, the darkening of the pre-irradiated area disappears. This was considered to be due to the elimination of the contamination layer containing abundant gas molecules by sputtering.
Le Gressus et al. [5C-1] discussed as follows.

Surface Contamination [5C-1]
The electron beam darkening does not correspond to surface contamination from residual gas as in ultrahigh vacuum, but from the "cleaning" of the sample. This sample modification goes as deep as the primary electrons, which can be verified in the following simple way.

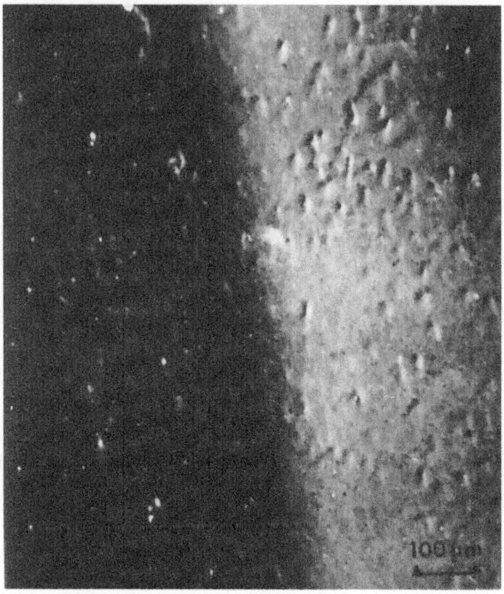

Fig. 5C.1 The darker area corresponds to the zone ion-etched before SEI observation (Le Gressus et al., 1979) [5C-1]

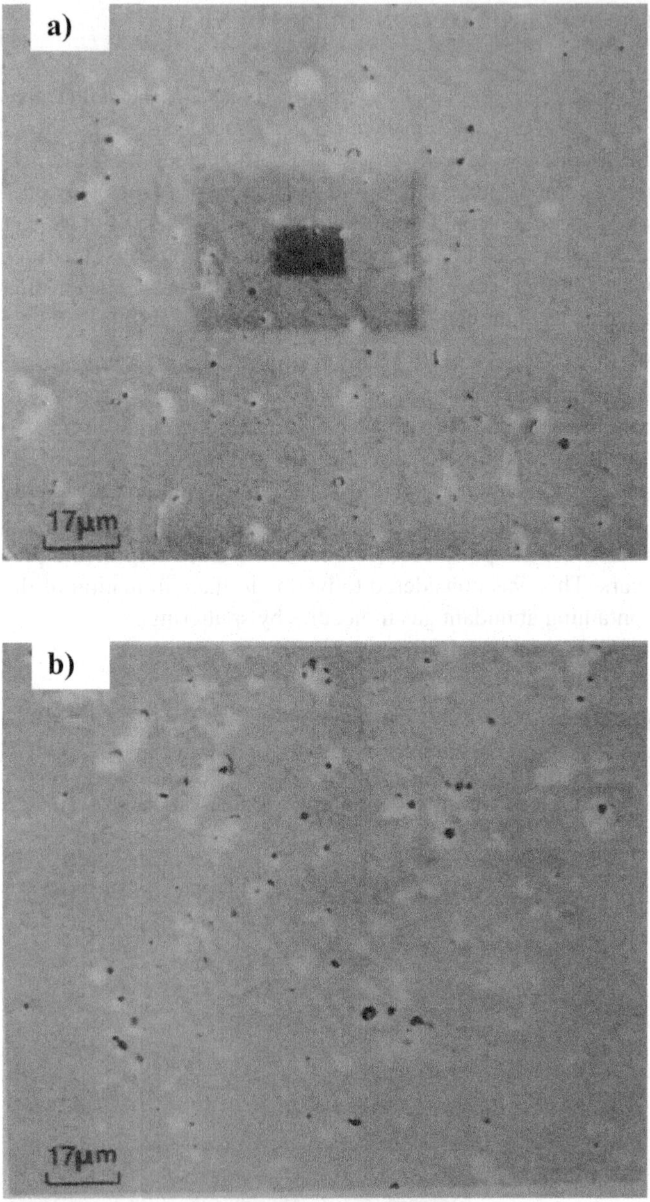

Fig. 5C.2 Electron beam darkening on aluminum alloy. **a**) The surface has been observed in a JEOL JSM-35 at 10^{-6} Torr. A dark area is seen. **b**) The sample is sputtered in a JEOL JFC-1100 unit, under argon for 5 minutes at 1 kV and 4 mA, and then examined in the same SEM. The gain of the electron multiplier is increased to the same brightness level. The darkening effect disappears (Le Gressus et al., 1979) [5C-1]

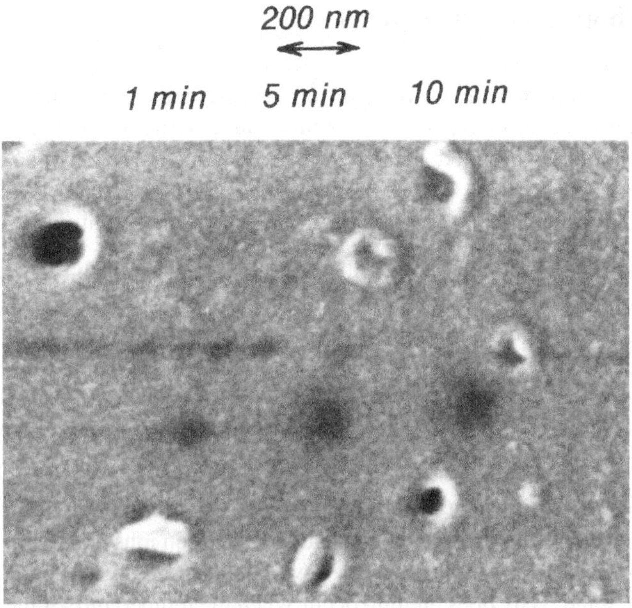

Fig. 5C.3 Three darkening images after different irradiation times, 1, 5 and 10 min. The fine-probe irradiation was conducted using the type-1 ACD (Yoshimura et al., 1983) [5B-11]

(i) Argon ion etching of a part of a contaminated surface leads to a darkening of the cleaned part (Fig. 5C.1), resulting from the increase in SE yield from the contamination layer. The cleaned area is also observable in optical microscopy.
(ii) After ion sputtering of the surface, the darkening effect becomes much less prominent, or disappears, as a consequence of the removal of the contamination layer. This technique can be applied also to conventional SEM (under 10^{-4} or 10^{-5} Torr) (Fig. 5C.2)

Experiment by Yoshimura et al. [5B-11]
Using the type-1 ACD presented in **Fig. 5B.3**, three spots were sequentially irradiated for 1, 5 and 10 min, respectively, directly after the EBS treatment. An SEI of the irradiated area is shown in **Fig. 5C.3**. The darkening images are seen at the irradiated spots and the darkening image area seems to enlarge with the irradiation time, while the darkness of images scarcely changes. The darkening line above the three darkening spots is the footprint of a manual spot scan.

The same area of **Fig. 5C.3** was observed in TEI (transmission electron image) and in STEI (scanning transmission electron image). Their images showed no discernible changes, compared with the virgin area's images before fine-probe irradiation. Therefore, it is reasonable to consider that this darkening observed in SEI shows a decrease in secondary-electron emission at the irradiated spots but not a change in the mass of the specimen itself.

The darkening images at the irradiated spots gradually became less dark, and then were no longer visible about 3 h after fine-probe irradiation [5B-11].

5D Etching of Carbonaceous Specimens

In electron microscopy, etching of a carbon-film specimen sometimes occurs during irradiation with an electron beam under some environmental conditions for specimens. This etching phenomenon is called "decontamination," a phenomenon opposite to contamination build-up.

Etching of a carbon-film specimen may occur under the following conditions.

1) When the oxygen partial pressure in the vicinity of the specimen is much higher due to air leakage, compared to the pressures of other gases.
2) When the water-vapor pressure in the vicinity of the specimen is much higher due to large outgassing of water vapor (H_2O) from the component materials located near the specimen.
3) In the cooling-down process of the ACD fins, which are cooled by conduction of copper wires connecting to a liquid-nitrogen tank wall.

It is considered that oxygen molecules or water-vapor molecules are ionized to oxygen ions by the bombardment of an electron beam, and such oxygen ions react with the carbon specimen to generate CO molecules. However, the etching phenomenon of 3) is difficult to understand. Why does etching occur when the temperature of the trap becomes lower? The lower the temperature of the trap is, the lower the pressure of water vapor becomes. Why?

Heide (1962) [5D-1] gave an understandable answer to the above question on the etching phenomenon of 3).

Experiment by Heide [5D-1]
The processes of contamination which take place when a cool chamber surrounds the object were more closely studied in the Elmiskop I (electron microscope). For this purpose, a new form of cooling apparatus was used. The temperature of the cool chamber could be reduced as far as –190 °C and the specimen either cooled as well or kept at room temperature. When the object is cooled there is always carbon removal at low temperature (**Fig. 5D.1 (a)**). **Figure 5D.1 (b)** shows results for an object at room temperature. One sees that the rapid removal of carbon, which usually occurs at fairly low temperatures, stops when the temperature is still further reduced to between –100 °C and –120 °C.

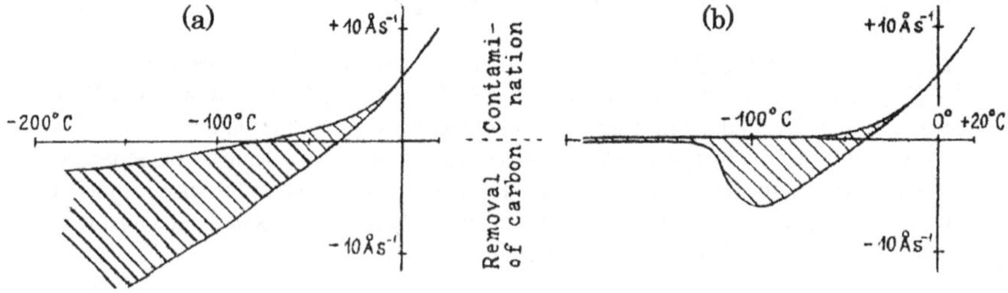

Fig. 5D.1 Rates of contamination build-up and carbon removal as functions of the cool chamber temperature. Beam-current density at object 0.4 A/cm^2, diameter of irradiated area about 1.8 μm, total pressure in microscope 2 to 5 $\times$ 10^{-5} Torr. (**a**) Specimen temperature ≅ Cool chamber temperature. (**b**) Specimen temperature ≅ Room temperature (Heide, 1962) [5D-1]

Table 5D.1 Vapor pressures of ice at low temperatures. μ m Hg = 10^{-3} Torr ["Scientific Foundation of Vacuum Technique" (second edition) by Dushman] [5D-2]

T[°C]	T[K]	P[μmHg]	t[°C]	T[K]	P[μmHg]
0	273.2	···	–80	193.2	4.13×10^{-1}
–10	263.2	1960	–90	183.2	7.45×10^{-2}
–20	253.2	779	–100	173.2	1.10×10^{-2}
–30	243.2	293.1	–110	163.2	1.25×10^{-3}
–40	233.2	99.4	–120	153.2	1.13×10^{-4}
–50	223.2	29.85	–130	143.2	6.98×10^{-6}
–60	213.2	8.40	–140	133.2	2.93×10^{-7}
–70	203.2	1.89	–150	123.2	7.4×10^{-12}
–78.5	194.7	0.53	–183	90.2	1.4×10^{-19}

Note: When the temperature is lowered from –140 °C to –150 °C, the vapor pressure of ice goes down by five orders.

To interpret these results we consider the partial pressures of the gases in the neighborhood of the specimen. For normal conditions we assume the following appropriate partial pressures: H_2O, 2×10^{-5} Torr; ΣCH, 5×10^{-6} Torr; CO, N_2, CO_2, H_2 each, about 1×10^{-6} Torr; O_2, 1×10^{-7} Torr. As the temperature is at first lowered the partial pressure of hydrocarbons falls, and the rate of specimen contamination falls also. At about –60 °C the partial pressure of hydrocarbons is so small that practically no further contamination occurs. (Comment by Yoshimura: The light hydrocarbon molecules such as CH_4 and C_2H_6 do not contribute to the contamination.) Between –60 °C and –100 °C rates of carbon removal up to 6 Å/s are observed, which corresponds to a removal of about 4×10^{15} carbon atoms·cm$^{-2} \cdot$ s^{-1} from the layer. In this temperature interval the partial pressure remains, with the exception of that of the hydrocarbons, the same as those given above. The object surface is covered with a monolayer of adsorbed gas molecules (probably water).

A calculation taking the ionization cross-section of these molecules and the intensity of the impinging electron beam into account shows that the number of ionization processes taking place per second in the surface layer is sufficient to explain the rate of carbon removal. For example, processes such as $H_2O \rightarrow H_2O^+ + e^-$ and $C + H_2O^+ \rightarrow H + H^+ + CO$ can occur, which require, naturally, that those positions on the surface which become empty are occupied quickly enough by new water molecules. This condition is fulfilled by an impact rate of 1×10^{16} water molecules·cm$^{-2} \cdot$ s^{-1} (corresponding 2×10^{-5} Torr). When the temperature of the chamber falls below –100 °C the partial pressure of water vapor falls very rapidly. At the same time the removal of carbon stops also if the object is not cooled. On the other hand, when the object is at the same low temperature as the cool chamber, the time spent on the surface by the unfrozen gas molecules is longer, and those H_2O molecules which pass through the cool chamber apertures and collide directly with the specimen condense upon it in a thicker-than-monomolecular layer.

Relating to the experiment by Heide (1962) [5D-1], the vapor pressures of ice at low temperatures are presented in **Table 5D.1**.

References

5-1. *"Handbook of Thin Film Technology"*, edited by Leon I. Maissel and Reinhard Glang, McGraw-Hill Book Company, 1970.

ESD/PSD

5A-1. Marie-Hélène Achard, R. Calder, and A. Mathewson, "The effect of bakeout temperature on the electron and ion induced gas desorption coefficients of some technological materials", *Vacuum* **29** (2), pp. 53–65 (1979).

5A-2. J. Gómez-Goñi and A. G. Mathewson, "Temperature dependence of the electron induced gas desorption yields on stainless steel, copper, and aluminum", *J. Vac. Sci. Technol. A* **15** (6), pp. 3093–3103 (1997).

5A-3. S. Ueda, M. Matsumoto, T. Kobari, T. Ikeguchi, M. Kobayashi, and Y. Hori, "Photodesorption from stainless steel, aluminum alloy and oxygen free copper test chambers", *Vacuum* **41** (7–9), pp. 1928–1930 (1990).

5A-4. G. Y. Hsiung, K. Y. Young, Y. J. Hsu, and J. R. Chen, "Study of the exposure-dose-dependent photon-stimulated-desorption phenomena", *J. Vac. Sci. Technol. A* **19** (4), pp. 1657–1661 (2001).

5A-5. J. R. Chen, K. Y. Yang, J. Y. Yang, and G. Y. Hsiung, "Photon-stimulated desorption from an aluminum surface after water vapor exposure", *J. Vac. Sci. Technol. A* **20** (3), pp. 857–860 (2002).

5A-6. D. Korzec, J. Rapp, D. Theirich, and J. Engemann, "Cleaning of metal parts in oxygen radio frequency plasma: Process study", *J. Vac. Sci. Technol. A* **12** (2), pp. 369–378 (1994).

Polymerization

5B-1. A. E. Ennos, "The origin of specimen contamination in the electron microscope", *Brit. J. Appl. Phys.* **4** April, pp. 101–106 (1953).

5B-2. R. W. Christy, "Formation of thin polymer films by electron bombardment", *J. Appl. Phys.* **31** (9), pp. 1680–1683 (1960).

5B-3. K. H. Müller, "Elektronen-Mikroschreiber mit geschwindigkeitsgesteuerter Strahlführung. I", *Optik* **33** (3), pp. 296–311 (1971) (in German).

5B-4. W. A. Knox, "Contamination formed around a very narrow electron beam", *Ultramicroscopy* **1,** pp. 175–180 (1976).

5B-5. L. Reimer and M. Wächter, "Contribution to the contamination problem in transmission electron microscopy", *Ultramicroscopy* **3**, pp. 169–174 (1978).

5B-6. J. T. Fourie, "Contamination phenomena in cryopumped TEM and ultrahigh vacuum field-emission STEM systems", *Scanning Electron Microscopy/1976/I*, pp. 53–60.

5B-7. J. T. Fourie, "The elimination of surface-originating contamination in electron microscopes", *Optik* **52** (5), pp. 421–426 (1978/79).

5B-8. J. T. Fourie, "A theory of surface-originating contamination and a method for its elimination", *Scanning Electron Microscopy/1979/II*, pp. 87–102.

5B-9. J. T. Fourie, "Electric effects in contamination and electron beam etching", *Scanning Electron Microscopy/1981/I*, pp. 127–134.

5B-10. J. S. Wall, "Contamination in the STEM at ultrahigh vacuum", *Scanning Electron Microscopy/1980/I*, pp. 99–106.

5B-11. N. Yoshimura, H. Hirano, and T. Etoh, "Mechanism of contamination build-up induced by fine electron probe irradiation", *Vacuum* **33** (7), pp. 391–395 (1983).

5B-12. A. E. Ennos, "The sources of electron-induced contamination in kinetic vacuum systems", *Brit. J. Appl. Phys.* **5** (1), pp. 27–31 (1954).

5B-13. N. Yoshimura and H. Oikawa, "Observation of polymerized films induced by irradiation of electron beams", *Shinku* (*J. Vac. Soc. Japan*) **13** (5), pp. 171–177 (1970) (in Japanese).

5B-14. B. K. Ambrose, L. Holland, and L. Laurenson, "Reduction of polymer growth in electron microscopes by use of a fluorocarbon oxide pump fluid", *J. Microscopy* **96**, Pt 3, pp. 389–391 (1972).

5B-15. L. Holland, L. Laurenson, R. E. Hurley, and K. Williams, "The behaviour of perfluoropolyether and other vacuum fluids under ion and electron bombardment", *Nuclear Instruments and Methods III*, pp. 555–560 (1973).

5B-16. B. Bauer und R. Speidel, "Herabsetzung der Kontaminationsrate im STEM bei einem Druck von 10^{-5} Torr", *Optik* **48** (2), pp. 237–246 (1977) (in German).
5B-17. D. E. Miller, "SEM vacuum techniques and contamination management", *Scanning Electron Microscopy/1978/I*, pp. 513–528.
5B-18. U. R. Bance, I. W. Drummond, D. Finbow, E. H. Harden, and P. Kenway, "Hydrocarbon contamination in vacuum dependent scientific instruments", *Vacuum* **28** (10/11), pp. 489–496 (1978).

Darkening

5C-1. C. Le Gressus, D. Massignon, A. Mogami, and H. Okuzumi, "Secondary electron emission dependence on electron beam density dose and surface interactions from AES and ELS in an ultrahigh vacuum SEM", *Scanning Electron Microscopy/1979/I*, pp. 161–172.

Etching

5D-1. H. G. Heide, "The prevention of contamination without beam damage to the specimen," *Fifth International Congress for Electron Microscopy*, A-4 (1962).
5D-2. *"Scientific Foundations of Vacuum Technique"* (second edition) by Saul Dushman: edited by J. M. Lafferty. John Wiley & Sons, Inc., New York, London, Sydney.

Other Articles

ESD/PSD

*5-1. M. Q. Ding and E. M. Williams, "Electron stimulated desorption of gases at technological surfaces of aluminium", *Vacuum* **39** (5), pp. 463–469 (1989).
*5-2. A. G. Mathewson, "Vacuum problems in particle accelerators due to interaction of synchrotron radiation, electrons and ions with surfaces", *Vacuum* **44** (5–7), pp. 479–483 (1993).
*5-3. T. Ohi and O. Konno, "Outgassing characteristics and surface analysis of pure alumimum plates under electron beam irradiation", *J. Vac. Sci. Technol. A* **12** (6), pp. 3186–3191 (1994).
*5-4. N. Ota, K. Kanazawa, M. Kobayashi, and H. Ishimaru, "Outgassing from aluminum surface layer induced by synchrotron radiation", *J. Vac. Sci. Technol. A* **14** (4), pp. 2641–2644 (1996).
*5-5. C. L. Foerster, C. Lanni, C. Perkins, M. Calderon, and W. Bartetta, "Photon stimulated desorption measurement of extruded copper and of welded copper beam chambers for the PEP-II asymmetric B factory", *J. Vac. Sci. Technol. A* **13** (3), pp. 581–584 (1995).
*5-6. C. L. Foerster, C. Lanni, J. R. Noonan, and R. A. Rosenberg, "Photon stimulated desorption measurement of an extruded aluminum beam chamber for the advanced photon source", *J. Vac. Sci. Technol. A* **14** (3), pp. 1273–1276 (1996).
*5-7. C. Herbeaux, P. Marin, V. Baglin, and O. Gröbner, "Photon stimulated desorption of an unbaked stainless steel chamber by 3.75 keV critical energy photons", *J. Vac. Sci. Technol. A* **17** (2), pp. 635–643 (1999).
*5-8. N. Ota, M. Saitoh, K. Kanazawa, T. Momose, and H. Ishimaru, "Reduction of photodesorption yield by oxygen discharge cleaning", *J. Vac. Sci. Technol. A* **12** (3), pp. 826–830 (1994).
*5-9. Minxu Li and H. F. Dylla, "Reduction of outgassing rate by glow discharge cleaning", *J. Vac. Sci. Technol. A* **13** (3), pp. 571–575 (1995).

*5-10. T. Momose, Y. Maeda, K. Asano, and H. Ishimaru, "Surface analysis of carbon on ozone treated metals", *J. Vac. Sci. Technol. A* **13** (3), pp. 515–519 (1995).

*5-11. C. L. Foerster, H. Halama, and C. Lanni, "Photon-stimulated desorption yields from stainless steel and copper-plated beam tubes with various pretreatments", *J. Vac. Sci. Technol. A* **8** (3), pp. 2856–2859 (1990).

*5-12. M. Saitoh, K. Kanazawa, T. Momose, H. Ishimaru, N. Ota, and J. Uramoto, "Oxygen discharge cleaning method for aluminum storage ring vacuum chambers", *J. Vac. Sci. Technol. A* **11** (5), pp. 2518–2524 (1993).

*5-13. D. Edwards, Jr., "Desorption of neutral molecules from Al(6010) by electron and ion bombardment", *J. Vac. Sci. Technol.* **15** (4), pp. 1586–1596 (1978).

*5-14. A. G. Mathewson, E. Alge, O. Gröbner, R. Souchet, and P. Strubin, "Comparison of the synchrotron radiation induced gas desorption in aluminum vacuum chambers after chemical and argon glow discharge cleaning", *J. Vac. Sci. Technol. A* **5** (4), pp. 2512–2515 (1987).

*5-15. B. R. Strohmeier, "The effects of O_2 plasma treatments on the surface composition and wettability of cold-rolled aluminum foil", *J. Vac. Sci. Technol. A* **7** (6), pp. 3238–3245 (1989).

*5-16. T. Kobari, M. Matumoto, N. Hirano, M. Katane, M. Matsuzaki, Y. Hori, M. Kobayashi, and M. Nagai, "Photodesorption from a copper chamber with a broached inner surface", *J. Vac. Sci. Technol. A* **13** (3), pp. 585–589 (1995).

*5-17. R. M. Lambert and C. M. Comrie, "A convenient electrical discharge method for eliminating hydrocarbon contamination from stainless steel UHV systems", *J. Vac. Sci. Technol.* **11** (2), pp. 530–531 (1974).

*5-18. R. P. Govier and G. M. McCracken, "Gas discharge cleaning of vacuum surfaces", *J. Vac. Sci. Technol.* **7** (5), pp. 552–556 (1970).

*5-19. A. W. Jones, E. Jones, and E. M. Williams, "Investigation by techniques of electron stimulated desorption of the merits of glow discharge cleaning of the surfaces of vacuum chambers at the CERN intersecting storage rings", *Vacuum* **23** (7), pp. 227–230 (1973).

*5-20. D. Blechschmidt, "In situ conditionig for proton storage ring vacuum systems", *J. Vac. Sci. Technol.* **15** (3), pp. 1175–1181 (1978).

*5-21. H. F. Dylla, K. Bol, S. A. Cohen, R. J. Hawryluk, E. B. Meservey, and S. M. Rossnagel, "Observations of changes in residual gas and surface composition with discharge cleaning in PLT", *J. Vac. Sci. Technol.* **16** (2), pp. 752–757 (1979).

*5-22. H. F. Dylla, S. A. Cohen, S. M. Rossnagel, G. M. McCracken, and Ph. Staib, "Glow discharge conditioning of the PDX vacuum vessel", *J. Vac. Sci. Technol.* **17** (1), pp. 286–290 (1980).

*5-23. F. Waelbroeck, J. Winter, and P. Wienhold, "Cleaning and conditioning of the walls of plasma devices by glow discharges in hydrogen", *J. Vac. Sci. Technol. A* **2** (4), pp. 1521–1536 (1984).

*5-24. H. F. Dylla, "Glow discharge techniques for conditioning high-vacuum systems", *J. Vac. Sci. Technol. A* **6** (3), pp. 1276–1287 (1988).

*5-25. N. R. Dean, E. W. Hoyt, M. T. Palrang, and B. G. Walker, "Glow discharge processing versus bakeout for aluminum storage ring vacuum chambers", *J. Vac. Sci. Technol.* **15** (2), pp. 758–760 (1978).

*5-26. M. Suemitsu, H. Shimoyamada, N. Matsuzaki, N. Miyamoto, and J. Ishibe, "Photoemission studies and outgassing-rate measurements on aluminum-alloy surfaces lathed with various alcohols", *J. Vac. Sci. Technol. A* **10** (1), pp. 188–192 (1992).

*5-27. T. Kobari and H. J. Halama, "Photon stimulated desorption from a vacuum chamber at the National Synchrotron Light Source", *J. Vac. Sci. Technol. A* **5** (4), pp. 2355–2358 (1987).

*5-28. T. E. Madey, Al. L. Johnson, and S. A. Joyce, "Electron and photon stimulated desorption: benefits and difficulties", *Vacuum* **38** (8–10), pp. 579–583 (1988).

*5-29. J. R. Chen, G. Y. Hsiung, Y. C. Liu, W. H. Lee, and C. C. Nee, "Secondary ion mass spectroscopy analysis for aluminum surfaces treated by glow discharge cleaning", *J. Vac. Sci. Technol. A* **13** (3), pp. 562–570 (1995).

*5-30. R. Calder, O. Gröbner, A. G. Mathewson, V. V. Anashin, A. Dranichnikov, and O. B. Malyshev, "Synchrotron radiation induced gas desorption from a Prototype Large Hadron Collider beam screen at cryogenic temperatures", *J. Vac. Sci. Technol. A* **14** (4), pp. 2618–2623 (1996).

Polymerization

*5-31. H. C. Poehlmann, R. R. Manning, and R. W. Jackman, "Contamination threats to critical surfaces from handling and storage practices", *J. Vac. Sci. Technol.* **9** (1), pp. 457–461 (1972).

*5-32. H. W. Conru and P. C. Laberge, "Oil contamination with the SEM operated in the spot scanning mode", *J. Phys. E: Scientific Instruments*, **8**, pp. 136–138 (1975).

*5-33. B. Bauer und R. Speidel, "Herabsetzung der Kontaminationsrate im STEM bei einem Druck von 10^{-5} Torr", *Optik* **48** (2), pp. 237–246 (1977).

*5-34. R. W. Springer and D. S. Catlett, "Rate and pressure dependence of contaminants in vacuum-deposited aluminum films", *J. Vac. Sci. Technol.* **15** (2), pp. 210–214 (1978).

*5-35. T. Tomita, Y. Harada, H. Watanabe, and T. Etoh, "Reduction of contamination in analytical electron microscopy", *Shinku* (*J. Vac. Soc. Japan*) **22** (4), pp. 158–164 (1979) (in Japanese).

*5-36. K. Yada, "Specimen contamination", *Electron Microscope* **16** (1), pp. 2–10 (1981) (in Japanese).

*5-37. A. Kumao, H. Hashimoto, and K. Shiraishi, "Studies on specimen contamination by transmission electron microscopy", *J. Electron Microsc.* **30** (3), pp. 161–170 (1981).

*5-38. Y. Harada, T. Tomita, T. Watabe, H. Watanabe, and T. Etoh, "Reduction of contamination in analytical electron microscopy", *Scanning Electron Microscopy/1979/II*, pp. 103–110.

Etching

*5-39. J. Kirz, "Specimen damage considerations in biological microprobe analysis", *Scanning Electron Microscopy/1980/II*, pp. 239–249.

Chapter 6
Vacuum Gauges

Vacuum gauges are necessary for controlling the evacuation system of scientific instruments. Such gauges must be rigid in structure and resistive against air leak accidents. The Pirani gauge of thermal conduction type is widely used for controlling evacuation systems, switching the evacuation mode from roughing to fine pumping, which is resistive against air leak. The Penning gauge is usually equipped for switching on the sputter ion pump after roughing, which is rigid in structure. When the sputter ion pump works, it gives the pressure indication in the range of high vacuum to ultrahigh vacuum. The sputter ion pump can control the field emission system in scientific instruments.

Some users sometimes require a vacuum gauge to check the pressures in scientific instruments. For this requirement a Bayard-Alpert gauge (BAG) or an extractor ionization gauge (EG) are often selected. Residual gas analyzers such as the quadrupole mass filter and the magnetic-sector-type mass spectrometer may be installed in scientific instruments. Such residual gas analyzers use an incandescent tungsten filament, accompanying outgassing from the hot filament and electron stimulated desorption (ESD), as same as the BAG.

For semiconductor-manufacturing equipment mechanical gauges such as the diaphragm capacitance gauge is widely used, which are resistive to many reactive gases.

First of all let us make a survey of almost all vacuum gauges which are commercially available, by referring an article by Lafferty 1972 [6-1].

Methods of Measuring Low Pressure [6-1]

Vacuum gauges may be categorized as follows, according to the way their response is affected by a physical property of the gas that is dependent on pressure.

Mechanical Gauges. The mechanical force exerted by the gas is used to lift a liquid column or deform a thin wall.

Thermal Gauges. The thermal conductivity of the gas is used to effect the heat loss from an element at elevated temperatures.

Viscosity Gauges. The viscosity of the gas is used to produce a drag on a rotating body or dampen the amplitude of oscillation of the suspended surface.

Momentum Gauges. Momentum is transferred from a hot surface to a cold surface by the gas molecules to produce a force on the cold surface.

Ionization Gauges. A fraction of the molecules in the gas are ionized, collected and measured. In a partial pressure gauge, the ions are separated and measured according to their mass.

Figure 6.1 shows the range over which the various types of vacuum gauges may be used to measure pressure. Mercury manometers have the advantage of providing an absolute method of measuring total gas pressure that is independent of the nature of the gas molecules. Since the McLeod gauge will measure absolute pressure below 10^{-3} Torr, it is frequently used to calibrate other gauges. Error may result in using the gauge if the gases do not obey Boyle's law or if they condense on compression.

Thermal conductivity gauges such as the Pirani, thermocouple and thermistor gauges are all limited to pressures above 10^{-4} Torr. The calibration of these gauges is dependent upon the molecular weight and specific heat of the gas. Surface contamination of the hot element can alter the gauge calibration by changing the thermal emissivity and the accommodation coefficient of the incident gas molecules.

An unusual form of viscosity gauge has been used to determine pressures as low as 5×10^{-10} Torr. This was accomplished by measuring the deceleration of a freely spinning high-speed rotor [1].

The Knudsen gauge has the outstanding advantage of having a response that is virtually independent of the nature of the gas. Designs have been made which permit operation from 10 to 10^{-8} Torr. This gauge has not been used widely because of its delicate suspension.

The ionization gauge, and in particular hot cathode ionization gauge, has proved to be the most popular gauge for measuring gas pressures below 10^{-4} Torr.

Notwithstanding its great popularity, the ionization gauge suffers from just about every conceivable fault one could imagine in a pressure measuring device. In the first place,

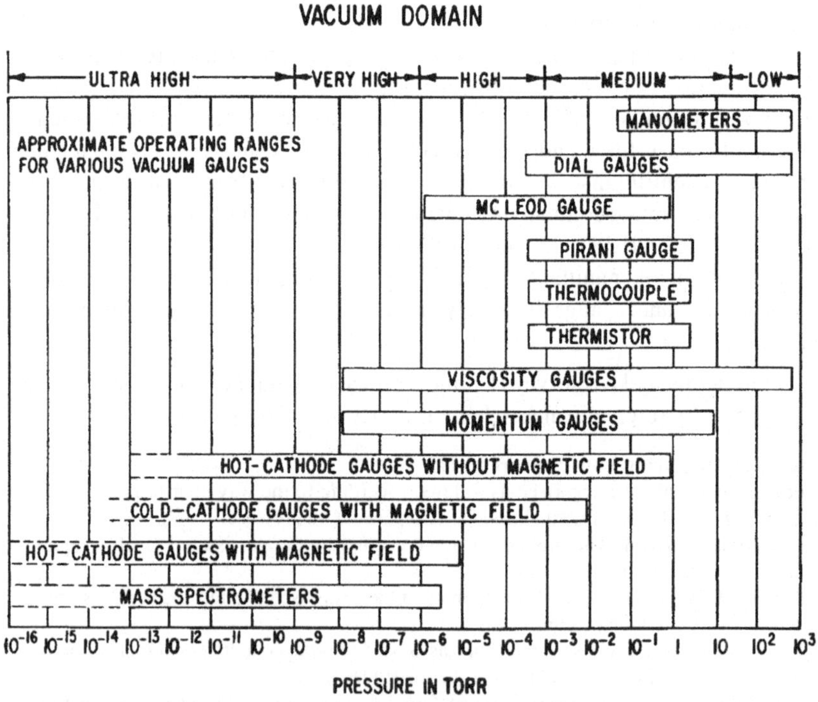

Fig. 6.1 Chart showing pressure ranges for different types of vacuum gauges (Lafferty, 1972) [6-1]

it doesn't measure pressure—it measures gas density. It does not provide an absolute method of measurement, and the calibration is dependent on the nature of the gas. The ionization gauge may give inaccurate readings, unless precautions are taken, because it may pump, dissociate, and react with the gas it is supposed to be measuring; it may generate gas of its own; it may produce false ion currents that are not related to gas pressure; it may generate gas of its own; it may produce false ion currents that are not related to gas pressure; it may produce other spurious currents that may be mistaken for ion current; and it may have inherent electronic instabilities. In spite of all these difficulties, when properly designed and operated the ionization gauge can be relied upon to give pressure readings that are accurate to better an order of magnitude or perhaps even a few percent under ideal conditions when carefully calibrated for a known gas composition.

Reference

1. J. M. Beams, D. M. Spitzer, and J. P. Wade, *Rev. Sci. Instrum*, 33, 151 (1962).

Mechanical Gauges

The diaphragm capacitance manometer is widely used in semiconductor processing because it is resistive to reactive gases. The capacitance manometer, the spinning rotor pressure gauge and crystal oscillator vacuum gauge belong to this category.

Capacitance Manometer

Utterback and Griffith (1966) [6-2] presented an article, "Reliable submicron pressure readings with capacitance manometer."

Abstract [6-2]: A cold trapped McLeod gauge, a capacitance manometer, and an ion gauge have been used simultaneously to measure gas pressure in the range 10^{-4}–3×10^{-3} Torr. Data were obtained for He, Ne, Ar, Xe, N_2, O_2, CH_4, CO, H_2, and CO_2. The capacitance manometer and pressure chamber were maintained at a constant room temperature to avoid thermal transpiration corrections. Comparison of data among gases clearly showed the systematic McLeod gauge error attributed to mercury streaming to the cold trap. The capacitance manometer and McLeod gauge readings agreed within 2% at 10^{-3} Torr for He, while the McLeod reading was 15% low for Xe at the same pressure. The errors for the other gases were intermediate. Spurious mercury capillary wetting effects, leading to serious errors in the McLeod gauge reading, were also noted for some of the molecular gases. The capacitance manometer was clearly superior to the McLeod gauge as a universal pressure sensor, and appears capable of reliable readings accurate to a few percent at 10^{-4} Torr.

Experiments (Utterback and Griffith, 1966) [6-2]
The following four objectives were set for this work:

(1) To check the calibration of the capacitance manometer against McLeod gauge using helium gas.

(2) To use the calibrated capacitance manometer to check McLeod gauge readings for ten other gases, in order to determine the magnitude of the McLeod gauge "mercury pumping error."
(3) To calibrate an ion gauge for all eleven gases.
(4) To explore the reliability of the capacitance manometer as a precise measuring instrument for the pressure range 10^{-4}–3×10^{-3} Torr.

Procedure [6-2]
McLeod gauge readings were made by two methods. For both methods it was necessary to determine first the difference in capillary mercury depression [1] for the well pumped gauge condition. It was found that the mercury in the closed capillary was always 5 ± 1 mm

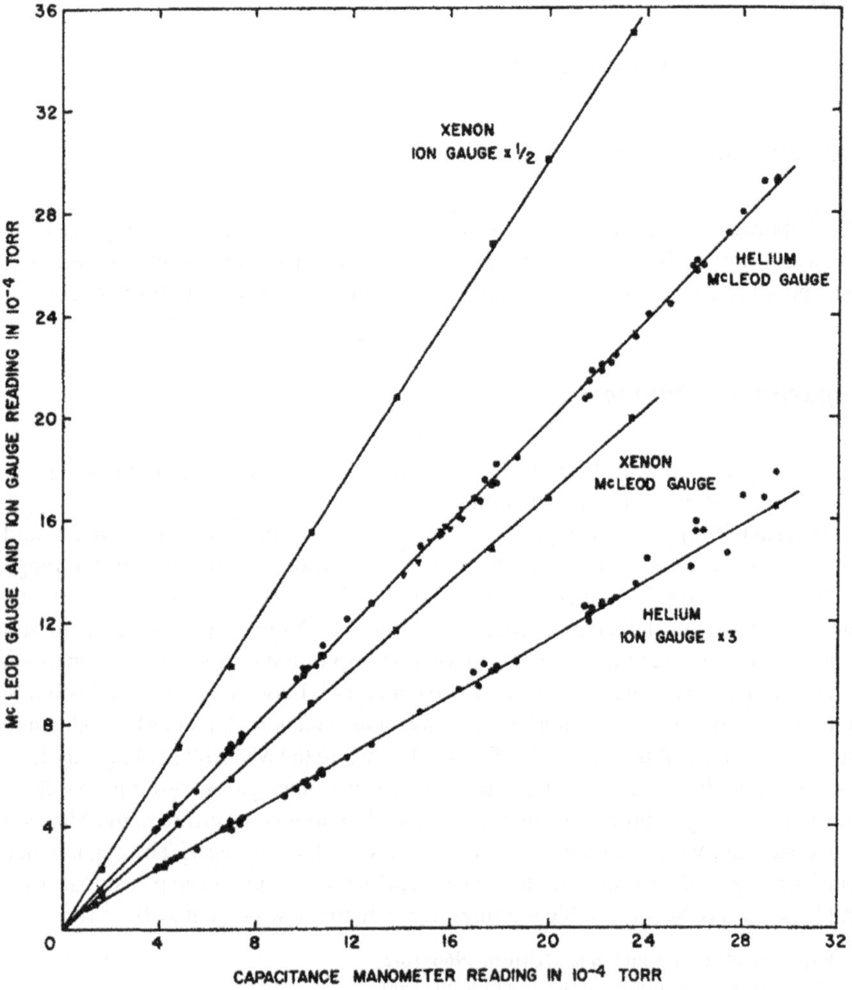

Fig. 6.2 McLeod gauge and ion gauge readings plotted against capacitance manometer readings for He and Xe (Utterback and Griffith, 1966) [6-2]

higher than in the open one, over the capillary length. This was reproducible over the 2-week period. Method I for reading involved bringing the mercury (open capillary) to 5 mm below the level of the closed end of the closed capillary and reading directly on the gauge scale (calibrated to the gauge constant). Method II involved a "variable compression" scheme [1] where the mercury columns were measured at four widely separated heights. The mercury depression difference was included. The first method was used on seven points of each run, and the second method was used as a final check at about 1.5×10^{-3} Torr for each run. The two methods were very consistent, with the "variable compression" scheme giving more reproducible results. (Individual values in the "variable compression" scheme, method II, seldom varied more than 2% from the average of the four individual values.) It was found necessary to jar and trap the instrument firmly to obtain reproducible results. The ion gauge readings were taken directly from the controller meter.

Results [6-2]
Figure 6.2 shows the readings obtained over the 2 week period for helium, and the xenon run for comparison. The McLeod gauge points near the capacitance manometer reading of 1.5×10^{-3} Torr were taken with method II (variable compression), and the rest with method I. The lowest He pressure points have been shown as their average on **Fig. 6.2**.

Reference

1. P. H. Carr, *Vacuum* **14**, 37 (1964).

Thermal Conductivity Gauges

The Pirani gauge, the thermo-couple gauge and the thermistor gauge belong to this category. Among these gauges the Pirani gauge using fine filaments may be most widely used due to its high sensitivity and high-speed response.

Pirani Gauge

The high-temperature-type Pirani gauge is widely used for controlling the evacuation system because it has sensitivity up to several Torr of pressure.

Leck (1952) [6-3] investigated the high temperature Pirani gauge in detail.

Abstract [6-3]: The operating conditions required for a Pirani gauge to operate in the pressure range 0–5 Torr have been investigated. A comparison between tungsten and platinum wire for use as the gauge element has shown that both have disadvantages when operated above 450 K.

Theoretical Consideration of the Calibration Curve [6-3]
List of symbols:
P pressure microns (10^{-3} Torr)
T_0 the temperature K of the hot wire at $P = 0$.

T_W the temperature of the hot wire $P \neq 0$. (In the constant temperature case $T_W = T_0$.)
T_g the temperature of the gauge wall (equal to room temperature, taken = 300 K throughout).
R_0 the resistance of the hot wire at $P = 0$.
R_2 the resistance of the wire at room temperature.
$R_W = R_2(1+\alpha t)$ the resistance of the hot wire at T_W where $t = T_W - T_g$.
K_R the heat radiation from the wire when $T_W^4 - T_g^4 = 1$ (for tungsten = 1.74×10^{-12} W).
K_C the heat conductivity from the wire at unit pressure (1μ), for $T_W - T_g = 1$, $K_C =$ 2.82×10^{-6} W for tungsten. (Values of K_R and K_C taken from paper by Dunoyer [1].)
V_0 the voltage across the wire at $P = 0$.
V the voltage across the wire when $P \neq 0$.
V_S the bridge supply voltage.
i_g the galvanometer current at any pressure P.
r_g the galvanometer resistance.

The following three basic methods of operating the Wheatstone bridge control circuit were first described by Pirani in 1905 [1]: (a) constant temperature bridge; (b) constant voltage bridge; (c) constant current bridge.

In (a) the bridge voltage V is adjusted to keep the bridge balanced, and consequently the wire temperature constant at all pressures.

For thermal equilibrium in the wire:

$$V^2/R_0 = K_P\left(T_W^4 - T_g^4\right) + K_C\left(T_W - T_g\right)P + \text{endlosses} \tag{6.1}$$

$$\therefore P \propto \left(V^2 - V_0^2\right) \tag{6.2}$$

For a thin wire Eq. (6.2) is approximately true up to pressures of 2000–3000 μ (2 to 3 Torr).

In (b) the bridge voltage is held constant, the out-of balance current being used as a measure of the pressure. It is usual, but by no means essential or even desirable, to make all four arms of the bridge approximately equal. If $R_1 >> R_0$ (**Fig. 6.3**) then the bridge operates with constant current characteristics to a close approximation. Both the absolute sensitivity and the shape of the calibration curve depend upon the ratio R_1/R_0. The following are the approximate thermal equilibrium equations for the simple cases where $R_0 >> R_1$ and $R_1 >> R_0$.

(1) $R_0 >> R_1$ ($V \cong$ constant)
neglecting end losses:

$$R_W I^2 \cong V_S^2/[R_2(1+\alpha t)] = K_R\left(T_W^4 - T_g^4\right) + K_C\left(T_W - T_g\right)P \tag{6.3}$$

$$\text{The out-of-balance current} = i_g \cong \left(V_S R_1/R_0 r_g\right)\left[(R_0 - R_2)/R_0\right] \tag{6.4}$$

(i_g made zero at $P = 0$)
(2) $R_1 >> R_0$ ($I \cong$ constant)

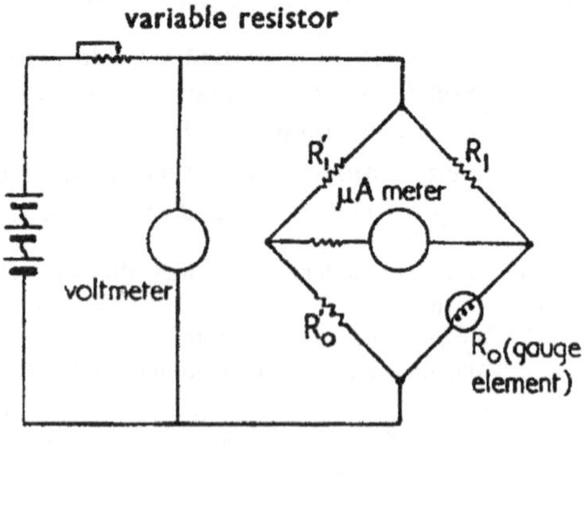

Fig. 6.3 Wheatstone bridge network. Out-of-balance current read by meter 6μA full scale deflection R_1' adjusted at zero pressure so that $R_1/R_0 = R_1'/R_0'$. By means of variable resistor and tappings on accumulator, bridge voltage could be varied between 0 and 18 V (Leck, 1952) [6-3]

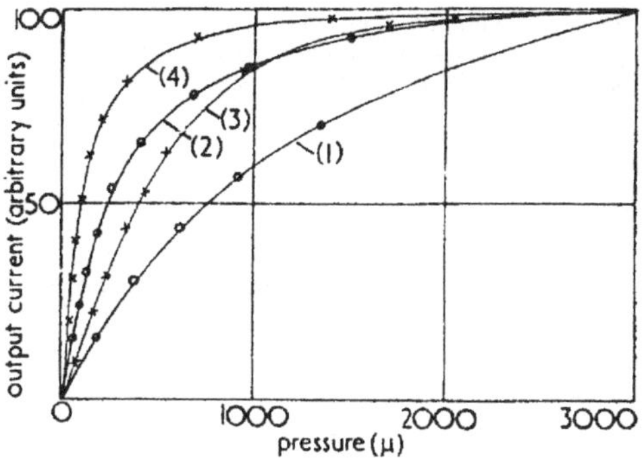

Fig. 6.4 Calibration curves for constant voltage and constant current operation (calculated) (Leck, 1952) [6-3]
(1) T_0=800 K, (2) T_0=400 K } constant voltage
(3) T_0=800 K, (4) T_0=400 K } constant current
(The "output current" scales adjusted so that all curves coincide at 3000μ.)

$$I^2R_2(1+\alpha t) \cong = K_R\left(T_W^4 - T_g^4\right) + K_C(T_W - T_g)P \tag{6.5}$$

$$\text{The out-of-balance current} = i_g \cong I(R_0 - R_2)/r_g \tag{6.6}$$

Using Eqs. (6.3) to (6.6) i_g has been calculated, and is plotted in **Fig. 6.4**, as a function of P for a tungsten wire 10 cm long and 0.001 in. diameter. These curves all saturate at high pressure because the fall in wire temperature reduces the heat loss/unit pressure. An increase in the fractional heat loss through the end supports reduces the rate of fall of temperature and consequently improves the linearity of the calibration curve. It can be seen from Eq. (6.1) that if this conductivity is increased then, for a given T_0, the wire temperature will be affected to a lesser degree by a given change in P. (In the limit if the end losses are made so large as to account for virtually all the heat dissipation, then the wire temperature will be independent of P.)

Viscosity Gauges

Spinning Rotor Gauge

The spinning rotor gauge is reliable in accuracy and becomes the reference gauge as a successor of the McLeod gauge.

Beams et al. (1962) [6-4] presented an article, "Spinning rotor pressure gauge," in which a magnetically suspended rotor pressure gauge is described for measuring pressures below 10^{-4} Torr.

Abstract [6-4]: The data are in good agreement with those obtained with a calibrated Alpert ionization gauge over the range 10^{-4}–5×10^{-8} Torr. When the gas in the vacuum system was frozen out with liquid helium, the rotor gauge recorded a residual pressure of about 5×10^{-10} Torr. A brief discussion is given of a number of effects which unless eliminated may introduce errors into the measurements at the lowest pressures.

Quoted from Beams et al., 1962 [6-4]:
In the process of measuring the pressure in a vacuum system, it is often important to avoid the introduction of high speed electrons and ions, hot filaments, and other surfaces which may produce contamination. Also, in many cases it is essential that the pressure gauge does

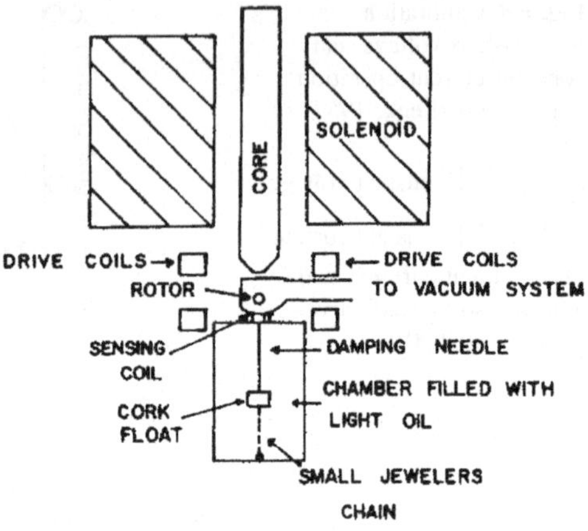

Fig. 6.5 Schematic diagram of rotor gauge (Beams et al., 1962) [6-4]

not change the composition of the gas, or introduce or remove gas from the system. The purpose of this paper is to describe a high vacuum gauge which, essentially, is free of the above difficulties and at the same time is capable of giving absolute values of the pressure when the composition of the gas is known. The principle of the method is old [1], but has been made usable for this purpose by the development of the magnetic suspension [2]. It consists in determining the frictional torque produced by the gas or vapor on a spinning rotor. The rotor is magnetically suspended inside of the vacuum chamber and spun to the desired speed. It is then allowed to coast freely and its deceleration measured. If the friction introduced by the magnetic support is negligible in comparison with the gaseous friction and if the mean free path of the molecules is longer than the dimensions of the chamber surrounding the rotor, then for the case of a spherical rotor, it can be shown that the pressure p in dyne/cm^2 is given by the relation

$$p = \frac{rd}{5C(t-t_0)} \left(\frac{2\pi RT}{M}\right)^{\frac{1}{2}} \log_e \frac{N}{N_0}, \tag{6.7}$$

where N_0 is the number of rev/sec at the time t_0, N is the number of rev/sec at time t, r is the radius of the rotor in cm, and d is the density of the rotor material, T is the absolute temperature, M is the molecular weight of the gas, R is the gas constant, and C is a constant which turns out to be approximately unity for a polycrystalline rotor surface [1, 3]. If the rotor is a solid right circular cylinder,

$$p = \frac{rd}{2C(t-t_0)} \left(\frac{2\pi RT}{M}\right)^{\frac{1}{2}} \left(\frac{1}{2+r/h}\right) \log_e \frac{N}{N_0}, \tag{6.8}$$

where r is the radius and h the length of the cylinder.

The apparatus is shown schematically in **Fig. 6.5**.

References

1. S. Dushman, *Scientific Foundations of High Vacuum Technique* (John Wiley & Sons, Inc. New York, 1949)
2. J. W. Beams, J. L. Young, III, and J. W. Moore, *J. Appl. Phys.* **17**, 886 (1946); J. W. Beams, *Science* **120**, 619 (1954); *Vacuum Symposium Trans. 7th Natl. Symp. on Vacuum Technology* (1960); *Phys. Today* **12**, 20 (1959).

3. M. Knudsen, *Kinetic Theory of Gases* (Methuen and Company, London, 1946), second edition.

Fremerey (1972) [6-5] presented an article, "High vacuum gas friction manometer."

Abstract [6-5]: The magnetic suspension is now achieved under a permanent magnet by means of a servo current which is applied to the rf sensing coil. The system involves negligible power dissipation. Instead of optical devices inductive pickup coils are used in the horizontal damping system as well as for the determination of the rotor frequency. The whole system thus avoids undesired heat and radiation transfer to the rotor and vacuum chamber. The gas pressure sensitivity is of the order 10^{-9} Torr.

Permanent Magnet Suspension [6-5]

The axial magnetic suspension of a small ferromagnetic rotor is usually attained under a magnetized cylindrical iron core (see **Fig. 6.6**). It is obvious that there exists a rotor position on the vertical axis under a permanent magnet where the rotor weight is exactly balanced by the upward magnetic attraction force. Unfortunately, the equilibrium at this

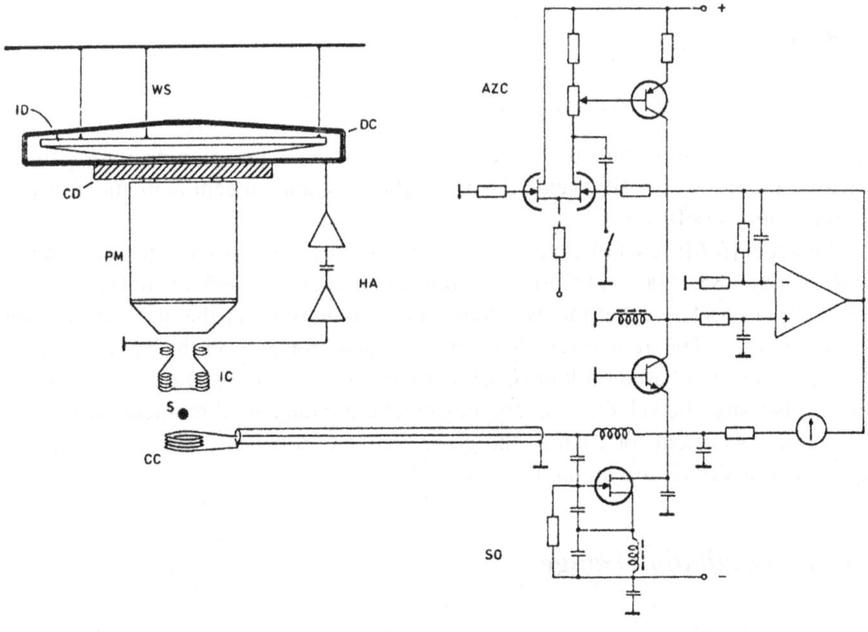

Fig. 6.6 Permanent magnet suspension with electronic circuitry: S — sphere rotor, PM — permanent magnet in rigid connection with iron disk ID. Vertical stabilization: CC — control coil for pickup and field correction, SO — sensing oscillator, AZC — automatic zero control circuit. Horizontal stabilization: IC — compensated differential coil set for pickup of lateral rotor movements parallel with the paper plane, HA — differentiating amplifier, DC — deflection coil (rotated into the drawing plane), WS — three wire suspension for the magnet system (PM, ID), turned to the natural frequency of the lateral rotor oscillation in order to provide proper phase relation within the servo loop; CD — copper disk for eddy current damping of the system (Fremerey, 1972) [6-5]

point is unstable. Additional means has to be supplied in order to provide a downward force above the equilibrium position, and an upward force below that point, i.e., to convert the unstable equilibrium into a stable one. This has been done so far by using a current controlled electromagnet, the field strength of which was regulated as a function of the vertical rotor positioning the proper sense. A simple theory of operation of the servo loop has been given in a preceding paper [1]. The use of an electromagnet for producing the whole dc support field requires stable power electronics, and is associated with the production of heat. When using a permanent magnet, essentially no power has to be wasted for field generation if the rotor is suspended at the natural equilibrium position, and only a weak variable field is needed if the rotor deviates from this position. The proper correction field is generated by means of an axial control coil (see **Fig. 6.6**) which is connected to the servo amplifier output, and simultaneously operates as an rf sensing coil in the usual way for pickup of the vertical rotor position.

Obviously, the correction field has to change its sign if the rotor moves across the equilibrium position. This feature offers the possibility of introducing an automatic zero control circuit which in the version of **Fig. 6.6** provides a mean long-term vertical rotor stability of about $\pm 1\,\mu m$ with respect to the ideal equilibrium position. The total power dissipation within the control coil (the rf power included) is of the order of only 10^{-4} W. The overall power consumption of the servo circuit due to the quiescent currents of the oscillator and amplifier stages is about 0.1 W at ± 4 V supply voltage. The electronic parts are easily available, inexpensive, and ensure long-life operation in this low-power circuit.

Reference

1. J. K. Fremerey, *Rev. Sci. Instrum.* **42**, 753 (1971).

Tekasakul et al. (1996) [6-6] presented an article, "The spinning rotor gauge: Measurements of viscosity, velocity slip coefficients, and tangential momentum accommodation coefficients."

Abstract [6-6]: A set of experimental measurements with a spinning rotor gauge (SRG) with 3.85, 4.00, and 4.50 mm nominal diameter steel spheres in He, Ar, and Kr is reported. The experiments covered the continuum and the slip regimes for all three gases. Theoretical results from a companion paper on the SRG, together with a calibration based on known viscosity for helium, are used to extract values of the viscosity, the velocity slip coefficient, and the tangential momentum accommodation coefficient for each of the gases. The measured viscosities are in good agreement with existing literature values.

Crystal Oscillation Gauge

The crystal oscillation gauge is strong in structure, small in size, and easy in operation. It indicates pressures in the range from 10^{-3} to 10^3 Torr.

Kokubun et al. (1984) [6-7] presented an article, "A bending and stretching mode crystal oscillator as a friction vacuum gauge."

Abstract [6-7]: The pressure dependences of the electric impedance of quartz oscillators are analyzed theoretically. A simple model of the quartz oscillator is proposed. Based on this model, the equations of fluid mechanics are analytically solved. The results calculated show good agreement with the experimental data. Especially, this theory succeeds in explaining the behavior that the impedance of the quartz oscillator is proportional to the square root of the pressure in the higher pressure region.

Theoretical Predictions and Their Experimental Verifications [6-7]

Based on the theoretical analysis described above, a few theoretical predictions are made about impedances for various kinds of gases. First, it is found that at the lower pressure, the impedance Z must be proportional to the square root of the molecular weight M. That is,

$$|Z| \propto \sqrt{M}. \tag{6.9}$$

This relation is well known.

Secondly, at the much higher pressure, the impedance must be proportional to the square root of the product of the coefficient of viscosity η and the density ρ:

$$|Z| \propto \sqrt{\eta\rho}. \tag{6.10}$$

Also, the density is proportional to the molecular weight, this relation is written in the form

$$|Z| \propto \sqrt{\eta M}. \tag{6.11}$$

Lastly, in the intermediate pressure region, the impedance must be proportional to the coefficient of viscosity:

$$|Z| \propto \eta. \tag{6.12}$$

It is difficult to verify the last prediction experimentally, since it is difficult to quantitatively separate the contribution of the impedance in the intermediate pressure region from the measured curve with high accuracy.

To verify these theoretical predictions, the pressure dependences of the impedance for various kinds of gases have been measured. The quartz oscillator used was tuning fork-shaped. Its resonant frequency is about 32 kHz. The gases used were argon, air, nitrogen, helium and hydrogen. The results are shown in **Fig. 6.7**. It is seen that the impedance value is different for each gas, but the shapes of the curves are almost similar for all gases.

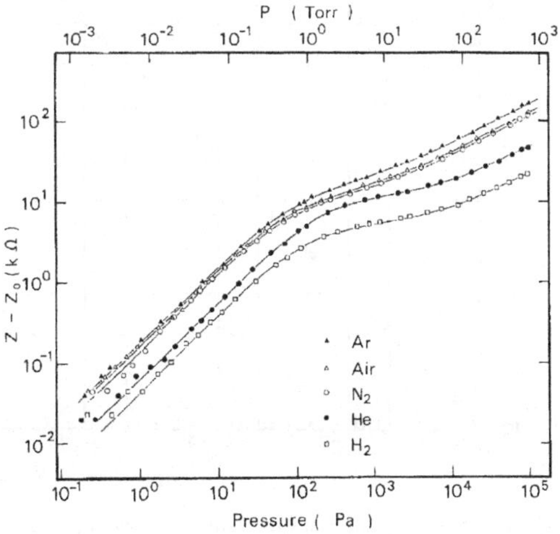

Fig. 6.7 Pressure dependence of the impedances for five gases (Kokubun et al., 1984) [6-7]

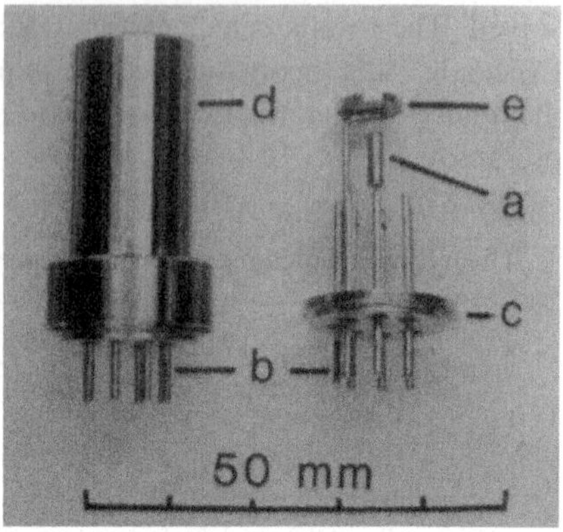

Fig. 6.8 Sensing head of the quartz friction (QF) gauge: (**a**) Metal case of quartz oscillator, (**b**) electrodes,(**c**) metal disk, (d) stainless steel tube, and (**e**) metal shield (Ono et al., 1985) [6-8]

Ono et al. (1985) [6-8] presented an article, "Design and performance of a qualtz oscillator vacuum gauge with a controller."

Abstract [6-8]: We designed , built, and tested the first model of a quartz friction (QF) vacuum gauge and its controller, using a tuning fork shaped miniature quartz

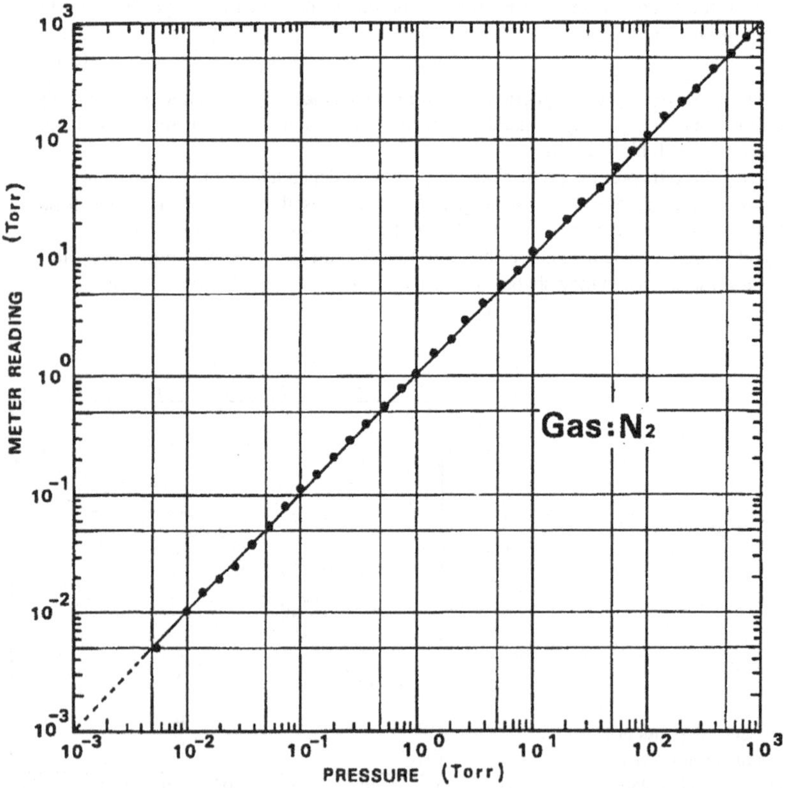

Fig. 6.9 Meter reading of QF gauge vs. pressure of nitrogen gas (Ono et al., 1985) [6-8]

oscillator of a wristwatch as a pressure sensor. This gauge made it possible to measure pressure from 0.01 Torr to more than 760 Torr with an accuracy of better than 10%. The oscillator was installed on a small-sized vacuum flange and driven at its resonance frequency by constant amplitude ac voltage fed by a phase-locked loop oscillation circuit. The resonance current of the oscillator was rectified and amplified to actuate a meter and/or a recorder. The pressure dependence of the readings of the gauge for various gases was measured, and found to be in agreement with theoretical predictions. The performance of the gauge was unaffected by hundreds of repetitions of a quick evacuation from atmospheric pressure to less than 0.001 Torr and an exposure to the air.

The sensing head and the meter reading of the QF gauge are presented in **Figs. 6.8** and **6.9**.

Ionization Gauges

Penning Gauge

The Penning gauge is widely used for controlling the vacuum systems of scientific instruments because it has rigid in structure.

Penning discharge is the basis of this gauge, as well as in the sputter ion pump (See Chap. 2). Comparing to the sputter ion pump, the Penning gauge has one anode cell, relatively low voltage and low magnetic flux, which may be due to reducing the manufacturing cost.

Conn and Daglish (1954) [6-9] presented an article "Cold cathode ionization gauges for the measurement of low pressure" in detail. He gave an attention on the generation of radio frequencies by cold cathode gauges.

Geometry and Arrangement of the Electrodes [6-9]

The three arrangements proposed by Penning [1] are shown in **Fig. 6.10** of which **(a)** illustrates the earliest form of gauge used. In this, the magnetic field was supplied by a permanent magnet, the cathodes were small rectangular or circular metal plates and the anode was a ring of wire. It has been suggested [2] that if the cathodes be extended until they 'overlap' the anode, the effect of slight variations of the magnetic field is reduced. In some of the commercial variants this has been elaborated by having cathodes in the form of a metal box surrounding the ring anode. The precise dimensions of the electrodes are not of great importance; the larger the electrodes, the larger the ion current. The ion current depends on the volume available for ionization; it is roughly proportional to the distance between the cathodes. It is also roughly proportional to the diameter of the anode ring. A typical gauge of this form would have circular cathodes about $1\frac{1}{2}$ in. in diameter, 1 in. apart, and anode-ring of 1 in. diameter.

A modification in which the ring anode was replaced by a metal cylinder of similar diameter was suggested by Penning and Nienhuis [3] and also by Hayashi and his co-workers [4]; it is illustrated in **Fig. 6.10 (b)**. The metal cylinder extends almost to the cathode end-plates so that the 'escape' of electrons is less than with a ring anode; consequently, at any

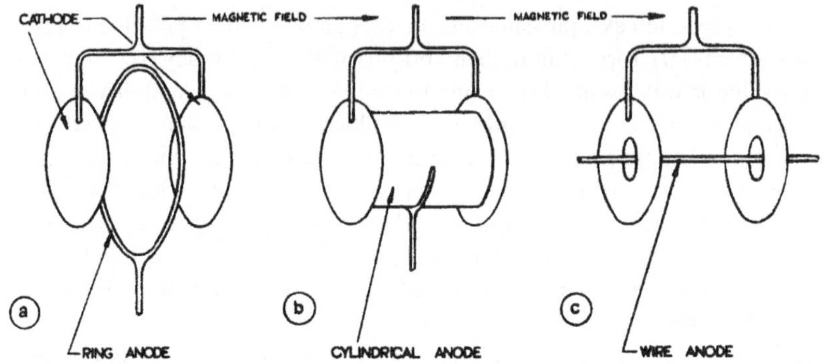

Fig. 6.10 Possible arrangements of electrodes. F. M. Penning [1] (Conn and Daglish, 1953) [6-9]

pressure, the ion current is larger. The narrow gap between the rims of the anode and the cathode plates leads to an increased electrostatic field and hence the primary current is larger. Consequently the critical or 'cutoff' pressure is lower than in the case of a similar gauge with a ring anode. Versions of this gauge with anode as large as 10 cm in diameter have been described by the Japanese [4] but the various possible modes of oscillation make gauges for this size somewhat unreliable. Leck at Liverpool University and Bloomer at the A.E.I. Laboratories, Aldermaston, have both concluded the model of Penning and Nienhuis was the best compromise between sensitivity and stability [5].

The Magnetic Field [6] [6-9]

It appears from **Fig. 6.11** that, under the conditions specified, the discharge will not strike if the field is less than about 300 oersted (0.03 T). With low fields the performance is erratic; oscillation may occur in more than one mode and transition from one mode to the other is irregular. For reliable operation, fields of the order of 1,000 oersted should be used.

Changes in the mode of oscillation, irregularities of the field or small changes in the voltage rarely cause changes in current which exceed a factor of two. In most applications vacuum gauges are used to indicate the pressure range and for this purpose such variations are not important. For the same reason the magnetic field is not critical, though it should clearly exceed the value which corresponds to the maximum ion current. This depends on the geometry of the gauge and differs slightly for the three types shown in **Fig. 6.10**. Greater accuracy can be obtained, if necessary, by stabilizing the electrical parameters.

The small irregularities revealed in **Fig. 6.11** have no significant bearing on the performance but they are remarkable in that the changes in slope, a, b, c, d, e occur at values of the field strength which are independent of pressure and of voltage. These changes occur at the same values using different gauges of the same form. Pressure, voltage and uniformity of field influence the extent of these variations, not their position.

Over a wide range of pressure the relationship between current, i, and pressure, p, is of the form

$$p = Ci^k,$$

where C is independent of pressure and current. This relation is illustrated by the logarithmic graph shown in **Fig. 6.12**. The departure from linearity at high pressures is determined by the magnitude of the ballast resistor. The variation given by the continuous line is obtained by reducing the pressure; when the pressure is raised this curve may not be retraced at pressures greater than 2.5×10^{-5} mm Hg (Torr), as indicated by the dotted curve. In this region there may be a change in the mode of oscillation, accompanied by a change in the

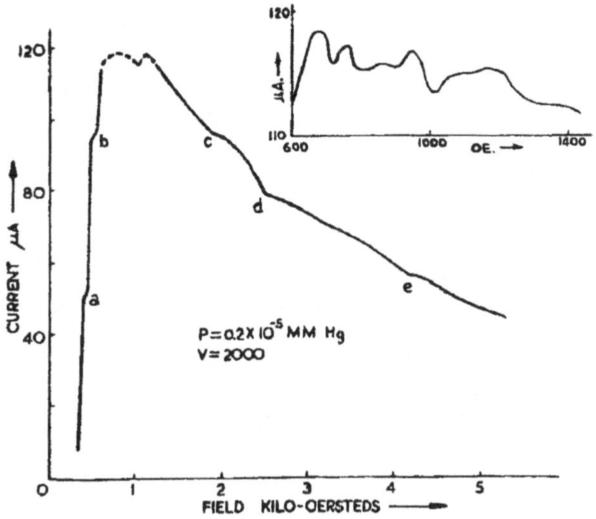

Fig. 6.11 Effect on the ion current of varying the magnetic field. The inset shows the shape near the maximum. The detail in this region is sensitive to changes of voltage and pressure. Cylindrical Anode (Conn and Daglish, 1953) [6-9]

sensitivity from one curve to the other. According to Beck and Brisbane [7], the index k depends on the magnetic field and approaches unity as this is increased. If the gauge is to have a scale which is linear with pressure, over a limited range, a magnetic field of the order of 2,500 oersted is desirable.

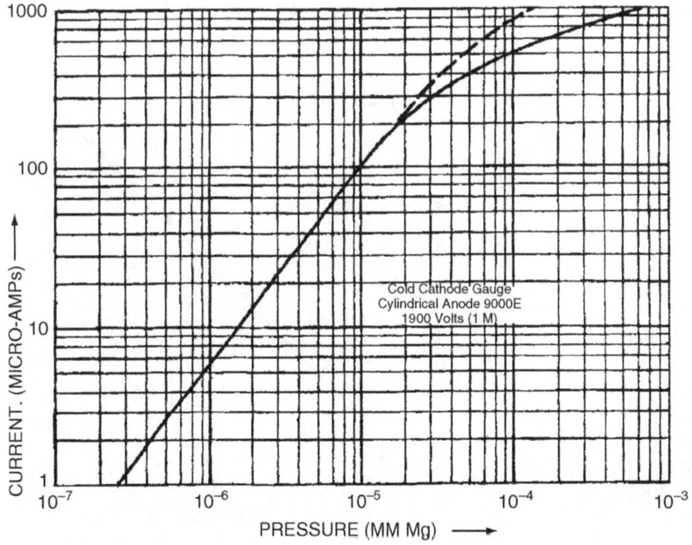

Fig. 6.12 A typical calibration curve for a Penning-type cold cathode gauge (Conn and Daglish, 1953) [6-9]

References

1. F. Arnot, "Collision Progresses in Gases", (Methuen & Co. Ltd., London, 1933), 89.
2. L. F. Wouters, *Univ. Calif. Rad. Lab. Rep.* No.: R.L. 20.6.18 (Jan. 1944). "The Characteristics of Philips-Type Gauges".
3. F. M. Penning and K. Nienhuis, *Philips Tech. Rev., II* (1949), 116. "Construction and Application of a New Design of the Philips Vacuum Gauge".
4. C. Hayashi et al. *Rev. Sci. Instrum.*, **20** (1949), 524. "Several Improvements on the Philips Gauge".
5. J. H. Leck and J. Bennett, Private Communication (1953).
6. F. M. Penning, *Physica*, **3** (1936), 873. "The Glow Discharge at Low Pressure between Coaxial Cylinders in an Axial Magnetic Field".
   J. D. Cobine and C. J. Gallagher, *J. Appl. Phys.* **18** (1047), 110. "Effects of Magnetic Field on Oscillations and Noise in Hot-Cathode Arcs".
   Jones F. Liewellyn, *Rep. Prog. Phys.*, **16** (1953), 216. "Electrical Discharges".
7. A. H. Beck, and A. D. Brisbane, *Vacuum*, **2** (1952), 137. "A Cylindrical Magnetron Ionization Gauge".

Sputter Ion Pump as a Pressure Indicator

Scientific instruments of ultrahigh vacuum (UHV), such as Auger electron spectrometers and electron microscopes, are usually evacuated with sputter ion pumps (SIPs). Practically, the SIPs act as a pressure indicator for UHV.

Well designed SIP (0.15 T, 24 mm diam. anode cells) has the linear ion-current characteristics down to pressures of ultrahigh vacuum, as shown in **Fig. 6.13** [2-29]. SIPs designed for XHV (extremely high vacuum) pumping, for instance the pump of 0.3 T and 29 mm-diam cells, have almost constant discharge intensity (current/pressure) down to 1×10^{-8} Pa, as shown in **Fig. 2.17(c)** [2-23] in Chap. 2.

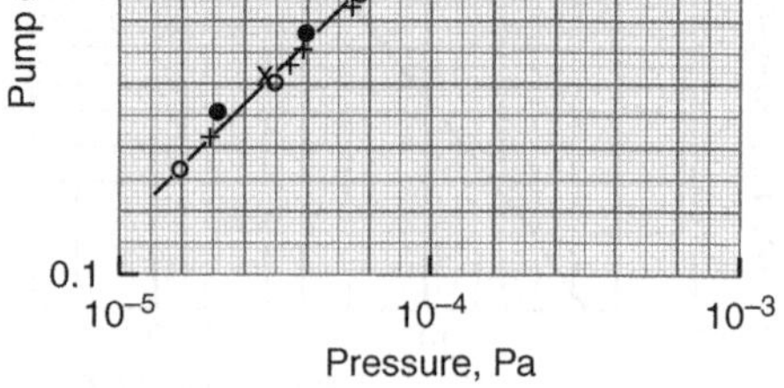

Fig. 6.13 Ion-current characteristics. -×-, "flat Ta/flat Ti" cathode pair; -•-, "holed Ta on flat Ti/flat Ti" pair; -o-, "slotted Ta on flat Ti/flat Ti" pair; -+ -, "slotted Ta on flat Ti/slotted Ti on flat Ti" pair (Yoshimura et al., 1992) [2-29]

Bayard-Alpert Ionization Gauge

The Bayard-Alpert ionization gauge (BAG) is often installed to check the pressure degree in scientific instruments. There are the glass-tube-type and nude-type available. The nude type gauge is generally selected for scientific instruments by the reasons that the glass tube is easy to break and the outgassing from the glass tube is considerably large.

Alpert (1953) [6-10] presented the famous article, "New developments in the production and measurement of ultrahigh vacuum," in which the developments of the Bayard-Alpert ionization gauge, that is, construction and manufacturing of the gauge head, ion gauge sensitivity, pumping action of the ionization gauge, ion gauge power supply (circuit) are described in detail.

Singleton (2001) [6-11] presented an article, "Practical guide to the use of Bayard-Alpert ionization gauges".

Abstract [6-11]: Sections in boldface type provide a practical guide and are followed in each instance by supplementary, detailed discussion of the rationale behind the guide. The topics addressed include gas pumping, by generation of ions, and by chemical interactions at the electron emitter; the change in the gas composition by interaction at the electron emitter; and errors in pressure measurement including Barkhausen-Kurtz oscillations, electron-stimulated desorption, and the x-ray effect. Factors which dictate the specific BAG selection, such as the method used for outgassing, and the selection of the electron emitter, are discussed."

The practical guide in boldface type of each section [6-11] are as follows.

Construction of the BAG [6-11]

Make sure that the gauge and supply are matched, and, when replacing a gauge, be sure that the correct model is used [6-10].

Three common examples of gauge structures are shown in **Fig. 6.14**.

Mounting an Ion Gauge on the Vacuum System [6-11]

The conductance of the connection to the system should be large enough that the pressure in the gauge reflects that in the vacuum chamber within the acceptable tolerance for the appli-

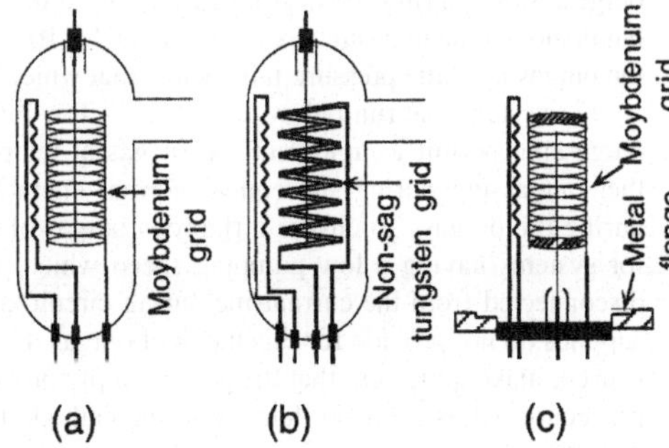

Fig. 6.14 Variations in BAG geometry. **(a)** The original BA gauge. **(b)** A version most suited to measurements in the 10^{-7} Torr range, and higher. **(c)** A nude gauge specifically designed for use in the UHV range. The ends of the grid are closed by grids, giving an improvement of a factor of 2 or more in sensitivity (Singleton, 2001) [6-11]

cation involved. When making electrical connections to the ion gauge use coaxial shielded cable for the ion collector electrode, and provide this with a reliable ground. It is customary to run the ion collector close to ground potential to minimize stray currents. Operation of a gauge in a strong magnetic field will change the gauge sensitivity [1]. Adequate magnetic shielding can minimize this effect.

Pumping by the Gauge as a Consequence of the Generation of Ions [6-11]
A gauge operating at 10 mA electron emission current and an accelerating potential of around 150 V, has a pumping speed as high as 0.1 L/s. Note that ion pumping speed is a linear function of emission current, so that it can be reduced by a factor of 10 or 100, by simply operating the gauge at 1 mA or 100 μA emission current, respectively.

Pumping by the Gauge as a Consequence of Reactions at the Hot-Electron Source [6-11]
An electron emitter in an ionization gauge, operating at an emission level of 10 mA will reach about 2100 K with a tungsten electron emitter, and about 1725 K with a thoria-coated electron emitter. Such high-temperature surfaces can have a very significant effect upon some gases. The rate of chemical reaction and, therefore, the pumping speed, is reduced by operating at a lower temperature, for example, by selecting a thoria-coated emitter, and further, by operating at a lower emission current. The choice of an emitter on an iridium support is also preferred to one on a tungsten support when oxygen or water vapor are significant part of the gas ambient, at pressures exceeding, roughly, 10^{-5} Torr: iridium does not react rapidly with these gases.

Safety Precautions [6-11]
An operating gauge requires a potential of around 185 V when measuring pressure, and as high as 1000 V if degassing by electron bombardment. Further, if such degassing is carried out at too high a pressure, say $> 10^{-2}$ Torr, a glow discharge develops which may result in potential buildup on electrodes insides the vacuum chamber quite remote from the gauge. To avoid shocking developments, the gauge tube, metal vacuum system, and any unused feedthroughs which connect into the system volume should be effectively grounded at all times.

Outgassing the BA Ionization Gauge [6-11]
When a system is pumped down after exposure to atmospheric pressure, do not run the ionization gauge continuously until the pressure falls below 1×10^{-4} Torr, and do not try to outgas the gauge until the pressure falls well below 1×10^{-5} Torr. In processing for ultrahigh vacuum (UHV), the entire system is outgassed at 200–400 °C before gauge outgas, which is started when the pressure is 1×10^{-6} Torr or lower, while the system is still warm.

For systems which operate in the 1×10^{-7} Torr range and higher, gauges with Ohmic heating outgas are quite satisfactory. The gauge should be used to monitor pressure during outgas. Slowly bring the heating current up to the maximum level specified for the gauge, but do not let the pressure rise above about 1×10^{-4} Torr for extended periods, and continue the outgas until the pressure falls below that which existed prior to starting outgas.

For system that run below the 1×10^{-7} Torr range, gauges with electron bombardment outgassing permit a rapid removal of gas at temperatures high enough to minimize further outgassing during subsequent normal operation. The gauge is not usually operated during the outgas. To ensure a rigorous outgas of the ion collector electrode, particularly for systems having a low pumping speed which operate in the UHV range, it should be disconnected from the current-measuring circuit, and connected to the grid; some modern supplies do not provide for this mode of outgas. If a thoria- or yttria-coated electron source is used, make quite sure that the power supply provides dc rather than ac for the outgassing procedure, so as to avoid damage to the cathode by electron bombardment. If a separate gauge is available for monitoring pressure, the same guidelines given for Ohmic heating can be followed, but do not allow the pressure to rise above 1×10^{-5} Torr to avoid damage to the oxide coating by ion bombardment. If there is no independent means of monitoring the pressure, as is often the case, the following guidelines can be used. The aim is to raise the temperature of the grid structure to around 1800 K [2]. For example, the original Westighouse BAG was designed for outgassing at 500 V and 200 mA (100 W, maximum). For such a gauge the following procedure is appropriate: with the grid at 500 V relative to the electron emitter, slowly raise the electron emission until 200 mA is reached and maintain this level for 30–60 min, depending on the gauge condition, and on the pumping speed of the vacuum system. During the early stages of outgas, watch carefully for the development of a discharge between the filament and grid, indicated by a blue glow, and immediately back off on the emission current if one is detected.

The outgas removes adsorbed gas from all gauge surfaces, and as the gauge cools to ambient temperature gas readsorption will continue until equilibrium is again achieved at the pressure in the system. This pumping action causes a drop in the pressure in the gauge. The problem will be quite negligible at pressures of 10^{-6} Torr and higher, lasting perhaps a few minutes, but the time taken to reach equilibrium is inversely proportional to pressure, and at 1×10^{-10} Torr could extend for several hours.

Pressure Measurement Using the Bayard-Alpert Ionization Gauge [6-11]

Absolute Pressure and "Nitrogen-Equivalent" Pressure [6-11]

Most ion gauge supplies are calibrated on the assumption that the gas is pure nitrogen. Hence, the custom of citing pressure readings as "nitrogen-equivalent pressure." Gauge response depends upon the specific gases present, **Table 6.1** shows the ratio of the sensitivity for several gases relative to that for nitrogen [3]. It is composite of data from several authors. In order to determine the absolute pressure in a system, both the gas composition and the sensitivity ratios for the particular gauge in use must be known.

The pressure indication given by a gauge controller may be subjected to considerable uncertainty. Calibrations of nominally identical gauges have shown differences of 20% in sensitivity [3]. Variation in the spacing of electrodes has a major influence on stability, so the use of gauges designed to minimize such changes will result in improved stability. Sensitivity also varies with electron energy, so when trouble shooting, check that the correct potentials are applied to the gauge electrodes. When using nude gauges, note that the geometry of the housing can seriously affect sensitivity [4, 5].

Selection of the Electron Emitter [6-11]

Judicious selection of the type of electron emitter and of emission current can provide for optimum operation in most applications.

Thoria- or yttria-coated emitters have the advantage of lower-temperature operation, which reduces the rate of many chemical reactions at the emitter, and also allows the gauge to run at a lower overall temperature, so that outgassing is minimized. The coatings are fragile and life is limited by the continuous evaporation of thorium and yttrium, respectively. The use of an iridium substrate, rather than tungsten, has the major advantage that exposure to atmosphere does not inevitably result in the burn out of the emitter. Such emitters are the

Table 6.1 BA gauge sensitivity for various gases relative to N_2.[a] (Singleton, 2001) [6-11]

Gas	He	Ar	H_2	CO	O_2	CO_2	H_2O
K/K_{N_2}	0.16	1.45	0.42	1.2	0.9	1.4	0.9

[a] See [3].

best choice whenever the gas atmosphere includes reactive gases, such as oxygen, hydrogen, and water. Oxide-coated cathodes must be reactivated after exposure to the atmosphere. Operate in the normal mode for a few minutes to allow this to proceed before switching to the outgas mode.

Setting the Emission Current [6-11]

The normal maximum level is about 10 mA. This provides the highest ion current, easing the problems of measuring low currents at the lower measurement limit of a gauge (around 10^{-11} A in the 10^{-10} Torr pressure range). It has the great advantage of maintaining the adsorbed gas coverage on the grid of the gauge to a lower level than exists at lower emission currents, so that one major, and virtually unpredictable error at pressures below 10^{-7} Torr, that caused by ESD (electron stimulated desorption), is kept to a minimum.

Since ion pumping and some thermally sensitive chemical pumping is maximized at high emitter temperatures, be sure that the tabulation to the gauge has a reasonable conductance to minimize a pressure drop.

Occasionally, high-frequency electronic (Barkhausen-Kurtz) oscillations will develop in a gauge, leading to incorrect pressure measurement. They appear more frequently in glass gauges, at pressures of 10^{-9} Torr and less, when running at 10 mA emission. In some instances, an obvious indication of their presence is that the ion current suddenly goes totally negative. The oscillations can often be quenched by bringing a small magnet (50 G) [3] up to the envelope of the gauge, but this reduces the accuracy of measurement [1]. Decreasing the emission current by a factor of 10 will often eliminate the oscillations. They can be largely avoided in a glass gauge by the use of a grounded conductive coating on the wall, or by a metal grid surrounding the gauge electrodes [6].

Electron-stimulated Desorption Problems [6-11]

Ions which are generated by electron impact on gases adsorbed on the grid can release ions which reach the collector, giving an erroneous reading. Errors of 10% at pressures in the 10^{-7} Torr range have been observed. At low pressures they can be much more serious. This problem has been documented for a number of "pure" gases, but the not unexpected finding that it can also be serious in the case of mixtures of gases makes it imperative to check for the problem whenever accurate pressure measurements are required. The use of a gauge fitted with a Redhead modulator electrode [7] is invaluable, but such gauges are not commonly available. A mass analyzer with a stability matching the degree of accuracy required is, in fact, the only viable option for serious quantitative work at low pressures. Such stability may be difficult to obtain.

Impurity Gas Generation Due to Ion Gauge Operation [6-11]

Serious perturbation of the gas composition in a system can result from gauge operation. The problem is greatest with very small, slow pumping vacuum systems. On a large vacuum chamber the effect may well be marginal.

References

1. A. R. Filippelli, *J. Vac. Sci. Technol. A* **5**, 249 (1987).
2. P. A. Redhead, J. P. Hobson, and E. V. Kornelson, *The Physical Basis of Ultrahigh Vacuum* (Chapman and Hall, London, U.K., 1968) p. 308.
3. W. J. Lange, *Phys. Today* **25**, 40 (1972).
4. L. G. Pittaway, *J. Phys. D* **3**, 1113 (1970).
5. A. R. Philipelli, *J. Vac. Sci. Technol. A* **14**, 2953 (1996).
6. P. A. Redhead, J. P. Hobson, and E. V. Kornelson, *The Physical Basis of Ultrahigh Vacuum* (Chapman and Hall, London, U.K., 1968) p. 309.
7. P. A. Redhead, *Rev. Sci. Instrum.* **31**, 343 (1960).

Yoshimura et al. (1970) [6-12] measured the pumping speeds of a BA gauge.

Pumping Speed of a B-A Gauge [6-12]
Figure 6.15 shows the apparatus for measuring pumping speeds of a BAG.
Measuring: Evacuate the chamber to a high vacuum with the greaseless cocks GC_1 and GC_2 being opened. Flash the filament tube FT for a few minutes under high vacuum until an equilibrium high vacuum is obtained. Then, stop flashing the FT and close the GC_1 to adsorb the gas molecules in the chamber on the filament surfaces sufficiently. Then, flash the FT a few second to increase the pressure rapidly, and just after flashing the cocks GC_1 and GC_2 are closed to eliminate the pumping function of the FT and the pump HP. The pressure goes down by the pumping functions of the BAG and chamber walls between two cocks.

The pumping-down curves are measured. Using the measured pumping-down curves, the pumping speeds of the closed chamber with the BAG (0.55 mA) are calculated from the initial gradient of pumping-down curves, which are presented in **Fig. 6.16**.

As seen in **Fig. 6.16**, pumping speeds vary depending on the condition of the chamber walls, which is due to the variation of the pumping speed of the chamber walls. On the other hand, the pumping speed of the gauge operating a constant emission current of 0.55 mA is considered to be constant. As a result, the pumping speed of a BAG with 0.55 mA is about 0.01 L/s, which is the base speed for three different conditions.

Comment: The pumping speed of the glass-tube-type BAG with an emission current of 1 mA may be about 0.02 L/s.
Hobson (1961) [6-13] measured the pumping speed for nitrogen of a Bayard-Alpert ionization gauge in an ultrahigh vacuum system with pressures between 10^{-9} and 10^{-7} mmHg.

Abstract [6-12]: The highest pumping speed measured was 2 L/s, mainly chemical or non-electronic. After about 10^{15} molecules of nitrogen had been pumped, the chemical pump had decayed leaving the pumping speed about 0.25 L/sec, almost entirely electronic, at 8 mA electron current. This speed was approximately constant until nearly 10^{-17} molecules had been pumped, when a major decline in electronic pumping speed was observed. There was no measurable re-emission of chemically

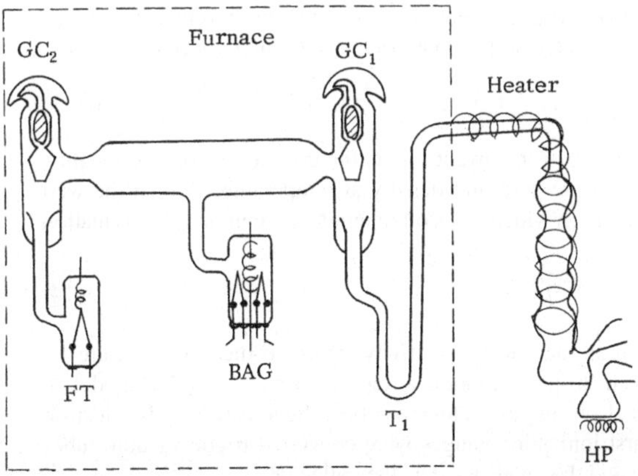

Fig. 6.15 Apparatus for measuring pumping speeds of a BAG. HP, Hickman pump; T_1, U-type liquid nitrogen trap; GC_1, Greaseless cock 1; GC_2, Greaseless cock 2; BAG, Bayard-Alpert ionization gauge (0.55 mA); FT, Filament tube. The apparatus is made of glass (Yoshimura et al., 1970) [6-12]

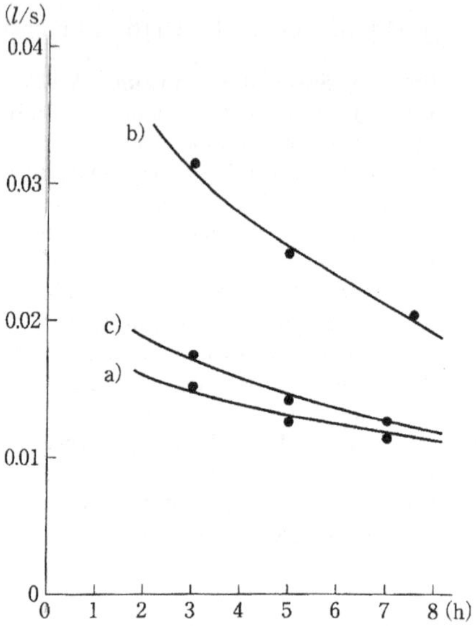

Fig. 6.16 Pumping speeds of the closed chamber with the BAG (0.55 mA) of **Fig. 6.15**. **a**) After bakeout at 100 °C for 3 h, **b**) after bakeout at 300 °C for 3 h, **c**) after bakeout at 300 °C for 3 h and subsequent exposure to the air introduced through a charcoal trap for 30 min (Yoshimura et al., 1970) [6-12]

pumped nitrogen, and the re-emission probability of electronically pumped nitrogen was 10^{-6}–10^{-5}/sec about 1 hr after completion of pumping. It is concluded that the main mechanism contributing to chemical pumping is the formation of a second adsorbed layer on the metal parts of the gauge, in particular the grid.

Weston (1979) [6-14] presented an article, "Measurement of ultra-high vacuum. Part I. Total pressure measurements", where various ionization gauges, the Bayard-Alpert gauge, the modulated Bayard-Alpert gauge, and the extractor gauge, the magnetron and similar gauges are described in details.

Ionization Gauges [6-14]

If a gas is ionized by electron impact from an electron current passing through it, then the number of positive ions formed is directly proportional to the molecular concentration ρ

$$i_+ = C\rho i_-, \tag{6.13}$$

where i_+ is the ion current, i_- the electron current and C the constant of proportionality. When there is equilibrium between pressure and density at temperature T then the well known gas-kinetic equation $p = \rho kT$, in which k is Boltzmann's constant, is valid. Equation (6.13) can then be written as

$$p = \frac{1}{K}\frac{i_+}{i_-}, \tag{6.14}$$

where $K = 1/CkT$ and is known as the gauge sensitivity. Thus, to measure pressure by ionization requires a source of electrons (cathode), an accelerating electrode (anode) to draw electron current and a third electrode (collector) to collect the ions formed by electron impact in the gas. Indeed, the first ionization gauges were converted triode vacuum tubes with the grid acting as the anode and the anode used as the collector.

K depends on temperature, from the gas kinetic equation, but is also depends on the gas species, the electron energy and the gauge geometry. The effect of the gas species and electron energy is illustrated in **Fig. 6.17** where the number of ions formed per electron per cm path as a function of electron energy is given for several gases commonly encountered in

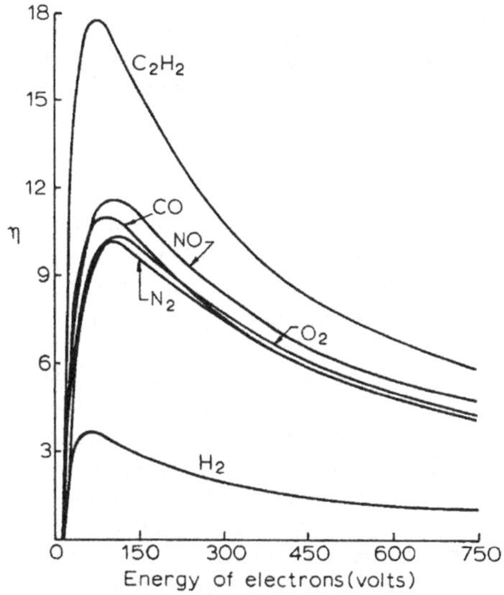

Fig. 6.17 The efficiency of ionization of N_2, CO, O_2, NO, H_2 and C_2H_2 according to Tate and Smith (J. P. Tate and P. T. Smith, *Phys. Rev.* **39**, 1932, 270), $\eta = ions/(electron \cdot cm)$ at $10^2 N/m^2$ and 0 °C (Weston, 1979) [6-14]

vacuum systems at $10^2 N \cdot m^{-2}$ and at 0 °C. These represent typical ionization efficiency curves applicable to other pressures, with ionization commencing at a threshold energy (ionization potential) rising to a maximum and falling again at higher energies. The geometry of the gauge determines the electron path length, the field distribution and ion collection efficiency. A number of workers over the last 50 years have attempted to derive K for particular gauges from fundamental data, but although reasonable agreement between the derived values and measured values has been claimed, in general it is necessary to obtain K from a calibration against a 'standard'.

The linear relationship of Eq. (6.14) holds from zero pressure up to the pressures where the ion current level affects the electron energy and current by space charge. However, the upper pressure limit is well above the ultrahigh vacuum range and is not concern of this review. Although theoretically there is no low pressure limitation on ion formation, there is the practical problem of measuring these extreme small ion currents. To give an adequate supply of ions according to Eq. (6.14) a large electron current and/or a high value of the sensitivity, K, is required. The electron current is limited by power dissipation in the cathode and there are other constrains which will be discussed later. So it is the sensitivity which becomes the important parameter for low pressure measurements. In the ultrahigh vacuum region the mean free path λ of the electrons is very much greater than the linear dimensions of the gauge envelop ($\lambda \approx 5\,\text{km}$ at $10^{-6}\,\text{N/m}^2$) and the probability of an ionizing collision by an electron traversing between cathode and anode is very small, resulting in a low value of K. To increase the ionization efficiency the electron path length would have to be considerably extended and this is the main criterion in the basic design of a vacuum ionization gauge. There are several configurations of electrostatic fields, or combinations of electrostatic and magnetic fields, which will cause electrons to oscillate or spiral within a limited volume. Many have been proposed for gauge use, but only two or three have evolved into practical ultrahigh vacuum gauges.

A more serious problem for low pressure measurements is the observed presence of extraneous currents in the gauge which are independent of pressure. These result in pressure reading errors since:

$$p' = \frac{1}{K}\left(\frac{i_+ + i_s}{i_-}\right) = p\left(1 + \frac{i_s}{i_+}\right), \tag{6.15}$$

where i_s is the total extraneous current, p' the measured pressure and p the true pressure. When $i_+ >> i_s$, the error is small but as the pressure and, thus i_+ decreases the error becomes increasingly important and eventually limits the lowest pressure measurements, as

$$p' \rightarrow \frac{1}{K}\left(\frac{i_s}{i_-}\right).$$

The main cause of this extraneous or residual current in hot cathode ionization gauges is the so-called x-ray effect. Soft x-rays are produced by electron bombardment of the positive electrode and some of these impinge on the ion collector to produce photo-electrons. The photo-electron current adds to the ion current and cannot easily be separated. In the conventional triode ionization gauge, where the ion collector is a metal cylinder surrounding the anode in form of a grid the x-ray effect restricts the pressure measurements to about $10^{-5}\,N/m^2$. Extraneous currents can also be caused by electrical leakage, ion desorption from the anode, again as a result of electron bombardment, photo-emission caused by radiation from the filament and field emission.

Thus an ultrahigh vacuum ionization gauge must be designed to give as high sensitivity as possible subject to having a low value of the residual current i_s. The ionization gauge that have been developed along these lines for ultrahigh vacuum application can be divided into two groups, hot filament (thermionic emission) ionization gauges and cold cathode (field emission or secondary emission) ionization gauges. In practice the groups are more specifically defined since most of the hot filament gauges are based on the Bayard-Alpert gauge and the cold cathode gauges on the magnetron gauge.

Nakao (1975) [6-15] determined the sensitivity of an ionization gauge for many inorganic gases and hydrocarbon gases up to C_{10} compounds using the relative ionization cross-section.

Abstract [6-15]: Results obtained indicate a linear relationship with a slope of n = 1 between the gauge sensitivity and the ionization cross-section. The ionization cross-section relative to nitrogen for both the inorganic and hydrocarbon gases shows an almost linear increase with the number of electrons per molecule as well as the gauge sensitivity relative to nitrogen. In the absence of better information the gauge sensitivity of a given gas can be determined by the relative ionization cross-section.

Extractor Gauge

The extractor ionization gauge of nude type (EG) is exclusively used for measuring ultrahigh vacuum of very low pressure.

Redhead (1966) [6-16] described the design and performance of a new type of hot-filament ionization gauge with low residual current.

Abstract: The ions are extracted from the ionization region, focused through an aperture in a shield electrode, and then collected on a short, fine wire. The sensitivity factor of this gauge is about $13\,\text{Torr}^{-1}$ for nitrogen. The ratio of the soft x-ray flux reaching the collector of the new gauge to that of a Bayard-Alpert gauge is calculated, and the x-ray limit of the new gauge estimated to be 3×10^{-13} Torr. Measurements at low pressures indicate that the x-ray limit cannot be greater than 7×10^{-13} Torr and may be considerably less. False pressure indications; caused by the electronic desorption of positive ions from chemisorbed gases on the grid (in particular, oxygen), are at least 500 times smaller in the new gauge than in the Bayard-Alpert gauge.

Construction [6-16]

Figure 6.18 is a cutaway drawing of the extractor gauge (EG) mounted on its glass header, before being sealed into its glass bulb. The glass bulb (not shown) is coated with a transparent conducting film on tin oxide. The filament is thoria-coated tungsten, supported by a hook at its midpoint. The top of the grid is closed, whereas the bottom is open. Below the shield is a hemispherical electrode, the ion reflector, at the same potential as the grid, and supported from the grid side rods. The collector is a short wire projecting through a small hole in the center of the ion-reflector. The modulator is a short wire projecting into the center of the grid from the top. By switching the modulator from grid potential to ground, the ion current to the collector can be modulated by 50% or more. All component parts of the gauges, except the grid and filament, were vacuum-fired before assembly. The grid, shield electrode, and ion-reflector were outgassed by the electron bombardment, 250 mA at 1 kV.

Positive ions are formed by electron impact within the grid cage and are attracted towards the negative shield electrode. Most of the ions pass through the aperture in the shield and are focused, by the positive potential on the ion reflector, onto the collector.

The x-ray flux striking the ion collector is reduced below that in a BAG because of two factors: (a) the collector is partially shielded from the grid by the shield plate, and (b) the collector is a short wire of small diameter, so that the solid angle subtended by the collector at the grid is very small.

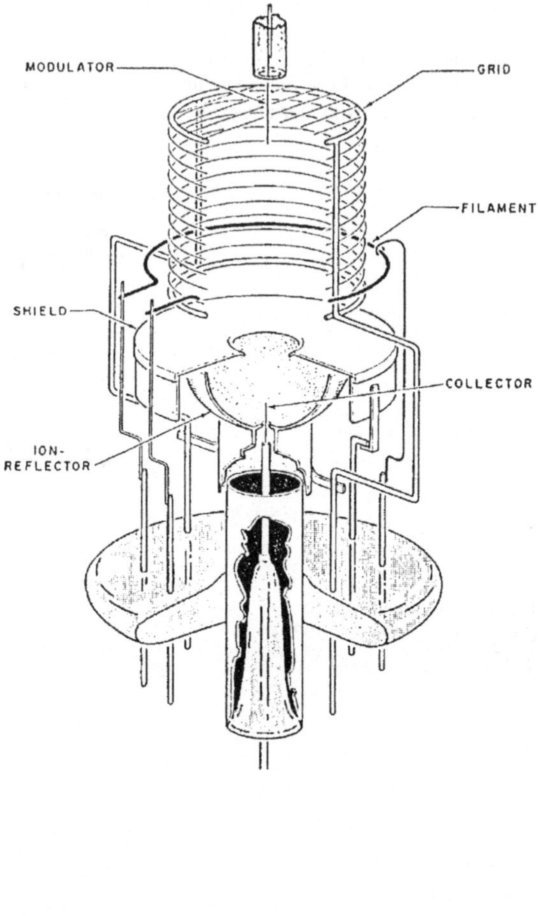

Fig. 6.18 Cutaway diagram of the extractor gauge; the envelope is omitted for clarity (Redhead, 1966) [6-16]

Beeck and Reich (1972) [6-17] presented an article, "Comparison of the pressure indication of a Bayard-Alpert and an extractor gauge."

Abstract [6-17]: Detailed comparative measurements were made and have shown: (1) When pumping a UHV system for the first time, or switching off for a period and re-energizing, the pressure indication by the EG is 3 to 5 times lower than that indicated by the BAG. (2) After degassing the gauges the pressure readings are almost the same, but higher because of increased desorption. The influence of the electronic desorption effect has been quantitatively investigated. The measurements confirmed that the reading of the EG is lower and more correct than the BAG reading under similar conditions.

Extractor Gauge [6-17]

Figure 6.19 shows diagrammatically a nude extractor gauge [1, 2, 3, 4, 5]. The ions produced by electron impact within the anode space are focused through the aperture to the ion collector and then detected as a current. However, ions which originate from electronic desorption at the anode and which are also focused in the direction of the ion collector will reach the ion reflector (in contrast to the ions created within the anode space) because this reflector is kept at the same potential as the anode. One has to expect, therefore, that only a relatively small fraction of the desorbed ions will arrive at the ion collector, so the error in the pressure indication caused by electronic desorption will also be small.

The connection between the shield and the flange is of high thermal conductivity, thus heating of the shield by radiation from the cathode causes only a slight increase of temperature and degassing of the diaphragm is negligible.

The cathode is a tungsten wire of 0.1 mm diameter coated with thorium oxide. The heating power was 1.5 W. As a result of this exceptionally low value the heating of the anode and the wall was much less than for systems using pure tungsten cathodes or iridium cathodes coated with thorium oxide.

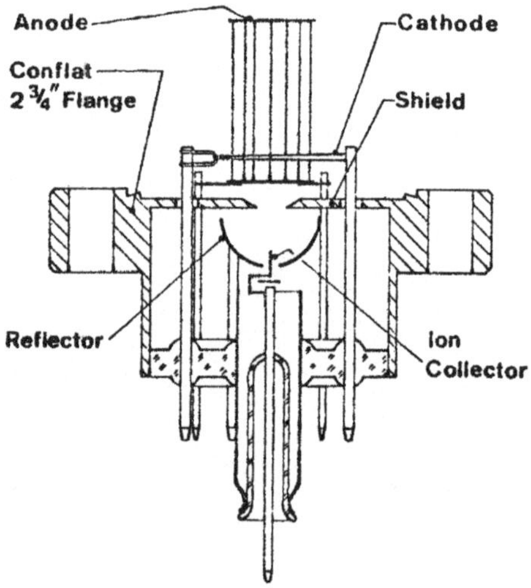

Fig. 6.19 Extractor gauge (EG) (Beeck and Reich, 1972) [6-17]

References

1. P. A. Redhead, *J. Vac. Sci. Technol.* **3**, 173 (1966).
2. P. A. Redhead, *J. Vac. Sci. Technol.* **7**, 182 (1970).
3. J. Groszkowski, *Bull. Acad. Polon. Ser. Sci. Tech.* **14**, 23 (1966).
4. F. P. Clay and L. T. Melfi, *J. Vac. Sci. Technol.* **3**, 167 (1966).
5. A. Barz and P. Kocian. *J. Vac. Sci. Technol.* **7**, 200 (1970).

Hot-Cathode Magnetron Ionization Gauge

Lafferty (1961) [6-18] presented an article, "Hot-cathode magnetron ionization gauge for the measurement of ultrahigh-vacua."

Abstract [6-18]: To accomplish increasing the sensitivity of the ionization gauge a magnetron with a cylindrical anode was selected for the ionization gauge and operated in a magnetic field with an intensity of 2.5 times the cutoff value. Two end plates maintained at a negative potential relative to the cathode collect the ion current generated in the magnetron and prevent the escape of electrons. Positive ion emission from the cathode is suppressed by mounting a hairpin filament on the axis of the cylindrical anode and well removed from the region of the negative ion collector. A very low level of electron emission is used to prevent instabilities in operation and give a maximum ratio of ion current to x-ray photocurrent. Measurements of sensitivity and x-ray photocurrent indicate that the magnetron gauge is linear down to a pressure of 4×10^{-14} mm Hg. Ability to measure low pressures with the gauge appears to be limited by the sensitivity of the circuit measuring the low ion currents. The ion pumping speed was found to be 0.003 L/s under normal operating conditions. Magnetron gauges have been built with ceramic metal envelopes.

The hot-cathode magnetron ionization gauge [6-18] is shown in **Fig. 6.20**.

Lafferty (1963) [6-19] presented an article, "Hot-cathode magnetron ionization gauge with an electron multiplier ion detector for the measurement of extreme ultrahigh-vacua."

Abstract [6-19]: The low-pressure limit of this gauge has been extended by the addition of an electron multiplier ion detector. The ions generated in the magnetron gauge are focused onto the first dynode of the multiplier by an electrostatic lens system. The multiplier gain is high enough to permit the use of an ion counting technique for the measurement of very low pressures. The lens system focuses the ions from the gauge into a beam that passes through two limiting apertures before striking the first dynode. This arrangement prevents the soft x-rays generated at the magnetron gauge anode from falling directly on the first dynode. The x-ray photoemission from the first dynode has been reduced by a factor of 50 000 below that observed for the magnetron gauge alone. Measurements of sensitivity and x-ray photocurrent indicate that this device should have a linear response down to pressures of 2×10^{-17} Torr. Ability to measure low pressures with the gauge appears to

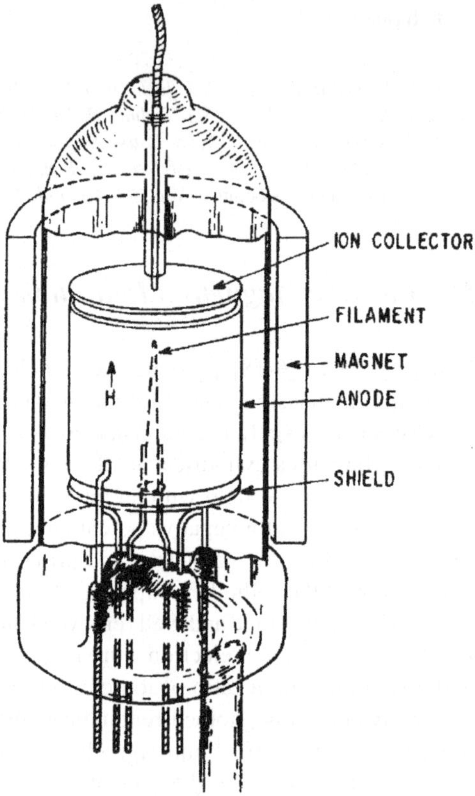

Fig. 6.20 Hot-cathode magnetron ionization gauge (Lafferty, 1961) [6-18]

be limited by the inherent dark current of the multiplier. The gauge may be used to advantage for the measurement of higher pressures by operating at reduced emission current levels to minimize pumping and degassing effects. Photoemission from the first dynode caused by incident light from the magnetron gauge filament has been reduced to values below the x-ray photoemission by the use of a lanthanum hexaboride cathode which operates at only 675 °C. This low operating temperature reduces the problem of gas reactions with the filament. Radiation from gas molecules excited by electron impact in the magnetron gauge has been observed in the 10^{-9} Torr pressure range.

Cold-Cathode Ionization Gauge for UHV

Peacock et al. (1991) [6-20] presented an article, "Comparison of hot cathode and cold cathode ionization gauges," where the performance of typical cold cathode gauges is described in detail.

Abstract [6-20]: Relative sensitivities for different gases are not the same for different HCGs or CCGs, or for HCGs compared to CCGs. Both types of gauge are susceptible to contamination. The filament heating power of the HCG is sufficient to cause local outgassing of adsorbed material, and possibly large pressure errors. Each has certain spurious effects. The lower limit of the HCG is set by x-ray photocurrent and electron impact desorption. The CCG is subject to discontinuities in the current versus pressure characteristic. Starting of the discharge in a CCG may be delayed at ultrahigh vacuum (UHV). Pumping speeds of both types of gauge are of magnitude 10^{-2} L/s, so that gauge pumping is usually negligible.

Cold Cathode Ionization Gauge (CCG) [6-20]

Figure 6.21 is a drawing of an inverted magnetron CCG [1]. All CCGs utilize crossed electric and magnetic fields to trap electrons. The high voltage ranges from 2 to 6 kV, and the magnetic field 1–2 kG. Starting from the first stray charged particle, a discharge slowly builds by cumulative processes to the point where the entry of further electrons is limited by space charge depression. At pressures below 10^{-4} Torr the discharge is an almost pure electron plasma [2]. **Figure 6.22** shows possible paths and events for electrons in the inverted magnetron. They move in approximately cycloidal jumps, circling about the anode. During part of each jump the energy of electron is sufficient to cause ionization. When the energy of electron "1" is reduced by an ionizing collision, it moves inward slightly. Electron "2" and an ion result from the collision. Upon striking the cathode, the ion frees electron "3." The trapping of the electrons is so efficient that at a pressure of 10^{-10} Torr it can require 20 min for the transit of an electron from the cathode to anode [2]. Ions, because of their greater mass, move in much larger jumps, and are quickly collected at the cathode.

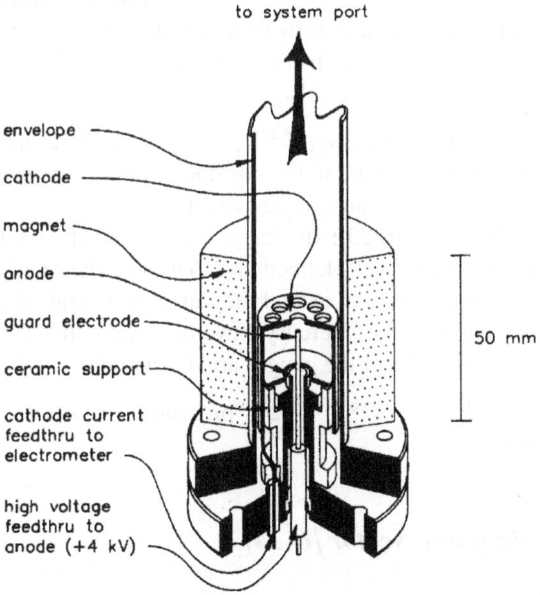

Fig. 6.21 Section of an inverted magnetron CCG showing construction. This gauge has separate feedthroughs for anode and cathode (Peacock et al., 1991) [6-20]

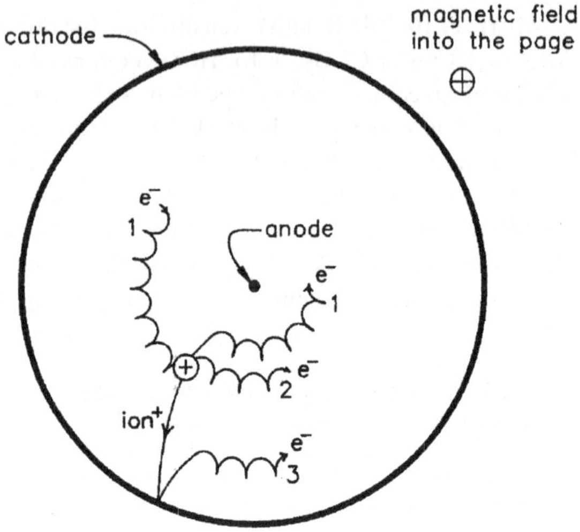

Fig. 6.22 Electron and ion trajectories in the CCG. The ion production process is described in the text (Peacock et al., 1991) [6-20]

References

1. Type 421 from HPD Division of MKS Instruments, Inc., Boulder, Colorado 80301.
2. P. A. Redhead, *Vacuum* **38**, 901 (1988).

Starting of the discharge in cold cathode gauge may be delayed at ultrahigh vacuum (UHV), which is a problem to be solved. The problem of starting the cold cathode gauge can be resolved by using a radioactive isotopes such as an americium source, as presented by Welch et al. (1996) [6-21] and Kendall and Drubetsky (1996) [6-22]. A radioactive source also works well for starting a contaminated Penning gauge.

Schulz and Phelps (1957) [6-23] presented an article, "Ionization gauges for measuring pressures up to the millimeter range", in which the development of a unique gauge, Schulz gauge, is described.

Abstract [6-23]: By designing an ionization gauge such that the electron paths are short and well-defined and that the collection efficiency of the ion collector is near unity, one can extend the range of ionization gauges to pressures of the order of a millimeter of mercury. Several specific tube constructions are proposed and the characteristics of the gauges are discussed.

Lichtman (1984) [6-24] presented an article, "Perspectives on residual gas analysis."

Magnetic Sector [6-24]

The magnetic sector employs the well known physical fact that moving, charged particles, traversing a magnetic field, follow a circular path (see **Fig. 6.23**). The radius (r) of the path is given by

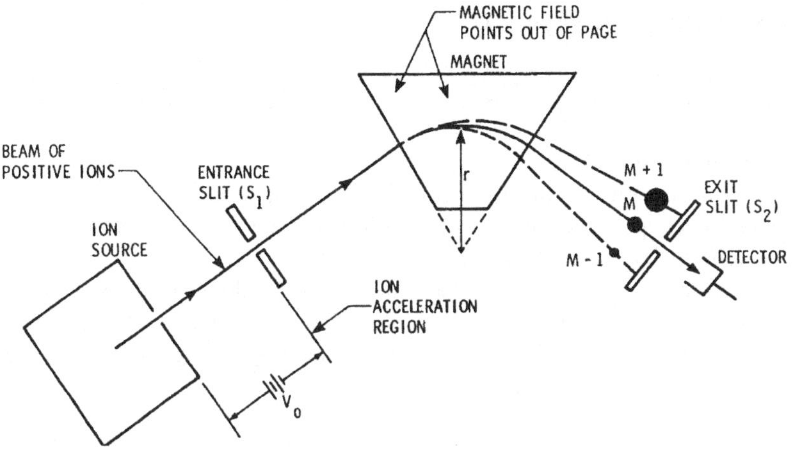

Fig. 6.23 Magnetic sector analyzer (Lichtman, 1984) [6-24]

$$r = mv/Bq. \tag{6.16}$$

From Eq. (6.16) it is apparent that the magnetic sector PPA (partial pressure analyzer) separates singly charged particles by momentum (mv) – not just mass. For nonrelativistic velocities, Eq. (6.16) can be rearranged and written in a more useful form as

$$M/q = r^2B^2/2V_0, \tag{6.17}$$

where V_0 =ion accelerating voltage. A double charged ion will follow the same path as one that is singly charged having half the mass. Magnetic sector PPAs normally deflect the ions through either 60°, 90°, or 180°. For monoenergetic ions in a perfectly uniform magnetic field the resolving power $M/\Delta M$ can be written as

$$M/\Delta M \cong r/(S_1 + S_2), \tag{6.18}$$

where r is the radius of ion path, S_1 is the entrance slit width, and S_2 is the exit slit width. Magnetic PPAs typically have a resolving power of approximately 100. Sensitivity to a partial pressures of 10^{-15}Torr has been reported for instruments using multiplier detectors.

The PPA mass spectrometers detect but one m/q ratio at a time; therefore a method of sweeping, in some order, different m/q across the detector is required. This may be accomplished by varying either the accelerating voltage or the magnetic field Eq. (6.17).

The magnetic PPAs have the advantage that there exists a large amount of useful reference information on their operating characteristics. In addition, they require the minimum of electronic equipment, all of which is basically dc or slowly varying dc. Generally, changes in resolution or sensitivity require removal of the analyzer from its envelope to adjust slits.

Further information on the magnetic type of PPA may be found in [1,2, 3, 4, 5, 6]

Quadrupole Mass-Filter [6-24]

The quadrupole mass filter (QMF) is, by far, the most popular PPA in use today. It was originally developed by Paul and co-workers in 1955 [7, 8, 9]. A schematic of a QMF is shown in **Fig. 6.24**.

The analyzing region is basically a set of four cylindrical rods. Ideally, these rods should have a hyperbolic cross-section. However, circular cross-section rods of radius $r = 1.16r_0$ work quite satisfactorily.

Ions are formed in the ion source and injected at A, along the z axis of the quadrupole (**Fig. 6.24**). The rods of the quadrupole are energized with a constant dc potential U and a rf potential $V \cos \omega t$. In the quadrupole region the ions are subjected to a potential of the form

$$\phi = (U + V \cos \omega t)\left(x^2 - y^2\right)/r_0^2. \tag{6.19}$$

The equation of motion for the ions can then be written:

$$\begin{aligned} m\ddot{x} + 2q(U + V \cos \omega t)x/r_0^2 &= 0, \\ m\ddot{y} - 2q(U + V \cos \omega t)y/r_0^2 &= 0, \\ m\ddot{z} &= 0. \end{aligned} \tag{6.20}$$

(Mathieu's differential equations)

The solution to the differential equation in z is trivial and fields z =constant. Thus, the axial velocity of the ions does not change as the ions traverse the quadrupole region. The solutions of the x and y equations are infinite series, of two types: (1) the stable solutions for which x and y remain finite for all values of time t

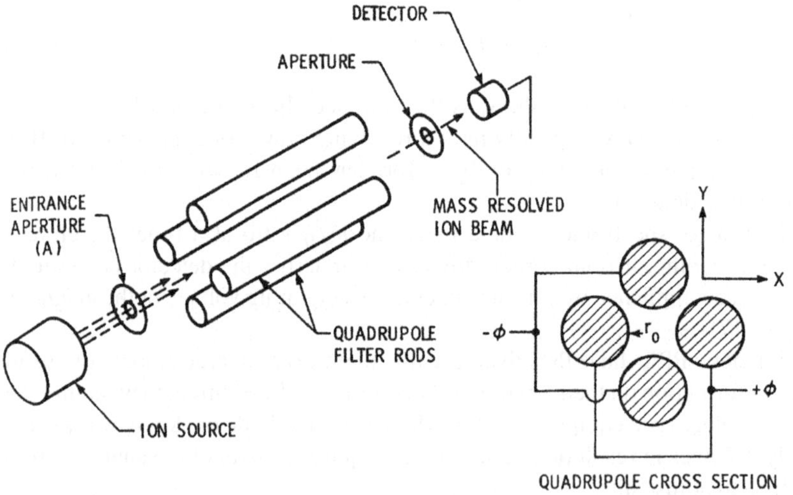

Fig. 6.24 Quadrupole mass filter (Lichtman, 1984) [6-24]

and (2) the unstable solutions for which x and y become infinite as $t \longrightarrow \infty$. The mass separation characteristics of the QMF arise from the existence of these stable and unstable solutions.

The solutions for which x and y remain finite will be called the resonant solutions. The conditions necessary for simultaneous x and y resonance are shown in **Fig. 6.25**. As shown for fixed U, V, ω, q, and r_0 there can be a range of masses for which there exist resonant solutions. The value of $\beta = U/V$ defines how wide that mass range can be. At the apex of the resonant regions where $\beta \cong 0.168$ there exists only one resonant mass value,

$$m = 5.65\left(Vq/r_0^2\omega^2\right). \tag{6.21}$$

Under these operating conditions only those particles having a mass specified by Eq. (6.21) will pass through the QMF; all other particles oscillate with increasing amplitude in the x and y directions; thus, they eventually strike the filter rods and are neutralized. It is important that the ions enter the QMF with a small enough velocity so that the rf potential will act on them for at least several cycles during their travel through the analyzer. The maximum allowable ion energies will depend on the particular instrument but will generally be in the range of 10's of eV.

Masses may be swept either by (1) varying the frequency while keeping the rf and dc voltages constant, or by (2) varying the rf and dc voltage level (keeping their ratio β constant) while maintaining a constant frequency. As shown in Eq. (6.21), method (2) provides a linear mass sweep. This is the method currently used in all commercial instruments.

Since the rf to dc voltage ratio β is kept constant in both methods of mass sweeping, it can be shown that the resolving power $R = M/\Delta M$ is constant regardless of mass. R depends only on β. However, very few commercial QMFs operate with constant resolving power. Because the mass sweep is linear (using voltage scan-

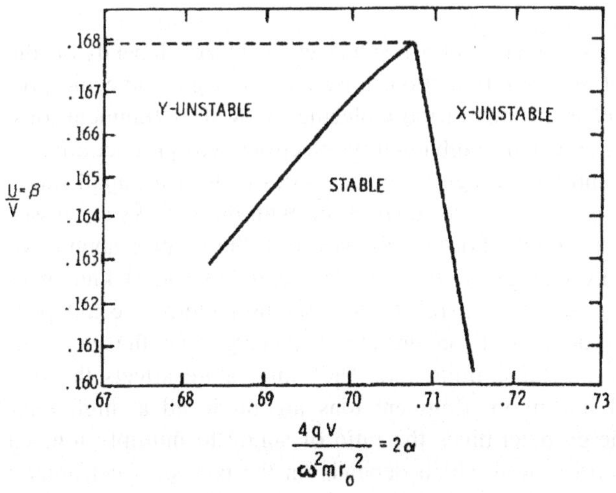

Fig. 6.25 Stability diagram for quadrupole mass filter (Lichtman, 1984) [6-24]

ning) it is convenient to have ΔM, the absolute resolution, remain constant; this is accomplished by continuously varying β electronically in just the right fashion while sweeping mass to establish constant ΔM. One should note that when operating in this mode, the sensitivity decreases with increasing M, thus making detection of more massive ions difficult. This problem is handled by suitable calibration procedures.

References

1. J. H. Leck, *Pressure Measurement in VaccumSystems,* 2nd ed. (Chapman and Hall, London, 1964)
2. W. D. Davis, *Trans, 9th Natl. Vac. Sym.* AVS 363 (1962).
3. W. D. Davis and T. A. Vanderslice, *Trans. 9th Natl. Vac. Sym.* AVS 417 (1962).
4. H. E. Duckworth, *Mass Spectroscopy* (Cambridge University, Cambridge, 1958).
5. C. A. McDowell, *Mass Spectroscopy* (McGraw-Hill, New York, 1963).
6. M. G. Inghram and R. J. Hayden, *Nuclear Science Series* (NationalAcademy of Science-National Research Council, Washington, D.C., 1954), Report Number 14.
7. W. Paul and H. Steinwedel, Z. *Naturforsch,* Teil A 8, 448 (1953).
8. W. Paul and M. Raether, Z. *Physik,* 140, 242 (1955).
9. W. Paul H. P. Reinhard, and U. Von Zahn, Z. *Physik*, 152, 143 (1958).

Mass Spectra Cracking Patterns

Weston (1980) [6-25] presented an article, "Measurement of ultrahigh vacuum. Part 2. Partial pressure measurements." Weston gave mass spectra cracking patterns in the table.

Cracking Pattern [6-25]

The usefulness of spectra obtained from a residual gas analyzer depends on the calibration. As mentioned in the introduction, ionization of a gas molecule produces singly and doubly charged ions of the molecules as well as fragment ions. For example, carbon dioxide when bombarded by electrons will produce ions of CO_2^+, CO^+, CO_2^{++}, O^+, and C^+ in significant quantities as well as small quantities of isotope ions, giving a characteristic spectrum with mass peaks at masses 44, 28, 22, 16 and 12, respectively. Fortunately, however, the relative abundance of the various ions for any one gas is more or less constant and is known as the cracking pattern of the gas. In general, these cracking patterns are independent of pressure but they do depend on the electron energy and the gas temperature and also on the mass discrimination. The temperature affects the dissociation probability and thus more fragment ions are produced at high temperature. The electron energy determines the ratio of singly to multiply ionized molecules. The mass discrimination which depends on the type of spectrometer and method of scanning, etc. will obviously affect the peak heights for the various masses. Because of this latter effect, different types of mass spectrometer can exhibit significant differences in cracking pattern for the same gas but instruments of the same type tend to show only minor variations.

Cracking patterns for most gases have been investigated for the magnetic-sector type instruments and can be found tabulated in any mass spectrometer data book. Similar information is being built up for the quadrupole. Conventionally the largest peak is scaled as 100, and is used as the base peak for sensitivity measurements, etc. **Table 6.2** gives some typical cracking pattern data in a magnetic-sector instrument for the gases likely to be met in the vacuum systems. Knowing the cracking pattern allows the spectrum to be analyzed. For example, if there is a mixture of CO_2 and N_2 present in the system, both will contribute to the peak at mass 28. However, by measuring the peak at mass 44, which is unique to the CO_2, the contribution of CO^+ to the mass 28 peak can be deduced and thus the relative height of the N_2^+ peak at 28 determined.

Outgassing from Ionization Gauges with Incandescent Filaments

Yoshimura et al. (1991) [6-26] reported the outgassing from incandescent filaments of the BAG and the EG in an article, "Outgassing characteristics of an electropolished stainless-steel pipe with an operating extractor ionization gauge."

Outgassing Characteristics [6-26]

The outgassing characteristics of a BAG (tungsten filament) and EG (iridium filament coated with thorium oxide) were investigated in detail in an isolation test. The isolation test was conducted for the EP SS304 pipe with a BAG (tungsten filament, glass-tube type, 1 mA emission current), which was evacuated by a turbomolecular pump through an all-metal valve. The experimental setup is shown in **Fig. 6.26**. The volume and surface area of the pipe are 0.75 L and 820 cm^2, respectively. The pipe has been in situ baked (150 °C, a few times) in the foregoing experiment [1].

The pipe was baked again in situ (150 °C, 20 h), while the metal valve was baked at about 100 °C, after exposure to air. The BAG was degassed by electron bombardment (21 W, 15 min), and then operated with 1 mA emission current. On the next day, the pressure just before the first isolation test was 3.6×10^{-7} Pa (N_2 equivalent pressure). An isolation test for 20 min was repeated eight times in sequence. The evacuation period between successive isolation tests was about 30 seconds each. The base pressure just before each isolation test increased as the isolation tests were repeated. The pressure-rise curves are presented in **Fig. 6.27**.

The pressure-rise characteristics in **Fig. 6.27** were surprising in that the pressure-rise rate of curve "8" was two orders higher than that of curve "1." And as seen from curves "4," "5," and "6," the rates increased with the pressure. This was believed to occur for the following reasons:

(1) The outgassing rate of the operating BAG increased as the pressure increased.
(2) Gas molecules emitted from the BAG during the isolation period were accumulated on the surface of the pipe wall, and as a result, the outgassing rate of the pipe wall increased as the isolation tests were repeated.
(3) Gas molecules diffusing out of the pipe wall impinged on the wall surface at high frequencies and were sorbed on the wall surface during the isolation period. As a result, the outgassing rate of the pipe wall increased as the isolation tests were repeated.

Table 6.2 Mass spectra cracking patterns (Weston, 1980) [6-25]

	Molecular weight / Formula																							
Mass	16	17	18	20	26	28	28	28	30	30	32	32	34	40	42	44	44	44	44	46	46	46	58	58
	CH_4	NH_3	H_2O	Ne	C_2H_2	C_2H_4	CO	N_2	C_2H_6	NO	CH_4O	O_2	H_2S	A	C_3H_6	C_3H_8	CO_2	N_2O	C_2H_4O	C_2H_6O	NO_2	CH_2O_2	C_4H_{10}	C_3H_6O
2	3.0		0.7										0.2											
12	2.4				2.5	2.1	4.5									0.4	6.0					3.3		
13	7.7				5.6	3.5										0.5	0.1					2.9		
14	18.6	2.2			0.2	6.3	0.6	7.2	3.4	7.5					3.9	2.5		12.9			9.6			
15	35.8	7.5							4.6	2.4					5.9	3.9		0.1					5.3	
16	100	80.0	1.1				0.9			1.5		11.4					8.5	5.0			22.3	5.2		
17	1.2	100	23.0																			17.1		
18		0.4	100								1.9									5.5				
19			0.1																	2.3				
20			0.3	100										10.7										
21				0.3																				
22				9.9													1.2							
24					5.6	3.7													1.6					
25					20.1	11.7			4.2							0.7			4.8					
26					100	62.3			23.0						11.3	7.6			9.1	8.3			6.3	8.3
27					2.8	64.8			33.3						38.4	37.9			4.5	23.9			37.1	8.0
28					0.2	100	100	100	100		6.4					59.1	11.4	10.8	2.7	8.9		17.2	32.6	
29						2.2	1.1	0.8	21.7		64.7					100	0.1	0.1	100	23.4		100	44.2	4.3
30							0.2		26.2	100	0.8					2.1		31.1		6.0	100	1.6		

Table 6.2 (continued)

	Molecular weight / Formula																							
Mass	16 CH_4	17 NH_3	18 H_2O	20 Ne	26 C_2H_2	28 C_2H_4	28 CO	28 N_2	30 C_2H_6	30 NO	32 CH_4O	32 O_2	34 H_2S	40 A	42 C_3H_6	44 C_3H_8	44 CO_2	44 N_2O	44 C_2H_4O	46 C_2H_6O	46 NO_2	46 CH_2O_2	58 C_4H_{10}	58 C_3H_6O
31										0.4	100							0.1		100				
32										0.2	66.7	100	44.4											
33											1.0	0.1	42.0											
34												0.4	100											
35													2.5											
36													4.2	0.3		0.4								
37															13.4	3.1								2.1
38														0.1	20.3	4.9								2.3
39															74.0	16.2							12.5	3.0
40														100	29.0	2.8								
41															100	12.4			3.9				27.8	2.1
42															69.6	5.1			9.2	2.9			12.2	7.0
43																22.3			26.7	7.6			100	100
44															26.2	100	100	45.7				10.0		
45															0.8	1.3	0.7		34.4			47.6		
46																0.4	0.2		16.5		37.0	60.9		

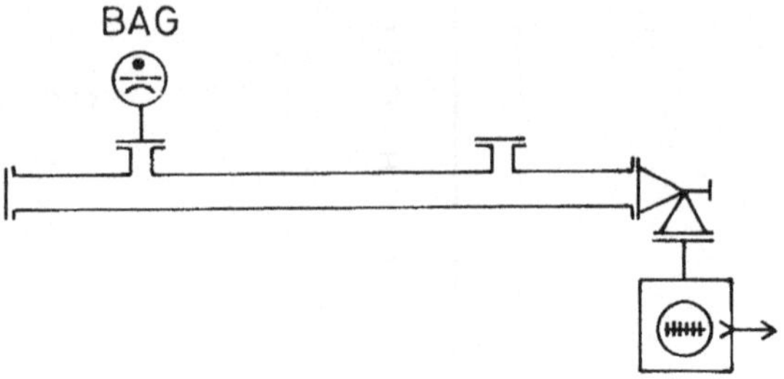

Fig. 6.26 Experimental setup for an isolation test with a BAG. The BAG was later replaced by an EG inserted in a water-cooled adapter for an isolation test with an EG (Yoshimura et al., 1991) [6-26]

Reason (1) must be responsible for the phenomenon that the pressure-rise rate increased with the pressure and that is attributed to the increasing temperature of the filament, resulting from the pressure rise.

The filament temperature of the operating BAG was measured over a wide pressure range by using a radiation thermometer. The metal valve was closed at a base pressure of 5.5×10^{-7} Pa, and then the pressure gradually increased. The filament temperature vs. pressure (N_2 equivalent) characteristics of the BAG are presented in **Fig. 6.28**.

As seen in **Fig. 6.28**, the filament temperature remained constant at 1575 °C up to 2.0×10^{-5} Pa, and then it rose gradually with the pressure rise. The temperature rise was as large as 50 °C for a one-order pressure rise above 4.0×10^{-5} Pa. It can be said that the BAG showed an increasing outgassing rate with the gradual pressure rise, due to the gradual increase of the filament temperature.

The extractor ionization gauge (EG) has been reported to be a reliable vacuum gauge for measuring ultrahigh vacuum (UHV) pressure [2]. The pressure-rise characteristics for

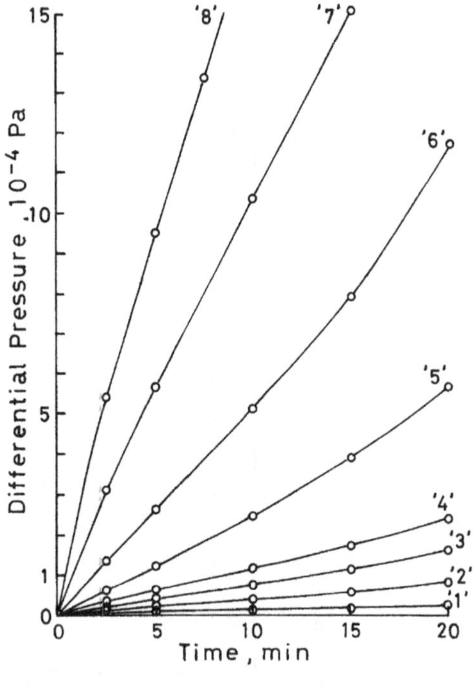

Fig. 6.27 Pressure-rise curves for the isolated pipe which had been in situ baked (150 °C, 20 h). Pressures were measured with the BAG. (Yoshimura et al., 1991) [6-26]

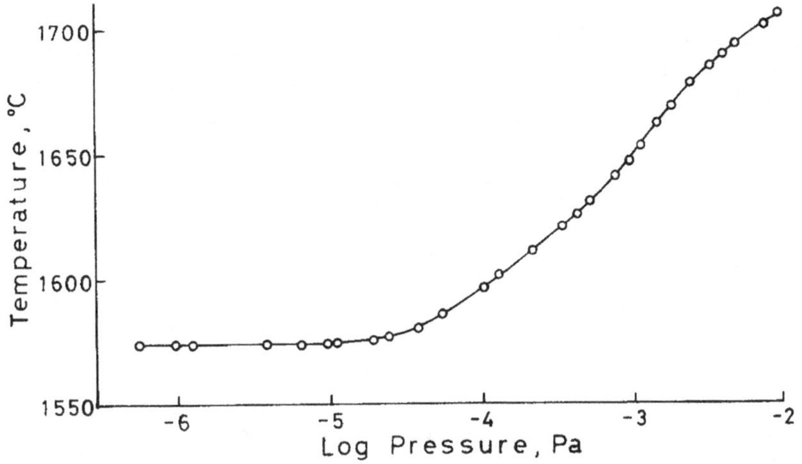

Fig. 6.28 Filament temperature of the BAG, varying with the pressure (Yoshimura et al., 1991) [6-26]

the closed pipe were repeated using an EG (iridium filament coated with thorium oxide, nude type, 0.59 mA emission). In this case the EG was inserted into a water-cooled SS304 adapter.

The process of isolation tests with the EG was similar to that with the BAG. The resulting pressure-rise curves are presented in **Fig. 4.7** of Chap. 4.

The pressure-rise curves in **Fig. 4.7** are quite different in form from those in **Fig. 6.27**. The pressure-rise rates in this case were very low in comparison. The lower outgassing of the EG must be due to its lower temperature in comparison with the tungsten filament. Low outgassing may also be related to a lower CO gas content in the filament or a smaller value of the diffusivity of the gas in the filament material. Apparently, the filament temperature of the EG remained almost constant during the pressure-rise periods of 60 or 90 minutes. Since the pressure-rise rate did increase as the isolation tests were repeated, the foregoing reasons (2) and (3) were also contributing factors. That is, gas molecules evolved from the pipe wall and the EG were accumulated on the wall surface during the isolation period, resulting in the increased outgassing rate of the pipe wall in the following isolation period.

Next, an isolation test was conducted on the same pipe evacuated for 2 h, following exposure to air (15 min), using the EG. The pressure just before the first isolation test was 5.0×10^{-6} Pa (N_2 equivalent pressure). The isolation test was repeated four times in sequence. For the first two tests the valve was closed for 5 min each, and for the last two tests, the valve was closed for 10 min each. The evacuation period between each successive isolation test was 30 seconds each. The resulting pressure-rise curves are presented in **Fig. 6.29**.

As seen in **Fig. 6.29**, the pressure-rise rates were fairly high. And, as seen from curves "3" and "4," the rates increased as the pressure increased. It was considered that when the pressure rose above 5×10^{-4} Pa, the filament temperature of the EG gradually increased, resulting in an increasing outgassing rate.

Fig. 6.29 Pressure-rise curves for the isolated pipe evacuated 2 h after exposure to air (15 min). Pressures were measured with the EG (Yoshimura et al., 1991) [6-26]

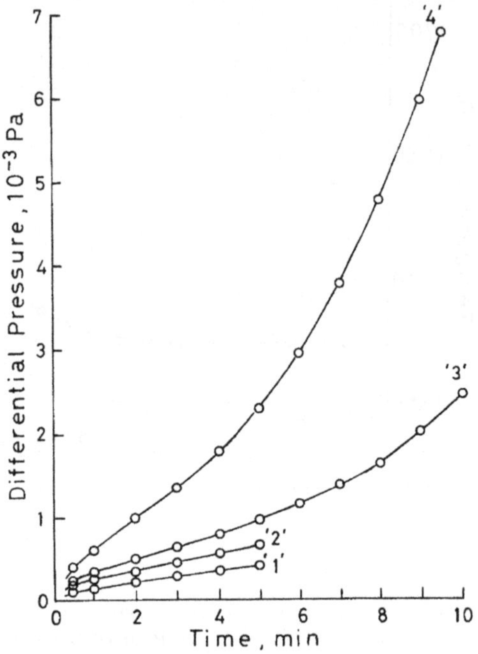

Gas Species Emitted from Ionization Gauges with Incandescent Filaments

Yoshimura et al. (1990) [6-27] reported the outgassing from a BAG (tungsten filament) and a QMF (tungsten filament) in an article, "Outgassing characteristics and microstructure of an electropolished stainless-steel surface"

Gas Species Emitted from Operating BAG and QMF [6-27]
The experimental setup is given in **Fig. 6.30**. The inner wall of the pipe (820 cm^2 inner surface area, 1 L volume) had been electropolished to 0.7 S. The pipe had already been degassed a few times (100 °C, 5 h of unit baking time) in a previous experiment.

Fig. 6.30 Experimental setup for the isolation test. The inner wall of the pipe was electropolished (0.7 S) (Yoshimura et al., 1990) [6-27]

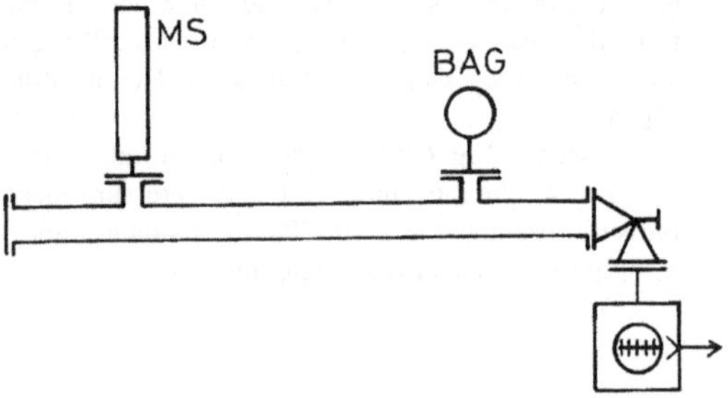

After an in situ Bakeout [6-27]

The pipe and the quadrupole mass spectrometer (MS) were in situ bakeout (150 °C, 20 h), while the metal valve was baked at about 100 °C. The MS with a W filament was first operated at 5 mA of emission current for 1 h, and then operated at 0.5 mA continuously. The BAG (glass-tube type, W filament) was degassed by an electron bombardment (30 mA/700 V, 20 min), and then operated at 1 mA.

Mass numbers of residual gases analyzed at 2.6×10^{-7} Pa of the base pressure after about 50 h pumping following baking, were 2 (H_2), 18(H_2O), 28 (CO and N_2), 44 (CO_2), and 16 (CH_4), in order of concentration.

First, the metal valve was closed at 2.6×10^{-7} Pa, thus the outgassing molecules being built-up inside the isolated pipe. The mass spectrum at 2.6×10^{-5} Pa in the isolation period is presented in **Fig. 6.31**. This pattern was quite different from that analyzed under evacuation. Most of the H_2O molecules detected under evacuation must be evolved from the TMP (blades and body walls), which had not been degassed at high temperature. Most of the CO (28), CO_2 (44), C (12), and CH_4 (16) molecules might be emitted from the operating MS (0.5 mA) and BAG (1 mA) through chemical changes by incandescent filaments [1] and electron stimulated desorption (ESD) [2].

Next, partial pressure-rise characteristics for individual masses were measured. The BAG was switched off at 2.6×10^{-7} Pa just before isolation, and again switched on 3 min after isolation. Building-up characteristics of masses 28 and 12 are presented in **Fig. 6.32 (a)**, and those of masses 16, 2, 18, and 44 are presented in **(b)**. The characteristics of masses 28 and 12 in the earlier 3 min after isolation are again presented in **Fig. 6.32 (b)** by chain lines for comparison.

Comments on individual masses are as follows [6-27].

2 (H_2). The peak rose steadily with time and jumped up appreciably when the BAG was switched on. Then, it soon lowered in a short time, and again rose slowly. Most of the H_2 molecules must be evolved from the EP pipe wall.

12 (C). The peak amplitude kept constant after isolation. The peak jumped up greatly when the BAG was switched on, and then it rose steadily with further time. The C molecules must be produced due to dissociation of CO, CO_2, and CH_4 in the MS and BAG with an incandescent filament.

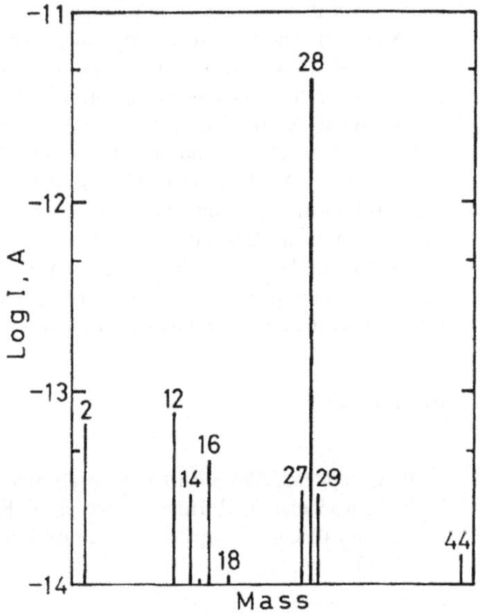

Fig. 6.31 Mass spectrum analyzed at 2.6×10^{-5} Pa in the isolated pipe. The valve was closed at 2.6×10^{-7} Pa (Yoshimura et al., 1990)[6-27]

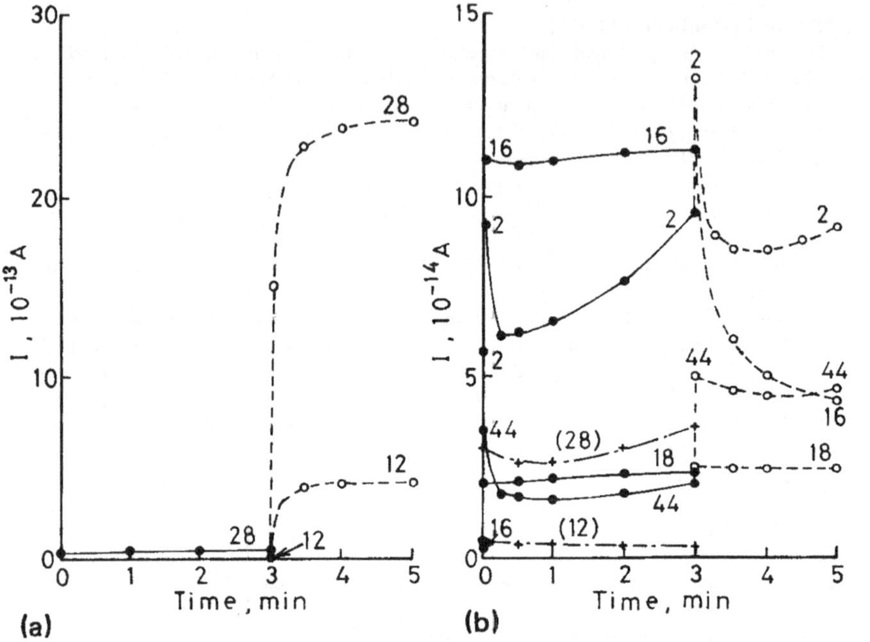

Fig. 6.32 Building-up characteristics of individual masses after the in situ bakeout (150 °C, 20 h). **(a)** Masses 28 (mostly CO) and 12 (C); and **(b)** masses 16 (CH_4), 2 (H_2), 18 (H_2O), and 44 (CO_2). The BAG was switched on 3 min after isolation (Yoshimura et al., 1990) [6-27]

16 (CH_4). The peak jumped up from 2.0×10^{-15} to 1.1×10^{-13} A when the metal valve was closed. It rose slowly until the BAG was switched on, and then it was lowered at a considerably high rate. CH_4 molecules may be partly dissociated into C and H_2 molecules in contact with an incandescent filament.

18 (H_2O). The peak rose very slowly. The variation of the peak amplitude, occurring when the BAG was switched on, was very small. The peak amplitude was kept almost constant after that. The outgassing rate was negligibly low.

28 (mostly CO). The peak jumped up from 7.0×10^{-15} to 2.5×10^{-14} A when the metal valve was closed, and then it rose steadily with time. It jumped up from 2.5×10^{-14} to 1.5×10^{-12} A when the BAG was switched on. The gas species must be mostly CO, which is known to be emitted through ESD [2]. Most of the CO molecules must be emitted from the operating MS and BAG.

44 (CO_2). The peak jumped up from 5.0×10^{-15} to 3.5×10^{-14} A when the metal valve was closed, and then it rose slowly. The peak jumped up to its double amplitude when the BAG was switched on, and then its amplitude was kept almost constant.

References

1. P. A. Redhead, *AVS 7th Vacuum Symposium Transactions*, 1960, p. 108
2. P. A. Redhead, J. P. Hobson, and E. V. Kornelson, The Physical Basis of Ultrahigh Vacuum (Electron-impact desorption) (Chapman and Hall Ltd., London EC4, 1968), p. 290.

Nakao (1975) [6-28] presented an article, "Simplified method for the calculation of partial pressure using the relative ionization cross-section."

Abstract [6-28]: Two simplified methods for the calculation of partial pressure are discussed theoretically in conjunction with the ionization probability of producing the most characteristic ion of molecules. Ionization probabilities for twenty molecules, usually encountered in many vacuum systems, are determined from the relative ionization cross-section by using a correction factor. Partial pressure values calculated from these methods are compared with those from the usual mass spectrometric method. The value can be evaluated within an accuracy of less than 10% for many molecules. The method of calculation using the ionization gauge correction factor is also developed. By these methods the partial pressure evaluation becomes possible without doing troublesome calibrations of the residual gas analyzer.

New Gauges for Measuring Extreme High Vacuum

Watanabe (1987) [6-29] presented an article, "Point collector ionization gauge with spherical grid for measuring pressures below 10^{-11} Pa."

Abstract [6-29]: A new hot-filament ionization gauge using a point collector and spherical grid (see **Fig. 6.33**) has been developed to reduce the low-pressure limit. If a very small point collector could be used in an ionization gauge, without reducing sensitivity, the small surface area of the collector could yield a very low x-ray limit.

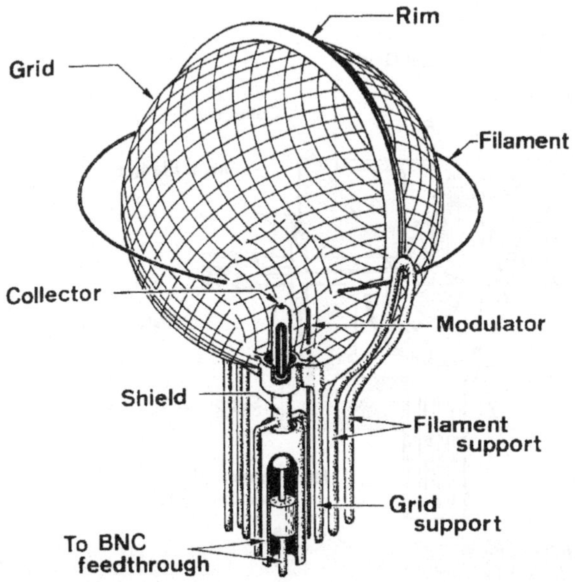

Fig. 6.33 Cutaway drawing of the point collector gauge with spherical grid and ring filament (Watanabe, 1987) [6-29]

In the present design a point collector is approximated using a needle tip ≈ 0.03 mm in diameter and ≈ 0.05 mm long projected from a thin tapered sleeve. A 26 mm-diam spherical grid is formed by joining two hemispheres made of finely woven molybdenum mesh. The collector is placed about 1.5 mm inside the pole of the spherical grid. A ring filament is stretched around the equator of the spherical grid. The x-ray limit of the new gauge, calculated theoretically from geometric considerations, is about 2.3×10^{-12} Pa; and its experimental result is about 2.5×10^{-11} Pa, with a sensitivity factor of 0.4/Pa (53/Torr).

Watanabe (1992) [6-30] presented an article, "Ion spectroscopy gauge: Total pressure measurements down to 10^{-12} Pa with discrimination against electron-stimulated-desorption ions."

Abstract [6-30]: A new hot-cathode total pressure gauge, called an ion spectroscopy gauge (see **Fig. 6.34**), which substantially avoids errors caused by ESD ions down to 10^{-12} Pa, has been developed. Discrimination against ESD ions is accomplished by combining three technical elements: (1) a spherical grid ion source; (2) a spherical 180° ion energy analyzer; and (3) burying the ion source in an aluminum alloy flange. The ion source plays an important role in the generation of a large kinetic energy difference between gas-phase ions and ESD ions, by using electron space charge. The 180° analyzer, which has a high energy resolution, promotes ion focusing and tends to reduce the x-ray limit. The performance of the platinum-grid gauge was confirmed down to 2×10^{-11} Pa, essentially free from errors caused by ESD ions. The x-ray limits and sensitivities of the new gauges were: below 2×10^{-12} Pa with a sensitivity of 4.5×10^{-4} A/Pa for the molybdenum grid and 5.6×10^{-12} Pa with a sensitivity of 1.8×10^{-4} A/Pa for the platinum grid.

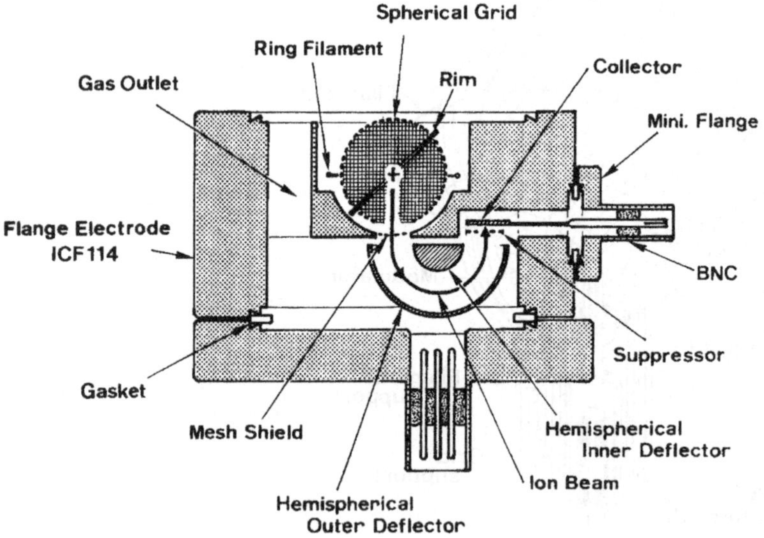

Fig. 6.34 Cutaway drawing of the ion spectroscopy gauge (Watanabe, 1992) [6-30]

Watanabe (1993) [6-31] presented an article, "Total pressure measurements down to 10^{-12} Pa without electron stimulated desorption ion errors."

Abstract [6-31]: The most advanced ion spectroscopy gauge was constructed with a fine platinum alloy spherical grid ion source, immersed in a copper block with stainless steel flanges (see **Fig. 6.35**). Gas phase ions were effectively separated from ESD ions by the actions of electron space charge in the grid and a hemispherical 180° ion energy analyzer. The x-ray limit of the advanced gauge was approximately 2.5×10^{-13} Pa, and the outgassing rate of 2×10^{-9} Pa · L/s (about 1/100 that of an ordinary nude Bayard-Alpert gauge) was achieved.

Oshima and Otuka (1994) [6-32] presented an article, "Performance of an ionization gauge with a large-angle ion deflector. I. Total pressure measurement in extreme high vacuum."

Abstract [6-32]: Performance of the ionization gauge with a large-angle ion deflector (254.6°) (see **Fig. 6.36**) has been examined in extreme high vacuum. We have estimated 4×10^{-13} Pa to be the lowest measurable limit of the gauge due to soft x-ray noise (1 min period of measurement) and to be less 2×10^{-13} Pa for electron stimulated desorption ion noise. The sensitivity of the gauge was 0.018 /Pa for hydrogen. By use of a channeltron multiplier for ion detection, the sensitivity of the gauge is improved: hence, a pressure of $\sim 3 \times 10^{-11}$ Pa is easily measured at an emission current < 0.1 mA. Deviation from the linear relation of ion current versus emission, allows outgassing from the gauge to be detected.

Akimichi et al. (1997) [6-33] determined the sensitivity factor of an axial symmetric transmission gauge (AT gauge) (see **Fig. 6.37**) by the conductance modulation method (CMM) in the pressure range from 10^{-10} to 10^{-6} Pa.

Abstract [6-33]: The gauge has a Bessel-box type energy filter between a Bayard-Alpert gauge type ionizer and an ion detector to eliminate the limiting fac-

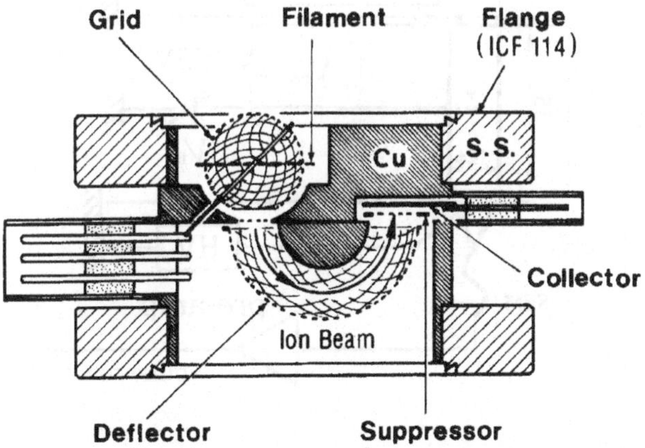

Fig. 6.35 Schematic diagram of the copper ion spectroscopy gauge. Grid material is Pt (80%)-Ir (20%) alloy. The copper body and stainless steel flanges were welded by electron beam (Watanabe, 1993) [6-31]

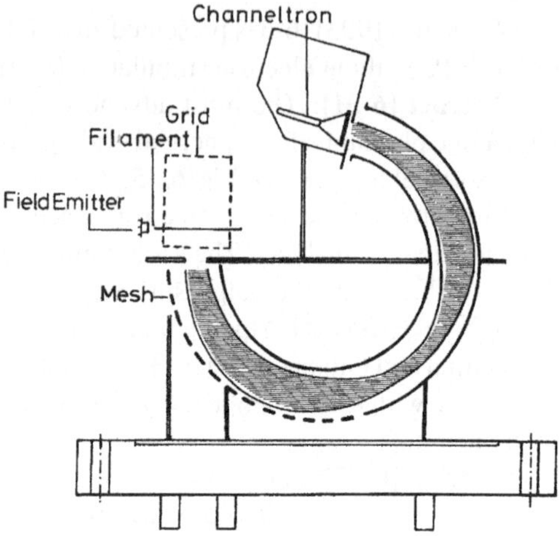

Fig. 6.36 Schematic diagram of the vacuum gauge with a large-angle deflector. There are two electron sources, a ring-type hot filament and a field emitter (Oshima and Otuka, 1994) [6-32]

tors of pressure measurement in extreme high vacuum such as the soft x-ray effect and the electron stimulated desorption of ions. The lower limit of the pressure measurement of the gauge is estimated to be about 10^{-12} Pa or lower. The sensitivity factor showed a slight pressure dependence caused mainly by outgassing from the

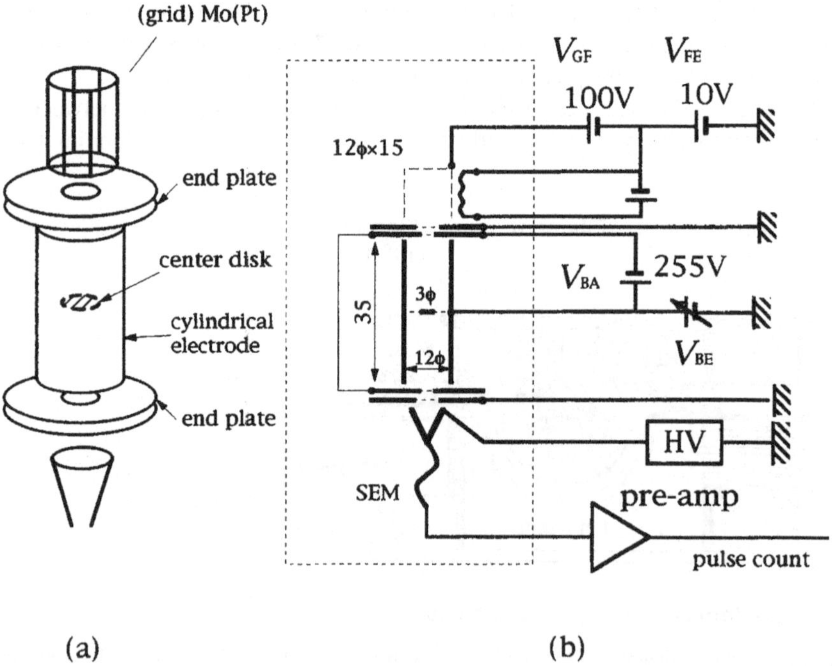

Fig. 6.37 Schematics of the AT gauge. (**a**) Bessel-box type energy filter, (**b**) network used under typical conditions (Akimichi et al. 1997) [6-33]

gauge itself. The sensitivity factor of the AT gauge, which was further corrected for the effect of outgassing by the CMM, is almost constant in the pressure range from 10^{-10} to 10^{-6} Pa and is $(2.3 \pm 0.1) \times 10^{-3}$/Pa for hydrogen.

Akimichi et al. (1999) [6-34] examined the long term characteristics of an axial symmetric transmission gauge, such as the outgassing rate and sensitivity factor, over a period of 3500 h.

Abstract [6-34]: It was found that the outgassing rate and the gain of the continuous-dynode-type electron multiplier, which was used for ion detection, were the main factors which determined the gauge characteristics. The lifetime of the electron multiplier estimated from the decay rate of the sensitivity factor during the acceleration test by hydrogen is about nine years for the operation of the gauge in the 10^{-10} Pa region.

Takahashi et al. (2005) [6-35] reported the basic characteristics of the axial-symmetric transmission gauge (AT gauge), that were modified for measurement over a wide pressure range.

Abstract [6-35]: A Faraday cup ion collector is used instead of a secondary electron multiplier to extend the measurable pressure range to 10^{-3} Pa. The sensitivity of the gauge was determined by a conductance modulation method, and by direct comparison against an extractor gauge and a spinning rotor gauge as well. When operating in O_2, there was systematic disagreement between pressure readings with the AT gauge and the extractor gauge. This is due to the difference in the elimination efficiency of the electron stimulated desorption (ESD) ion: (The schematic diagram of the AT gauge is shown in **Fig. 6.38**.).

Conclusion [6-35]

The ion energy spectra obtained by the Bessel-box-type energy analyzer show that the AT gauge successfully eliminates the influence of the ESD ions. The lower limit of the pressure measurement is determined by the fluctuations of the background current and gas phase signal intensity and is estimated to be 5×10^{-11} Pa. The sensitivity factor of the AT gauge, calibrated by the conductance modulation method, was almost constant in the pressure range from 10^{-10} to 10^{-6} Pa. The constant sensitivity factor of the gauge was also confirmed in the pressure range from 10^{-9} to 10^{-3} Pa by direct comparison with the extractor gauge and the spinning rotor gauge.

The comparison of the pressure readings between the AT gauge and the extractor gauge in oxygen showed that the difference of the readings between the two gauges increased with the oxygen pressure. This difference was caused by the ESD ion current in the extractor gauge. The disturbance caused by the ESD ion current in the extractor gauge is quantitatively estimated. The AT gauge, however, detects only the gas phase ion and is suitable for the pressure measurement of high to ultrahigh vacuum including high ESD ion yield gases such as oxygen, water vapor, carbon dioxide, etc.

Watanabe and Suemitsu (1999) [6-36] successfully separated the electron-stimulated-desorption neutrals from outgassing originating from the grid surface of emission-controlled gauges with the construction of a heated-grid (see **Fig. 6.39**), whose grid temperature can be varied independent of the electron emission

Abstract [6-36]: In XHV, these two processes are both dominated by hydrogen molecules, while their strengths depend strongly on the grid temperature and the

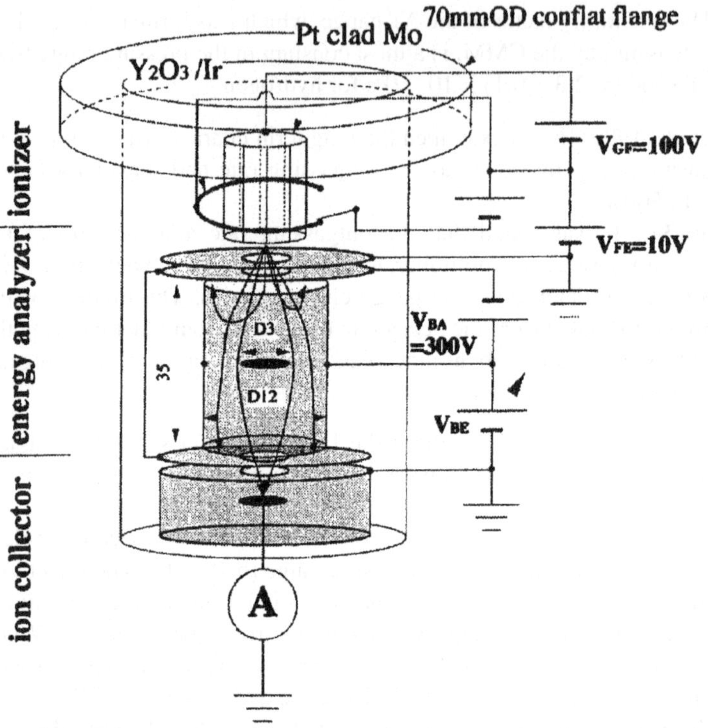

Fig. 6.38 Schematic diagram of the AT gauge with a Faraday cup ion detector. The potential applied to electrodes under typical condition is also shown (Takahashi et al., 2005) [6-35]

grid material. At elevated temperatures, outgassing caused by the out-diffusion of hydrogen atoms dissolved in the grid begin to dominate. This suggests use of a material with low hydrogen solubility for XHV-compatible grid material.

Watanabe and Kasai (1995) [6-37] presented an article, "Low outgassing residual gas analyzer with a beryllium-copper-alloy-flanged ion source."

Abstract [6-37]: By using a newly developed beryllium-copper (BeCu)-alloy ConFlat flange to house the hot-cathode ion source, a remarkable decrease in the outgassing from a quadrupole residual gas analyzer (RGA) has been achieved. The reduction in outgassing between the new BeCu-flanged RGA (see **Fig. 6.40**) and an ordinary stainless-steel RGA of otherwise similar design was a factor of 60 or more in the 10^{-9} Pa total pressure range. From these results, the possibility of high accuracy residual gas analysis below 10^{-9} Pa is introduced.

Satou and Oshima (1995) [6-38] presented an article, "Performance of an ionization gauge with a large-angle ion deflector. II. Mass analysis of residual gas and electron stimulated desorption ions."

Abstract [6-38]: Time of flight (TOF) mass analysis has been applied to residual gas in extreme high vacuum. By a combination of energy analysis with TOF mea-

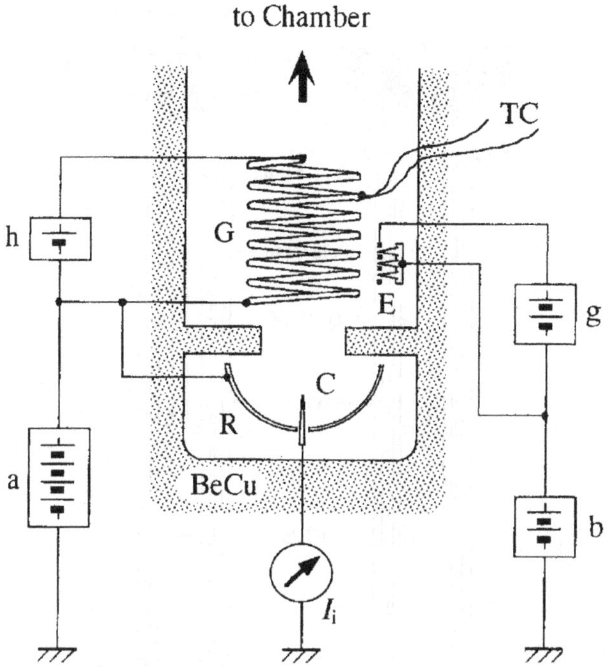

Fig. 6.39 Schematic diagram of the heated-grid gauge. G: Pt-clad Mo spiral grid, E: Spindt-type field emitter array, TC: R-type thermocouple, H: grid-heating power supply, A: grid-bias power supply, g: gate-bias power supply for emitter, B: emitter-bias power supply. All electrodes are immersed in a beryllium-copper-alloy envelope (Watanabe and Suemitsu, 1999) [6-36]

surement through the detector, we have separated the molecular species of gas-phase ions from the electron stimulated desorption (ESD) ions generated at the ionization grid. Only H_2 molecules are detected for the gas-phase ions at 10^{-10} Pa, while H^+, H_2^+, O^+, and other molecules are observed for the ESD ions. The outgassing from the instrument was negligible under the present experimental conditions. (A schematic diagram of the time of flight technique is shown in **Fig. 6.41**.)

Dong and Myneni (1999) [6-39] presented attempts at using the Spindt-type molybdenum field emitter arrays in the extractor gauges and a residual gas analyzer [**Fig. 6.42 (a)**, **(b)**].

Abstract [6-39]: The sensitivity of the field emitter gauge is as high as 11/Torr. The departure from linearity of the pressure versus ion current measurements did not exceed 10% over the pressure range of 10^{-10}–10^{-6} Torr. Stable sensitivities for nitrogen, helium, and hydrogen were achieved below 10^{-7} Torr with the field emitter residual gas analyzer. The slightly reduced emission current and sensitivity, after long-term operation, are of concern and need to be addressed. Residual gas spectra indicate that when using field emitters, the electron stimulated desorption ions (O^+, F^+, and Cl^+) are reduced as compared to those made using a hot filament source.

Fig. 6.40 Cross-section views of the improved RGA with the BeCu-alloy-flanged ion source (**left-hand side**) and of the ordinary standard RGA (**right-hand side**). The cross-hatched region indicates the BeCu alloy (Watanabe and Kasai, 1995) [6-37]

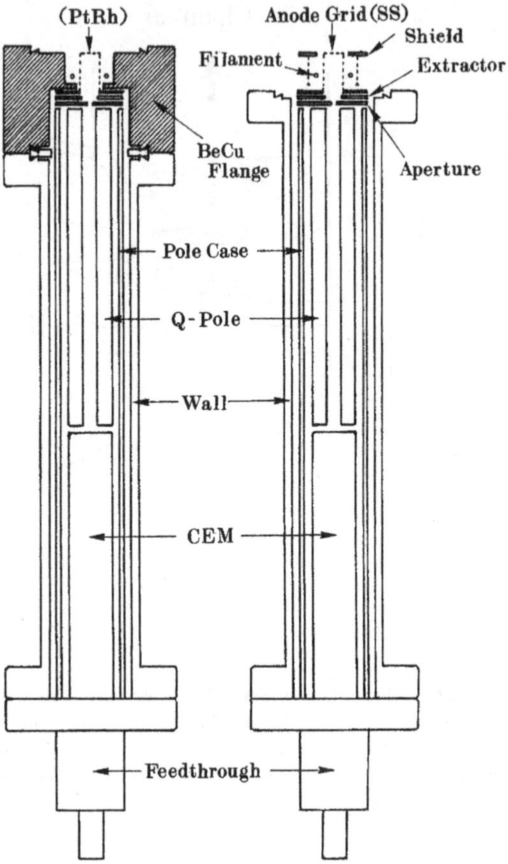

Takahashi et al. (2001) [6-40] developed the quadrupole mass spectrometer (QMS) with the Bessel-Box type energy analyzer (see **Fig. 6.43**)

Abstract [6-40]: The analyzer is placed between the ionizer and the quadrupole mass filter for the separation of the gas phase ions and the electron stimulated desorption (ESD) ions produced in the ionizer according to their kinetic energies. Three important results have been obtained: First, the elimination of the ESD ions gives the simple mass spectra of gas phase ions for the accurate analysis of the spectra. Second, the signal intensity profiles of the peaks of parent ions and the fragment ions with same mass to charge ratio (M/e) in the mass spectra are obtained and are discussed in relation to the mechanism of ion formation. Third, the sensitivity of the QMS for the very small amount of impurities in the gases of higher pressure (high vacuum region) is also improved down to a few tens of parts per billion of the total pressure due to the reduction of background noise. The noise is much reduced by the geometrical structure of the energy analyzer which prevents the ion collector, a secondary electron multiplier, from the radiations of vacuum ultraviolet photons and soft x rays generated in the ionizer.

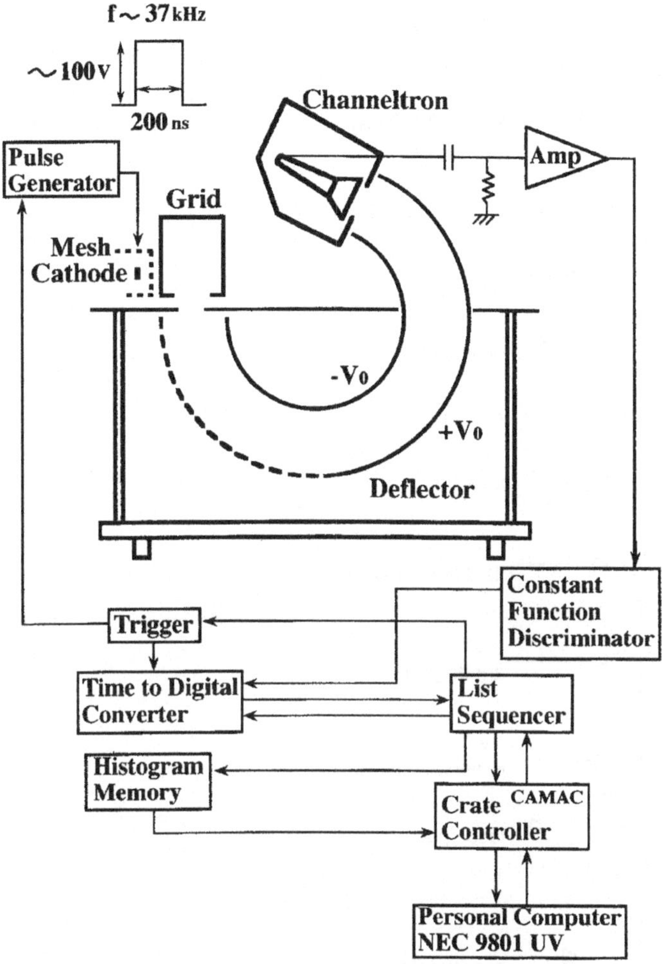

Fig. 6.41 A schematic diagram of the time of flight technique that is used to clarify molecular species in XHV (Satou and Oshima, 1995) [6-38]

Watanabe (2002) [6-41] presented an article, "Investigation and reduction of spurious peaks caused by electron-stimulated desorption and outgassing by means of a grid heating method in a hot-cathode quadrupole residual gas analyzer."

Abstract [6-41]: Once the grid surface is thoroughly cleaned at over 900 °C by ohmic heating or electron bombardment, the major ESD species observed for both neutrals and ions dramatically changes to mainly a spurious H peak with a yield of 10^{-11} (species/electron, with neutral/ion $\cong 2$). When the operational grid temperature is then increased to 500 °C, the spurious H peak is decreased by over 3 orders of magnitude and the spectrum becomes mainly H_2 molecules, reflecting the true pressure of the system; i.e., the relative error for the pressure is reduced to 8% from 614%. The lower limit caused by outgassing in these instruments is due to the

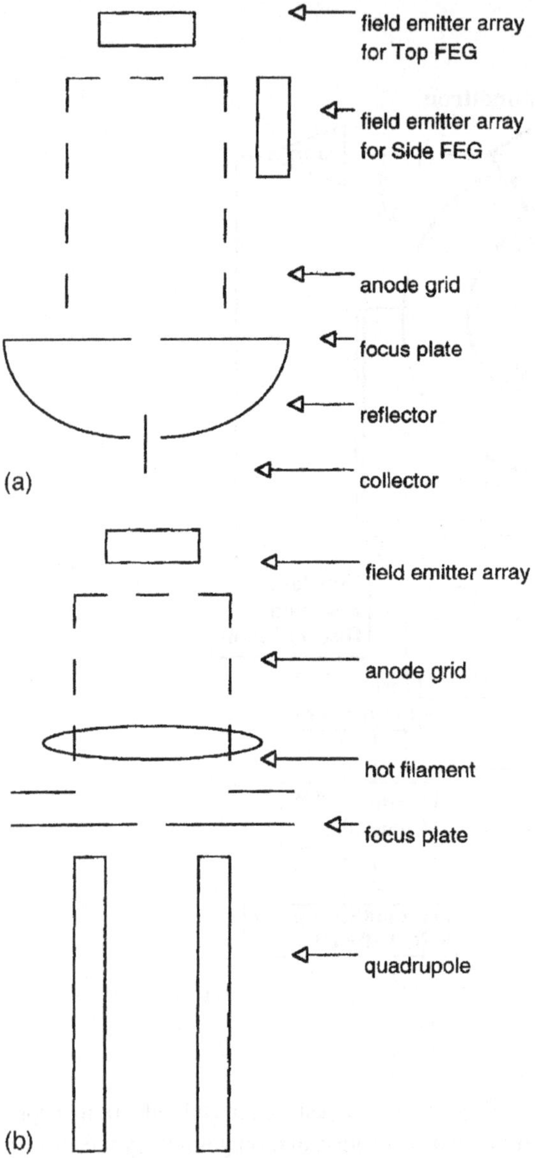

Fig. 6.42 Schematic arrangement of the (**a**) FEG and (**b**) FERGA (Dong and Myneni 1999) [6-39]

grid bulk content of hydrogen. Finding a lower-hydrogen solubility material for the grid may become the last work required for lower-limit XHV gauge development. In order to decrease the limits caused by ESD and outgassing in a hot-cathode gauge, employing ion energy filters and cold-cathode field emitters is unnecessary.

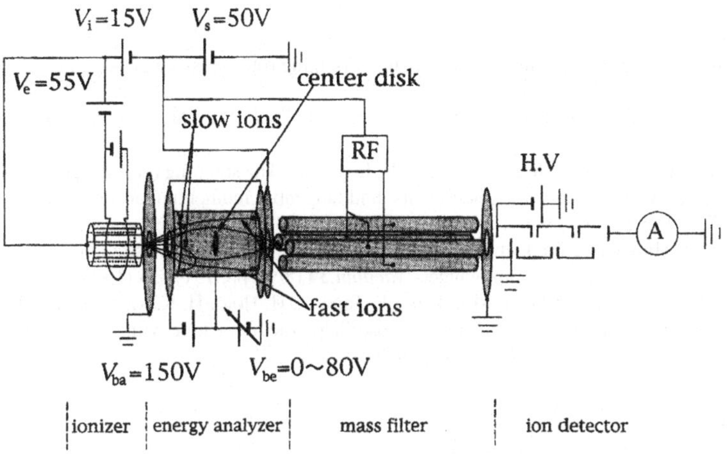

Fig. 6.43 Schematic view of the quadrupole mass spectrometer with the Bessel-Box type energy analyzer (Takahashi et al., 2001) [6-40]

Wilfert and Edelmann (2004) [6-42] presented an article, "Miniturized vacuum gauges."

Abstract [6-42]: Due to the rapid advancement in microsystem and micromachining technology in the past several years, the most important common pressure measuring principles have been successfully applied in nimiturized vacuum sensors. In addition to MicroPiranis®, microstructured gas friction gauges and miniaturized capacitance manometers have been developed recently too, whose dimensions are already in the millimeter and partly in the micrometer range. Exceptions in this development are ionization gauges which, due to their basic measuring principle, cannot be arbitrarily miniaturized without substantial restrictions in their operating performance. The article gives a rough summary of the present level of development of selected types of miniaturized vacuum gauges and points out limits and causes for the restricted miniaturizability of certain total pressure vacuum gauges.

References

6-1. J. M. Lafferty, "Review of pressure measurement techniques for ultrahigh vacua", *J. Vac. Sci. Technol.* **9** (1), pp. 101–107 (1972).

Mechanical Gauges

6-2. N. G. Utterback and T. Griffith, Jr., "Reliable submicron pressure readings with capacitance manometer", *Rev. Sci. Instrum.* **37** (7), pp. 866–870 (1966).

Thermal Conductivity Gauges

6-3. J. H. Leck, "The high temperature Pirani gauge", *J. Sci. Instrum.* **29**, pp. 258–263 (1952).

Viscosity Gauges

6-4. J. W. Beams, D. M. Spitzer, Jr., and J. P. Wade, Jr., "Spinning rotor pressure gauge", *Rev. Sci. Instrum.* **33** (2), pp.151–155 (1962).
6-5. J. K. Fremerey, "High vacuum gas friction manometer", *J. Vac. Sci. Technol.* **9** (1), pp. 108–111 (1972).
6-6. P. Takasakul,J. A. Bentz, R. V. Tompson, and S. K. Loyalka, "The spinning rotor gauge: Measurements of viscosity, velocity slip coefficients, and tangential momentum accommodation coefficients", *J. Vac. Sci. Technol. A* **14** (5), pp. 2946–2952 (1996).
6-7. K. Kokubun, M. Hirata, H. Murakami, Y. Toda, and M. Ono, "A bending and stretching mode crystal oscillator as a friction vacuum gauge", *Vacuum* **34** (8/9), pp. 731–735 (1984).
6-8. M. Ono, M. Hirata, K. Kokubun, H. Murakami, F. Tamura, H. Hojo, H. Kawashima, and H. Kyogoku, "Design and performance of a qualtz oscillator vacuum gauge with a controller", *J. Vac. Sci. Technol. A* **3** (3), pp. 1746–1749 (1985).

Ionization Gauges

6-9. G. K. T. Conn and H. N. Daglish, "Cold cathode ionization gauges for the measurement of low pressures", *Vacuum* **3** (1), pp.24–34 (1953).
6-10. D. Alpert, "New developments in the production and measurement of ultra high vacuum", *J. Appl. Phys.* **24** (7), pp. 860–876 (1953).
6-11. J. H. Singleton, "Practical guide to the use of Bayard-Alpert ionization gauges", *J. Vac. Sci. Technol. A* **19** (4), pp. 1712–1719 (2001).
6-12. N. Yoshimura, H. Oikawa, and O. Mikami, "Measurement of outgassing rates from materials by differential pressure rise method", *Shinku (J. Vac. Soc. Japan)* **13** (1), pp. 23–28 (1970) (in Japanese).
6-13. J. P. Hobson, "The pumping of nitrogen by a Bayard-Alpert ionization gauge in an ultra-high-vacuum system", *Vacuum* **11**, pp.16–25 (1961).
6-14. G. F. Weston, "Measurement of ultra-high vacuum. Part I. Total pressure measurements", *Vacuum* **29** (8/9), pp. 277–291 (1979).
6-15. F. Nakao, "Determination of the ionization gauge sensitivity using the relative ionization cross-section", *Vacuum* **25** (9/10), pp. 431–435 (1975).
6-16. P. A. Redhead, "New hot-filament ionization gauge with low residual current", *J. Vac. Sci. Technol.* **3** (4), pp. 173–180 (1966).
6-17. U. Beeck and G. Reich, "Comparison of the pressure indication of a Bayard-Alpert and an extractor gauge", *J. Vac. Sci. Technol.* **9** (1), pp. 126–132 (1972).
6-18. J. M. Lafferty, "Hot-cathode magnetron ionization gauge for the measurement of ultrahigh vacua", *J. Appl. Phys.* **32**, (3) pp. 424–434 (1961).
6-19. J. M. Lafferty, "Hot-cathode magnetron ionization gauge with an electron multiplier ion detector for the measurement of extreme ultra-high vacua", *Rev. Sci. Instrum.* **34** (5), pp. 467–476 (1963).
6-20. R. N. Peacock, N. T. Peacock, and D. S. Hauschulz, "Comparison of hot cathode and cold cathode ionization gauges", *J. Vac. Sci. Technol. A* **9** (3), pp. 1977–1985 (1991).
6-21. K. M. Welch, L. A. Smart, and R. J. Todd, "Enhanced ignition of cold cathode gauges through the use of radioactive Isotopes", *J. Vac. Sci. Technol. A* **14** (3), pp. 1288–1291 (1996).
6-22. B. R. Kendall and E. Drubetsky, "Starting delays in cold-cathode gauges at low pressures", *J. Vac. Sci. Technol. A* **14** (3), pp. 1292–1296 (1996).
6-23. G. J. Schulz and A. V. Phelps, "Ionization gauges for measuring pressures up to the millimeter range", *Rev. Sci. Instrum.* **28** (12), pp. 1051–1054 (1957).

Partial Pressure Gauges

6-24. D. Lichtman, "Perspectives on residual gas analysis", *J. Vac. Sci. Technol. A* **2** (2), pp. 200–205 (1984).

6-25. G. F. Weston, "Measurement of ultra-high vacuum. Part 2. Partial pressure measurements", *Vacuum* **30** (2), pp. 49–67 (1980).
6-26. N. Yoshimura, H. Hirano, K. Ohara, and I. Ando, "Outgassing characteristics of an electropolished stainless-steel pipe with an operating extractor ionization gauge", *J. Vac. Sci. Technol. A* **9** (4), pp. 2315–2318 (1991).
6-27. N. Yoshimura, T. Sato, S. Adachi, and T. Kanazawa, "Outgassing characteristics and microstructure of an electropolished stainless steel surface", *J. Vac. Sci. Technol. A* **8** (2), pp. 924–929 (1990).
6-28. F. Nakao, "Simplified methods for the calculation of partial pressure using the relative ionization cross-section", *Vacuum* **25** (5), pp. 201–209 (1975).

Development of New Gauges for XHV

6-29. F. Watanabe, "Point collector ionization gauge with spherical grid for measuring pressures below 10^{-11} Pa", *J. Vac. Sci. Technol. A* **5** (2), pp. 242–248 (1987).
6-30. F. Watanabe, "Ion spectroscopy gauge: Total pressure measurement down to 10^{-12} Pa with discrimination against electron-stimulated-desorption ions", *J. Vac. Sci. Technol. A* **10** (5), pp. 3333–3339 (1992).
6-31. F. Watanabe, "Total pressure measurement down to 10^{-12} Pa without electron stimulated desorption ion errors", *J. Vac. Sci. Technol. A* **11** (4), pp. 1620–1626 (1993).
6-32. C. Ohshima and A. Otuka, "Performance of an ionization gauge with a large-angle ion deflector. I. Total pressure measurement in extreme high vacuum", *J. Vac. Sci. Technol. A* **12** (6), pp. 3233–3238 (1994).
6-33. H. Akimichi, T. Arai, K. Takeuchi, Y. Tuzi, and I. Arakawa, "Calibration of an axial symmetric transmission gauge in ultrahigh and extreme high vacuum", *J. Vac. Sci. Technol. A* **15** (3), pp. 753–758 (1997).
6-34. H. Akimichi, K. Takeuchi, Y. Tuzi, and I. Arakawa, "Long term behavior of an axial-symmetric transmission gauge", *J. Vac. Sci. Technol. A* **17** (4), pp. 2021–2025 (1999).
6-35. N. Takahashi, J. Yuyama, Y. Tuzi, H. Akimichi, and I. Arakawa, "Axial-symmetric transmission gauge: Extension of its pressure measuring range and reduction of the electron stimulated desorption ion effect in ultrahigh vacuum", *J. Vac. Sci. Technol. A* **23** (3), pp. 554–558 (2005).
6-36. F. Watanabe and M. Suemitsu, "Separation of electron-stimulated-desorption neutrals from outgassing originating from the grid surface of emission-controlled gauges: Studies with a heated-grid gauge", *J. Vac. Sci. Technol. A* **17** (6), pp. 3467–3472 (1999).
6-37. F. Watanabe and A. Kasai, "Low outgassing residual gas analyzer with a beryllium-copper-alloy-flanged ion source", *J. Vac. Sci. Technol. A* **13** (2), pp. 497–500 (1995).
6-38. T. Satou and C. Oshima, "Performance of an ionization gauge with a large-angle ion deflector. II. Mass analysis of residual gas and electron stimulated desorption ions", *J. Vac. Sci. Technol. A* **13** (2), pp. 448–450 (1995).
6-39. C. Dong and G. R. Myneni, "Field emitter based extractor gauges and residual gas analyzers", *J. Vac. Sci. Technol. A* **17** (4), pp. 2026–2033 (1999).
6-40. N. Takahashi, T. Hayashi, H. Akimichi, and Y. Tuzi, "Development of the quadrupole mass spectrometer with the Bessel-Box type energy analyzer: Function of the energy analyzer in the partial pressure measurements", *J. Vac. Sci. Technol. A* **19** (4), pp. 1688–1692 (2001).
6-41. F. Watanabe, "Investigation and reduction of spurious peaks caused by electron-stimulated desorption and outgassing by means of a grid heating method in a hot-cathode quadrupole residual gas analyzer", *J. Vac. Sci. Technol. A* **20** (4), pp. 1222–1229 (2002).
6-42. St. Wilfert and Chr. Edelmann, "Miniaturized vacuum gauges", *J. Vac. Sci. Technol. A* **22** (2), pp. 309–320 (2004).

Other Articles

Mechanical Gauges

*6-1. J. Dimeff, J. W. Lane, and G. W. Coon, "New wid-range pressure transducer", *Rev. Sci. Instrum.* **33** (8), pp. 804–811 (1962).

*6-2. G. C. Baldwin and M. R. Gaerttner, "Thermal transpiration error in absolute pressure measurement with capacitance manometers", *J. Vac. Sci. Technol.* **10**(1), pp. 215–217 (1973).

Thermal Conductivity Gauges

*6-3. B. G. Dickins, "The effect of accommodation on heat conduction through gases", *Proc. R. Soc. London, Ser. A* **143**, pp. 517–540 (1933).

*6-4. J. H. Leck and C. S. Martin, "A Pirani gauge for operation up to a pressure of 10 mm of mercury", *J. Sci. Instrum.* **33**, May, pp. 181–183 (1956).

*6-5. J.-S. Shie, Bruce C. S. Chou, and Y.-M. Chen, "High performance Pirani vacuum gauge", *J. Vac. Sci. Technol. A* **13** (6), pp. 2972–2979 (1995).

*6-6. Vincent S. Smentkowski and John T. Yates, Jr., "Universal calibration of W5%Re vs. W26%Re (type-C) thermocouples in the temperature range 32–2588 K", *J. Vac. Sci. Technol. A* **14** (1), pp. 260–265 (1996).

*6-7. W. J. Alvesteffer, D. C. Jacobs, and D. H. Baker, "Miniaturized thin film thermal vacuum sensor", *J. Vac. Sci. Technol. A* **13** (6), pp. 2980–2985 (1995).

*6-8. O. Paul, O. Brand, R. Lenggenhager, and H. Baltes, "Vacuum gauging with complementary metal-oxide-semiconductor microsensors", *J. Vac. Sci. Technol. A* **13** (3), pp. 503–508 (1995).

*6-9. R. E. Ellefson and A. P. Miiller, "Recommended practice for calibrating vacuum gauges of the thermal conductivity type", *J. Vac. Sci. Technol. A* **18** (5), pp. 2568–2577 (2000).

Viscosity Gauges

*6-10. K. Kokubun, M. Hirata, M. Ono, H. Murakami, and Y. Toda, "Frequency dependence of a qualtz oscillator on gas pressure", *J. Vac. Sci. Technol. A* **3** (6), pp. 2184–2187 (1985).

*6-11. J. K. N. Sharma, P. Mohan, and D. R. Sharma, "Comparison of two primary pressure standards using spinning rotor gauges", *J. Vac. Sci. Technol. A* **8** (2), pp. 941–947 (1990).

*6-12. S.-H. Choi, Sharrill Dittmann, and Charles R. Tilford, "Stabilization techniques for spinning rotor gage residual drag", *J. Vac. Sci. Technol. A* **8** (6), pp. 4079–4085 (1990).

*6-13. A. Chambers, A. D. Chew, and A. P. Troup, "Rotating disk gauge for absolute total pressure measurement in high vacuum", *J. Vac. Sci. Technol. A* **10** (4), pp. 2655–2660 (1992).

*6-14. W. Jitschin, K. Jousten, and D. Wandrey, "Design and evaluation of a primary high-vacuum standard", *J. Vac. Sci. Technol. A* **10** (5), pp. 3344–3351 (1992).

*6-15. S. K. Loyalka, "Theory of the spinning rotor gauge in the slip regime", *J. Vac. Sci. Technol. A* **14** (5), pp. 2940–2945 (1996).

*6-16. P. D. Levine and J. R. Sweda, "A primary standard high vacuum calibration station for industrial applications", *J. Vac. Sci. Technol. A* **14** (3), pp. 1297–1302 (1996).

*6-17. C. Boffito, M. Moraja, and G. Pastore, "Spinning rotor gauge in the range from 10^{-4} Pa to atmospheric pressure", *J. Vac. Sci. Technol. A* **15** (4), pp. 2391–2394 (1997).

*6-18. K. Jousten, A. R. Filippelli, C. R. Tilford, and F. J. Redgrave, "Comparison of the standards for high and ultrahigh vacuum at three national standards laboratories", *J. Vac. Sci. Technol. A* **15** (4), pp. 2395–2406 (1997).

*6-19. J. Šetina, "Two point calibration scheme for the linearization of the spinning rotor gauge at transition regime pressures", *J. Vac. Sci. Technol. A* **17** (4), pp. 2086–2092 (1999).

Ionization Gauges

*6-20. P. A. Redhead, "Modulated Bayard-Alpert Gauge" *Rev. Sci. Instrum.* **31** (3), pp. 343–344 (1960).

*6-21. J. P. Hobson and P. A. Redhead, "Operation of an inverted-magnetron gauge in the pressure range 10^{-3}–10^{-12} mmHg", *Can. J. Phys.* **36**, pp. 271–288 (1958).

*6-22. P. A. Redhead, "Magnetron gauge: A cold-cathode vacuum gage", *Can. J. Phys.* **37**, 1260–1271 (1959).

*6-23. N. Yoshimura, H. Oikawa, and O. Mikami, "Measurement of outgassing rates from materials by differential pressure rise method", *Shinku* (*J. Vac. Soc. Japan*) **13** (1), pp. 23–28 (1970) (in Japanese).

*6-24. W. Jitschin, "Accuracy of vacuum gauges", *J. Vac. Sci. Technol. A* **8** (2), pp. 948–956 (1990).

*6-25. S. Ruthberg, "Standards and vacuum measurement accuracy", *J. Vac. Sci. Technol.* **9** (1), pp. 186–195 (1972).

*6-26. P. A. Redhead, "Errors in the measurement of pressure with ionization gauges", *1960 7th National Symposium on Vacuum Technology Transactions* (Pergamon Press, 1961), pp.108–111.

*6-27. G. Barnes, "Erroneous readings of large magnitude in a Bayard-Alpert ionization gauge and their probable cause", *Rev. Sci. Instrum.* **31** (10), pp. 1121–1127 (1960).

*6-28. T. E. Hartman, "Ultra-High vacuum use of Bayard-Alpert ionization gauges", *Rev. Sci. Instrum.* **34** (3), pp. 281–285 (1963).

*6-29. T. E. Hartman, "Anomalous residual currents in the ultrahigh vacuum use of Bayard-Alpert ionization gauges", *Rev. Sci. Instrum.* **34** (11), pp. 1190–1195 (1963).

*6-30. P. A. Redhead, "The sensitivity of Bayard-Alpert gauges", *J. Vac. Sci. Technol.* **6** (5), pp. 848–854 (1969).

*6-31. G. Comsa, "The behavior of ions in a BAG structure", *J. Vac. Sci. Technol.* **8** (4), pp. 582–589 (1971).

*6-32. G. Comsa, "Ion collection in Bayard-Alpert gauges", *J. Vac. Sci. Technol.* **9** (1), pp. 117–121 (1972).

*6-33. J. R. Young, "Measuring hydrocarbon gas pressure with an ionization gauge", *J. Vac. Sci. Technol.* **10** (1), pp. 212–214 (1973).

*6-34. R. Holanda, "Investigation of the sensitivity of ionization-type vacuum gauges", *J. Vac. Sci. Technol.* **10** (6), pp. 1133–1139 (1973).

*6-35. J. E. Bartmess and R. M. Georgiadis, "Empirical methods for determination of ionization gauge relative sensitivities for different gases", *Vacuum* **33** (3), pp. 149–153 (1983).

*6-36. C. R. Tilford, "Reliability of high vacuum measurements", *J. Vac. Sci. Technol. A* **1** (2), pp. 152–162 (1983).

*6-37. D. G. Bills, P. C. Arnord, S. L. Dodgen, and C. B. Van Cleve, "New ionization gauge geometries providing stable and reproducible sensitivities", *J. Vac. Sci. Technol. A* **2** (2), pp. 163–167 (1984).

*6-38. P. C. Arnold and D. G. Bills, "Causes of unstable and nonreproducible sensitivities in Bayard-Alpert ionization gauges", *J. Vac. Sci. Technol. A* **2** (2), pp. 159–162 (1984).

*6-39. H. Gentsch, J. Tewes, and G. Messer, "An improved ion gauge with gold coated electrodes for reliable operation in reactive gases and for use as reference standard", *Vacuum* **35** (3), pp. 137–140 (1985).

*6-40. N. T. Peacock, "Measurement of x-ray currents in Bayard-Alpert type gauges", *J. Vac. Sci. Technol. A* **10** (4), pp. 2674–2678 (1992).

*6-41. M. Saitoh, K. Shimura, T. Iwata, T. Momose, and H. Ishimaru, "Influence of vacuum gauges on outgassing rate measurements", *J. Vac. Sci. Technol. A* **11** (5), pp. 2816–2821 (1993).

*6-42. P. C. Arnold and S. C. Borichevsky, "Nonstable behavior of widely used ionization gauges", *J. Vac. Sci. Technol. A* **12** (2), pp. 568–573 (1994).

*6-43. P. C. Arnold, D. G. Bills, M. D. Borenstein, and S. C. Borichevsky, "Stable and reproducible Bayard-Alpert ionization gauge", *J. Vac. Sci. Technol. A* **12** (2), pp. 580–586 (1994).

*6-44. D. G. Bills, "Causes of nonstability and nonreproducibility in widely used Bayard-Alpert ionization gauges", *J. Vac. Sci. Technol.* A **12** (2), pp. 574–579 (1994).
*6-45. C. R. Tilford, A. R. Filippelli, and P. J. Abbott, "Comments on the stability of Bayard-Alpert ionization gages", *J. Vac. Sci. Technol. A* **13** (2), pp. 485–487 (1995).
*6-46. C. M. Spencer and D. Stäheli, "High-stability controller for Bayard-Alpert ionization gauges", *J. Vac. Sci. Technol.* **5** (4), pp. 105–108 (1968).
*6-47. A. R. Filippelli and P. J. Abbott, "Long-term stability of Bayard-Alpert gauge performance: Results obtained from repeated calibrations against the National Institute of Standards and Technology primary vacuum standard", *J. Vac. Sci. Technol. A* **13** (5), pp. 2582–2586 (1995).
*6-48. A. R. Filippelli, "Influence of envelope geometry on the sensitivity of "nude" ionization gauges", *J. Vac. Sci. Technol. A* **14** (5), pp. 2953–2957 (1996).
*6-49. B. R. F. Kendall and E. Drubetsky, "Stable cancellation of x-ray errors in Bayard-Alpert gauges", *J. Vac. Sci. Technol. A* **16** (3), pp. 1163–1168 (1998).
*6-50. P. A. Redhead and J. P. Hobson, "Total pressure measurements below 10^{-10} Torr with non-magnetic ionization gauges", *Brit. J. Appl. Phys.* **16**, pp. 1555–1566 (1965).
*6-51. A. Van Oostrom, "Modulation of Bayard-Alpert ionization gauges with grid end-caps", *J. Sci. Instrum.* **44**, pp. 927–930 (1967).
*6-52. P. J. Szwemin, "Behavior of the Bayard-Alpert gauge with cage modulator operating without electron current", *J. Vac. Sci. Technol.* **9** (1), pp. 122–125 (1972).
*6-53. L. G. Pittaway, "Modulation of the desorbed ion current in Bayard-Alpert gauges", *J. Vac. Sci. Technol.* **10** (4), pp. 507–512 (1973).
*6-54. A. Barz and P. Kocian, "Extractor gauge as a nude system", *J. Vac. Sci. Technol.* **7** (1), pp. 200–203 (1970).
*6-55. D. Blechschmidt, "A miniature extractor gauge for the UHV", *J. Vac. Sci. Technol.* **11** (6), pp. 1160–1165 (1974).
*6-56. H. Gentsch, "Inertes Ionisationsvakuummeter mit extrahierendem Kollektor (EXKOLL)", *Vakuum-Technik* **36** (3), pp. 67–74 (1987).
*6-57. L. T. Melfi, Jr., "Buried collector gauge for measurements in the 10^{-11} Torr pressure range", *J. Vac. Sci. Technol.* **6** (2), pp. 322–325 (1969).
*6-58. G. A. Beitel and C. M. Gosselin, "Performance of a modified buried collector gauge", *J. Vac. Sci. Technol.* **7** (6), pp. 580–585 (1970).
*6-59. J. H. Leck, "The measurement of total pressure in the range 10^{-8}–10^{-12} Torr", *Vacuum* **20** (9), pp. 369–372 (1970).
*6-60. K. Close and J. Yarwood, "The measurement of low gas pressures in terms of ion currents", *Vacuum* **20** (2), pp. 56–64 (1970).
*6-61. D. Blechschmidt, "An ionization gauge using a channel electron multiplier for pressures below 10^{-12} Torr", *J. Vac. Sci. Technol.* **10** (2), pp. 376–380 (1973).
*6-62. T. Satoh and C. Ohshima, "Characteristic of electron emission and gas desorption from the impregnated cathode of ionization gauge for an extremely high vacuum", *Shinku* (*J. Vac. Soc. Japan*) **37** (2), pp. 77–79 (1994).
*6-63. N. Gotoh, T. Momose, H. Ishimaru, and R. Paitich, "Residual current of the modified Lafferty gauge", *J. Vac. Sci. Technol. A* **13** (5), pp. 2574–2578 (1995).
*6-64. B. R. F. Kendall, "Ionization gauge errors at low pressures", *J. Vac. Sci. Technol. A* **17** (4), pp. 2041–2049 (1999).
*6-65. E. H. Chao, S. F. Paul, and R. C. Davidson, "Dynamics of the $m = 1$ diocotron mode in the electron diffusion gauge experiment", *J. Vac. Sci. Technol. A* **17** (4), pp. 2034–2040 (1999).
*6-66. J. S. Cleaver, "A new high pressure ionization gauge", *J. Sci. Instrum.* **44**, pp. 969–972 (1967).
*6-67. P. S. Choumoff and B. Iapteff, "High pressure ionization gauge for calibration in the 10^{-7}–10^{-1} Torr Range", Proc. 6th Housingernl. Vacuum Congr. 1974, Japan. *J. Appl. Phys.* Suppl. 2, Pt. 1, 1974, pp. 143–146.
*6-68. N. Ohsako, "A new wide-range B-A gauge from UHV to 10^{-1} Torr", *J. Vac. Sci. Technol.* **20** (4), pp. 1153–1155 (1982).

*6-69. F. M. Penning and K. Nienhuis, "Construction and applications of a new design of the philips vacuum gauge", *Philips Technical Review* **11** (4), pp.116–122 (1949).

*6-70. J. R. Young, "Pressure dependence of the axial ion current in a Penning discharge", *J. Vac. Sci. Technol.* **5** (4), pp. 102–104 (1968).

*6-71. L. De Chernatony and D. J. Crawley, "UHV measurements by cold cathode gauges", *Vacuum* **20** (9), pp. 389–391 (1970).

*6-72. I. Bello, S. Bederka, and L. Haworth, "Optical method for low pressure measurements", *J. Vac. Sci. Technol. A* **13** (3), pp. 509–514 (1995).

*6-73. M. G. Buehler, L. D. Bell, and M. H. Hecht, "Alpha-particle gas-pressure sensor", *J. Vac. Sci. Technol. A* **14** (3), pp. 1281–1287 (1996).

*6-74. B. R. Kendall and E. Drubetsky, "Cold cathode gauges for ultrahigh vacuum measurements", *J. Vac. Sci. Technol. A* **15** (3), pp. 740–746 (1997).

*6-75. M. Hirata, "Precise gauge calibration of means of comparison method using a personal computer", *Shinku* (*J. Vac. Soc. Japan*) **36** (4), pp. 420–423 (1993).

*6-76. I. Arakawa, M. Kim, and Y. Tuji, "Unidirectional vacuum gauge by means of detecting excited neutrals", *J. Vac. Sci. Technol.* A **2** (2), pp. 168–171 (1984).

*6-77. H. Saeki, T. Magome, and Y. Shoji, "Performance of a hot-cathode-ionization-gauge head with correcting electrode and shield tube, operated with an automated-pressure-compensating circuit in a synchrotron radiation environment", *J. Vac. Sci. Technol. A* **24** (4), pp. 1148–1150 (2006).

*6-78. J. T. Tate and P. T. Smith, "The efficiencies of ionization and ionization potentials of various gases under electron impact", *Phys. Rev.* **39**, pp. 270–276 (1932).

Residual Gas Analyzens

*6-79. P. della Porta, "Present knowledge on residual gases in electron tubes", *Advances in Electron Tube Techniques* **2**, pp. 55–69 (1962).

*6-80. R. D. Craig and E. H. Harden, "The interpretation of mass spectra in vacuum measurement", *Vacuum* **16** (2), pp. 67–73 (1966).

*6-81. H. Hoch, "Total- und Partialdruckmessungen bei Drücken zwischen 2×10^{-10} und 10^{-12} Torr", *Vakuum-Technik* **16** (1/2), pp. 8–13 (1967).

*6-82. M. Kobayashi and Y. Tuzi, "Performance of a directional detector of molecular density", *J. Vac. Sci. Technol.* **16** (2), pp. 685–688 (1979).

*6-83. K. F. Poulter, "Effect of gas composition on vacuum measurement", *J. Vac. Sci. Technol. A* **2** (2), pp. 150–158 (1984).

*6-84. F. Watanabe and H. Ishimaru, "Ion current modulation of a residual gas analyzer", *J. Vac. Sci. Technol. A* **3** (6), pp. 2192–2195 (1985).

*6-85. W. K. Huber, N. Müller, and G. Rettinghaus, "Total and partial pressure measurement in the low 10^{-12} mbar range", *Vacuum* **41** (7–9), pp. 2103–2105 (1990).

*6-86. P. A. Redhead, "Measurement of residual currents in ionization gauges and residual gas analyzers", *J. Vac. Sci. Technol. A* **10** (4), pp. 2665–2673 (1992).

*6-87. J. A. Basford, M. D. Boeckmann, R. E. Ellefson, A. R. Filippelli, D. H. Holkeboer, L. Lieszkovszky, and C. M. Stupak, "Recommended practice for the calibration of mass spectrometers for partial pressure analysis", *J. Vac. Sci. Technol. A* **11** (3), pp. A22–A40 (1993).

*6-88. R. J. Ferran and S. Boumsellek, "High-pressure effects in miniature arrays of quadrupole analyzers for residual gas analysis from 10^{-9} to 10^{-2} Torr", *J. Vac. Sci. Technol. A* **14** (3), pp. 1258–1265 (1996).

*6-89. S, Watanabe, M. Aono, and S. Kato,"Reduction of outgassing rate from residual gas analyzers for extreme high vacuum measurements", *J. Vac. Sci. Technol. A* **14** (6), pp. 3261–3266 (1996).

*6-90. N. Nagayama, K. Araki, and H. Mima, "The difference of the pattern coefficient of the quadrupole mass spectrometer", *Shinku* (*J. Vac. Soc. Japan*) **40** (8), pp. 650–654 (1997).

Chapter 7
Microdischarges in High Vacuum

The ion microscope generally equips the electrostatic lens system using high voltage electrodes, which often causes the problem concerning microdischarges in vacuum. Also, high-voltage transmission electron microscopes may meet problems concerning microdischarges in the electron beam acceleration tubes. When microdischarge, even a very small one, occurs in observing electron-microscope images, it causes inferior image quality.

Two kinds of microdischarge may occur, microdischarge over an insulator surface and one between high-voltage electrodes.

Keywords

Triple junction (cathode side): Junction of cathode, insulator, and vacuum.

Anode-initiation model: Anode-initiated microparticles approach the cathode, which becomes a trigger to initiate microdischarge.

Total voltage effect: In a multi-electrode high-voltage system, the breakdown voltage is limited by the total voltage applied between the top electrode and the bottom electrode rather than by the electric field in each electrode gap.

Ion-exchange process: Microdischarge occurs due to the regenerative exchange of positive ions and negative ions between the top electrode and the bottom electrode.

Conditioning: Aging of high-voltage system with application of a high voltage. After conditioning, a higher voltage could be applied without accompanying microdischarge.

AGC (Argon-glow conditioning): An argon glow discharge treatment for electrodes and insulators for aging. Typical conditions: 350 Vac, 180 mA, 5 min in about 10 Pa of argon.

HVC (high-voltage conditioning): A voltage higher than the rated voltage is applied in high vacuum for aging.

7A Microdischarges over Insulator Surfaces

Factors

Gleichauf (1951) [7A-1] presented an article, "Electrical breakdown over insulators in high vacuum."

Abstract [7A-1]: The breakdown voltage of vacuum gaps depends on the electrode material, but when the gap is bridged by an insulator it is independent of the electrode material and varies with the kind of insulator. Some indication is found that the breakdown voltage increases with increasing surface resistivity of the insulator, but no apparent correlation is found between breakdown voltage and dielectric constant, density, or vapor pressure. It is found that a roughening of the intervening surface of the insulator in the region adjacent to the cathode increases the breakdown voltage. For electrode separations of 1 mm or greater the breakdown voltage does not increase linearly with the length of the insulator. (The relationship $V_B \propto \sqrt{L}$ has been found, where L is the length of insulator and V_B is the breakdown voltage.) Experiments were made in which one of the electrodes was separated from the insulator. It seems that the critical gradient at breakdown in the vacuum space between the cathode and the insulator, calculated for the case of plane electrodes from the ratio of dielectric constants, is not as large as would be required in gaps without insulators. These gradients are almost the same for copper as for stainless steel electrodes. The breakdown voltage over an insulator is raised when the edge of the insulator close to either electrode is rounded. When a layer of glass, thin compared with the separation of electrodes, is fused to the cathode, breakdowns occur at lower voltages than for an identical vacuum gap.

Pillai and Hackam (1985) [7A-2] measured the surface flashover of Teflon, Plexiglas, quartz, Pyrex glass, Macor glass-ceramic, and sapphire solid insulators in vacuum ($\sim 10^{-8}$ Torr, $\sim 10^{-6}$ Pa) and in atmospheric air using dc, ac (60 Hz), and $1.2/50 - \mu$sec lightning impulse voltages.

Abstract [7A-2]: The dependence of the flashover voltage on the following parameters is investigated: (1) spacer material, (2) diameter of the spacer, (3) spacer length, (4) number of spacers stacked in series, (5) air pressure in the range 10^{-6} to 10^5 Pa, (6) electrode material, (7) spark conditioning, and (8) the external resistance in series with the gap. At a fixed insulator length the flashover voltage decreases with increasing spacer diameter. The withstand voltage of spacers stacked in series increases with increasing the number of spacers. The dc flashover voltage of different insulating materials is theoretically calculated in vacuum as a function of the length of the insulator and compared with the experimentally obtained results. Good agreement is obtained.

Charging of Dielectric Surfaces

Tourreil and Srivastava (1973) [7A-3] presented an article, "Mechanism of surface charging of high-voltage insulators in vacuum.

Abstract [7A-3]: In vacuum, the surface of insulators becomes electrically charged when subjected to high-voltage stresses. The charging mechanism is described. A model simulating the charging shows that the surface charge densities are proportional to the applied voltage and depend on the secondary electron emission of the dielectric surface. It is also time dependent. Surface charges are shown to explain many results obtained in studies of insulator surface breakdown.

Sudarshan and Cross (1973) [7A-4] presented an article, "DC electric-field modifications produced by solid insulators bridging a uniform-field vacuum gap."

Abstract [7A-4]: The results of measurements of the electric field near the surface of stainless-steel electrodes in vacuum are described. It is found that the field is modified by a solid insulator bridging the inter-electrode gap. The extrapolated cathode field is enhanced sufficiently to equal the value considered necessary to initiate breakdown in a plane vacuum gap. The measured fields are greater than the applied field and consistent with the accumulation of positive surface charge on the solid insulator. Analysis of the data suggests that the surface charge results from bombardment of the surface by electrons emitted at the insulator-cathode junction.

The characteristics of junction-field vs. applied-field measured for alumina are presented in **Fig. 7A.1** [7A-4].

Anderson and Brainard (1980) [7A-5] presented an article, "Mechanism of pulsed surface flashover involving electron-stimulated desorption."

Surface Flashover Model [7A-5]

Secondary Emission Avalanche [7A-5]:
A magnified view of the insulator surface early in the pre-breakdown time delay is shown in **Fig. 7A.2**. The layers of avalanche electrons, neutrals, and positive ions are identified, and typical trajectories of each of these particles are indicated. Electrons drift toward the anode through repeated secondary emission events of unity yield, as if hopping across the insulator surface; neutrals drift away from the surface; and ions are accelerated away from the surface and toward the cathode.

Observation of Avalanche Current [7A-5]:
Measurements of the current of electrons arriving at the anode electrode at the edge of a 20 mm-long rectangular insulator (**Fig. 7A.3**) are consistent with the description of discharge development summarized above. Electrons passing through a 1.5 mm-diam hole in the anode electrode were collected by a shielded Faraday cup. A small needle on the cathode electrode initiated a region of surface charging in line with the hole in the anode electrode. Typical simultaneous cathode voltage and Faraday-cup current waveforms obtained from an alumina ceramic insulator are displayed in **Fig. 7A.4**. Results for polymethyl methacrylate (Plexiglas) are similar.

The waveforms in **Fig. 7A.4** indicate that after an initial burst of current corresponding to a few tenths of an ampere per mm surface current, a smaller steady current persisted until it grew rapidly immediately prior to the collapse of insulator impedance. As the secondary emission avalanche propagated across the insulator, some of the emitted electrons apparently escaped the avalanche and were accelerated to the anode to be detected as the initial burst of current (which may be seen in **Fig. 7A.4** as a 1.5 ns-wide hump in the current waveforms occurring soon after the cathode voltage attains its maximum value). The steady current observed during the latent period prior to breakdown is roughly equal to the predicted current due to the slow drift of avalanche electrons toward the anode.

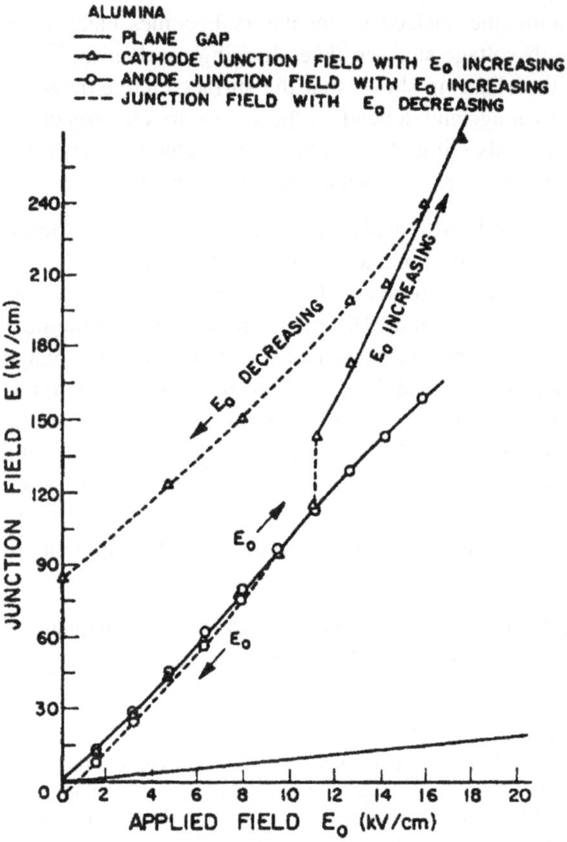

Fig. 7A.1 Junction field vs. applied field for alumina. Insulator making flat contact with electrodes (Sudarshan and Cross, 1973) [7A-4]

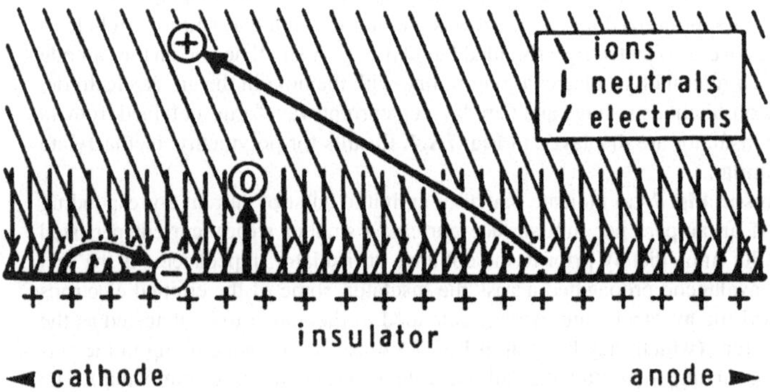

Fig. 7A.2 Layers and avalanche electrons, neutrals, and positive ions early in the prebreakdown delay time. Typical trajectories of each of these particles are indicated (Anderson and Brainard, 1980) [7A-5]

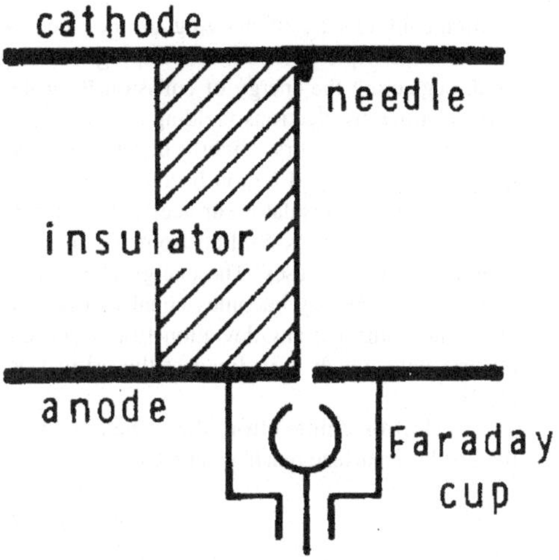

Fig. 7A.3 Arrangement for measuring the prebreakdown current. The hole in the anode electrode is aligned with a small needle attached to the cathode electrode and in contact with the insulator. The anode is at ground potential (Anderson and Brainard, 1980) [7A-5]

Pillai and Hackam (1982) [7A-6] presented an article, "Surface flashover of solid dielectric in vacuum."

Analysis of the Charging Mechanism [7A-6]
The secondary electron emission rate δ from an insulator surface initially increases with increasing impinging energy of the electrons until it reaches saturation and starts to decrease at high energy. There are two energies at which $\delta = 1$ [3].

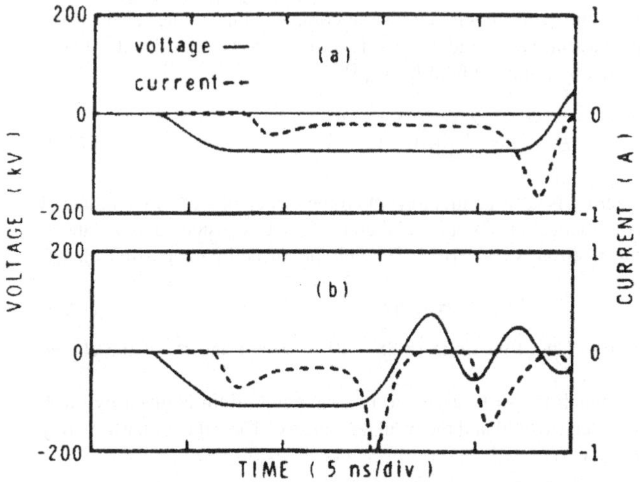

Fig. 7A.4 Simultaneous cathode voltage and Faraday-cup current waveforms, measured as in **Fig. 7A.3** (alumina ceramic insulator 20 mm long). Collapse of impedance at breakdown is evidenced by the onset of inductive-capacitive ringing in the cathode voltage. The applied voltage was 46% larger in (**b**) than in (**a**), which reduced the prebreakdown time delay by a factor of approximately 2 (Anderson and Brainard, 1980) [7A-5]

In the initial stage of charging, all the electrons striking the insulator's surface are emitted from the triple junction. The impinging energy is that gained from the inter-electrode field, neglecting the energy of emission from the cathode [2]. The region on the insulator surface struck by electrons yielding $\delta > 1$ becomes positively charged. Near the anode, the surface becomes at first slightly negatively charged. When the surface-charge density of the positively charged region becomes sufficiently high, many of the secondary electrons emitted from the insulator surface will be attracted back onto the surface. The trajectory range of the electrons emitted from the insulator's surface will decrease as the surface-charge density increases. The energy at impact will therefore decrease. It will do so until the impact energy becomes equal to that corresponding to $\delta = 1$ which is the stable operating point and any deviation from it is self-correcting. This also corresponds to the electric field which is inclined at the critical angle with respect to the insulator surface [1, 2].

An electron emitted from the surface with energy A_0 and having a direction normal to the insulator surface reaches height x

$$x = (A_0/eE_\perp), \tag{7A.1}$$

where x is cm, A_0 in eV and $E_\perp$ (in V/cm) is the field normal to the surface of the insulator resulting from the surface charge density.

The range of the trajectory is

$$y = (4A_0E_{//}/eE_\perp^2), \tag{7A.2}$$

where y is in cm and $E_{//}$ (in V/cm) is the field parallel to the surface of the insulator and equal to the average applied field in the case of a solid cylindrical insulator. $E_{//} = V/l$, where V is the applied voltage between the electrodes and l the gap spacing. Taking into account the cosine-law distribution in the direction of emission, the average value of impact energy A_i becomes [1]

$$A_i = A_0\left[1 + 2(E_{//}/E_\perp)^2\right]. \tag{7A.3}$$

When the steady state of charging is reached at unity yield, the impinging energy is A_1. It should be noted that the stability of the surface charge corresponding to the lower of the two energies A_1, at which the emission yield is unity, is sufficient to be considered [4]. The relationship between the field $E_{//}$ and the field $E_\perp$ is [5]

$$E_{//} = E_\perp\sqrt{\frac{1}{2}\left[(A_1/A_0) - 1\right]}. \tag{7A.4}$$

The height of the trajectory is very small, a few micrometers, compared to the radius of curvature of the surface of practical insulators, and therefore can be considered as a plane as far as the trajectory of the electron is concerned. The field produced by the positive charge in the plane geometry is

$$E_\perp = \sigma_+/2\varepsilon_0, \tag{7A.5}$$

where ε_0 is the permittivity of the free space (8.85×10^{-14} F/cm) and σ_+ is the charge density in C/cm^2.

The critical angle for which the surface charge is at an equilibrium is given by $\tan\theta$. θ is the angle between the electric field and the insulator surface. The value of $\tan\theta$ mainly depends on the impact energy A_1.

$$\tan\theta = (E_\perp/E_{//}) = \sqrt{2A_0/(A_1 - A_0)}. \tag{7A.6}$$

In general, the secondary-emission parameters A_0 and A_1 for solid insulators are a few eV and a few tens of eV, respectively. Using the data available on the secondary-electron emission yield from insulators [3], the value of θ obtained from Eq. (7A.6) is found to vary from about 10°–35°. The expression for positive surface charge density σ_+ can be written in terms of the interelectrode field $E_{//}$

$$\sigma_+ = 2\varepsilon_0 E_{//} \tan\theta. \tag{7A.7}$$

The current carried by the secondary emission avalanche depends on the density of negative space charge σ_- and the drift velocity of electrons in the avalanche. The density of the electrons' space-charge in the emitted secondary electron avalanche cannot exceed the surface charge σ_+ in magnitude without causing electrons to be repelled from the insulator. When the electrons drift towards the anode, the value of σ_- diminishes, but at the same time, the positive surface charge at the insulator-cathode junction enhances the field at the cathode triple junction. The increased field emission maintains σ_- equal to σ_+. Hence in our analysis, it is assumed that σ_- is equal to σ_+. The surface current carried by the secondary emission avalanche per centimeters can be written as

$$I_{//} = \sigma_- v_e, \tag{7A.8}$$

where $I_{//}$ is the current per unit length of the insulator surface (A/cm). The average velocity v_e is used as an approximation for the drift velocity.

$$v_e = \sqrt{2A_1 \times 10^4 / m_e}, \ (\text{cm/s}) \tag{7A.9}$$

where m_e and e are the electron mass and charge, respectively. A_1 is expressed in joule.

From the above relation, the average velocity of an electron in the avalanche having an energy A_1 eV is

$$v_e = 5.94 \times 10^7 . \sqrt{A_1} \tag{7A.10}$$

The surface current expression becomes

$$I_{//} = \sigma_- \times 5.94 \times 10^7 \times \sqrt{A_1}. \ (\text{A/cm}) \tag{7A.11}$$

The electrons return to the insulator after traveling an average distance of $s = A_1/E_{//}$. The value of s is of the order of 10^{-4}–10^{-3} cm, when σ_+ is at a saturated steady state. The current density perpendicular to the insulator surface can be obtained as

$$J_\perp = (\sigma_- v_e E_{//}/A_1). \ (\text{A/cm}^2) \tag{7A.12}$$

The desorption rate from the insulator surface depends on $J_\perp$.

Electron-Impact Gas Desorption and Discharge Formation [7A-6]

When electrons impinge on the surface of a solid insulator, the particles released, depending on the electrons' energy, include neutral molecules, atoms, excited neutrals, and positive and negative ions. The ratio of neutrals evolved is very much higher than other particles. In most cases, the electron-impact desorption cross-sections are smaller than the corresponding ionization cross-section in the gas phase. Electron impact cross-sections vary with electron energy in a similar fashion to gas-phase ionization cross-sections. The average desorption cross-section is assumed to be $10^{-16}\,\text{cm}^2$ [6], as the major components of the desorbed gases are carbon monoxide and hydrogen.

The electron-stimulated desorption (ESD) efficiency can be found from experiments on pre-breakdown currents by measuring the outgassing rates of gases [6, 7]. It is assumed that the pre-breakdown current is due to secondary-emitted electrons hopping across the surface of a fully charged insulator. For CO, H_2, N_2, H_2O, and other molecules which are known to be adhered to the surface, the number of molecule per square centimeter on the surface of the dielectric may be taken as the order 10^{16}–$10^{17}/\text{cm}^2$ [8, 9]. In order to estimate the desorption probability, the results found in a study of the gas evolution from insulators subjected to an electric field in the direction parallel to the surface are considered [7]. The rate of gas evolution depends on the insulating material as well as the desorption probability. A value of outgassing rate of the order of 2–5 $\times 10^{-7}$ Torr · L/s for different insulating materials (Plexiglass, Teflon, vinyl, ceramics) has been reported [7]. The desorption probability corresponding to this outgassing rate is in the range 5–15 molecules per impinging electron calculated from pre-breakdown currents of 2–6 $\times 10^{-7}$ A observed in our (Pillai and Hackam) experiments. A desorption probability of a few molecules per electron is consistent with the product of an electron-stimulated desorption cross-section of a small multiple of 10^{-16} cm^2 [6, 10] and the number of molecules per square centimeter on the surface of 10^{16}–10^{17} [8, 9].

The rate of electron-impact gas desorption J_d depends on the electron-current density $J_\perp$ (A/cm^2), the desorption cross-section Q_d (cm^2) and the surface-charge density σ_d (molecules/cm^2). The expression for J_d is

$$J_d = (\sigma_d Q_d J_\perp / e)(\text{molecules} \cdot \text{cm}^{-2} \cdot \text{s}^{-1}). \tag{7A.13}$$

In general the desorbed particle rate is written as

$$J_d = (\gamma J_\perp / e), \tag{7A.14}$$

where γ is the desorption probability equal to $\sigma_d Q_d$ molecules/electron.

The electron stimulated desorption probability of gases from a clean and well-outgassed insulator surface can be found by observing the pressure changes in the system housing it during electron bombardment of the sample [10]. The variation in the desorption rate for a particular insulator depends among other things on the sample temperature, the electron-impact energy and the electron-current density. The main components of the desorbed gases are hydrogen, nitrogen, carbon monoxide, and water vapor. The relative amount of these gases present depends on the structure, composition, and history of the insulator. The desorbed gases from materials like Teflon (polytetrafluoroethylene), Plexiglass (polymethyl methacrylate), ceramics, alumina (Al_2O_3) contain a higher percentage of hydrogen and water vapor particularly in the case of an unbaked vacuum system. In a well-baked system the gases normally consist of at least 30–50% CO. In order to find the density of desorbed gas, the velocity v_0 at which a desorbed neutral leaves the insulator surface must be known. However, only a few data are available on the value of v_0. The average velocity of the gas molecules may be assumed of the order of 1.5–2.0 $\times 10^5$ cm/s [3, 11, 12] The density of the desorbed molecules N_d can be estimated from the value of v_0 and the gas-desorption rate

$$N_d = (\gamma J_\perp / e v_0)\ (\text{molecules/cm}^3). \tag{7A.15}$$

The amount of the desorbed gas M per square centimeter from the insulator surface due to a pre-discharge field current is given by the product of N_d and the length of the insulator, l

$$M = (\gamma J_\perp / e v_0) l. \tag{7A.16}$$

Flashover Breakdown Criteria [7A-6]

The electrical field is enhanced at the cathode end of the insulator due to the effect of positive-charge accumulation, and consequently, the electron-emission current from the cathode increases. Increasing the applied electric field leads to a rise in the field-emission current, and consequently, the amount of desorbed gas increases. The ionization rate increases rapidly with increasing of the desorbed gas. When the pre-breakdown current is low, the gas desorption is also low. In this case, the condition for a self-sustained discharge can only be satisfied at a very high field. There is obviously some critical current I_{cr} and a corresponding critical desorbed amount of gas density M_{cr} at which the electron density required for breakdown is reached [13].

The breakdown field can be calculated from the value of M_{cr} corresponding to I_{cr}.

From Eqs. (7A.5), (7A.6), (7A.12), (7A.16) and assuming $M = M_{cr}$, the critical value at the onset of the surface flashover, the expression for the field $E_{//}$ required to cause flashover can be written as

$$E_{//} = \sqrt{\frac{M_{cr}A_1 v_0 e}{2\varepsilon_0 l \gamma v_e \tan\theta}}. \text{ (V/cm)} \tag{7A.17}$$

Equation (7A.17) indicates the dependence of $E_{//}$ on the secondary-electron impact energy, desorption probability and length of the insulator. It also gives a correct description of the decrease in the dielectric strength of the surface with increasing insulator length.

The breakdown voltage can also be calculated by using Eq. (7A.17). The breakdown voltage is almost independent of the electrode material and strongly dependent on the insulator material [7, 14]. The breakdown voltage V_B is

$$V_B = E_{//}l = \sqrt{\frac{M_{cr}A_1 v_0 e l}{2\varepsilon_0 \gamma v_e \tan\theta}}. \text{ (V)} \tag{7A.18}$$

Equation (7A.18) shows that the flashover voltage across a solid insulator held between two planar parallel electrodes increases in a nonlinear form with increasing length of the insulator. The reported results of [14, 15, 16, 17, 18] are plotted in **Fig. 7A.5** for various insulating materials where it can be seen that a dependence of the flashover voltage on length of the insulator to a power law of $\sqrt{l}$ is suggested. The predicted dependence of the flashover voltage given by Eq. (7A.18) on insulator length l is close to that observed experimentally (**Fig. 7A.5**).

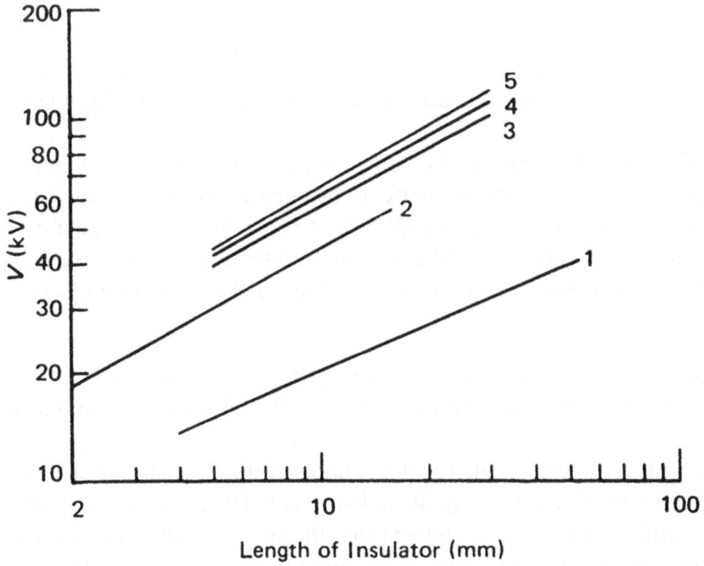

Fig. 7A.5 Flashover voltage as a function of insulator length for different materials. (1) Pyrex [17], (2) Alumina ceramic [18], (3) Polyethylene [16], (4) Teflon [16], (5) Plexiglass [16] (Pillai and Hackam, 1982) [7A-6]

References

1. H. Boersch, H. Hamisch, and W. Ehrlich, *Z. Angew Phys.* **15**, 518 (1963).
2. C. H. De Tourreil and K. D. Srivastava, *IEEE Trans. Elect. Insul.* **8**, 17 (1973).
3. S. C. Brown, *Basic Data of Plasma Physics* (MIT, Cambridge, 1959).
4. R. A. Anderson and J. B. Brainard, *J. Appl. Phys.* **51**, 1414 (1980).
5. J. P. Brainard and D. Jensen, *J. Appl. Phys.* **45**, 3260 (1974).
6. P. A. Redhead, J. P. Hobson, and E. V. Kornelson, *The Physical Basis of Ultrahigh Vacuum* (Chapman and Hall, London, 1968).
7. A. A. Avdienko and A. V. Kiselev, *Sov. Phys. Tech. Phys.* **12**, 381 (1967).
8. S. P. Bugaev, A. M. Iskoldskii, and G. A. Mesyats, *Sov. Phys. Tech. Phys.* **12**, 1358 (1968).
9. S. Dushman and J. M. Lafferty, *Scientific Foundations of Vacuum Technique* (Wiley, New York, 1962).
10. M. J. Drinkwine and D. Lichtman, *Prog. Surf. Sci.* **8**, 123 (1977).
11. J. H. Leck and B. P. Stimpson, *J. Vac. Sci. Technol.* **9**, 293 (1972).
12. A. A. Avidenko and M. D. Malev, *Sov. Phys. Tech. Phys.* **24**, 581 (1979).
13. H. Raether, *Electron Avalanches and Breakdown in Gases* (Butterworth, London,1964).
14. R. Hawley, *Vacuum* **18**, 383 (1968).
15. Hackam, IEE invited review for colloqium "*Electrical Phenomena on Insulating Surfaces in Gases and in Vacuum,*" IEE Digest No. 26 (1975).
16. S. Grzybowski and E. Kuffel, *IEEE Trans. Power Appar. Syst.* **99**, 1788 (1980).
17. P. Gleichauf, *J. Appl. Phys.* **22**, 766 (1951).
18. O. I. Kondratov, *Proceedings of the 5th International Symposium on Discharge and Electrical Insulation in Vacuum*, Poznan, Poland, 1972, p. 289.

Cross and Sudarshan (1974) [7A-7] experimentally investigated the effect of a cuprous oxide coating upon the surface flashover of high-density alumina in vacuum.

Abstract [7A-7]: It is shown that such coatings improve the impulse strength of the system and eliminate the conditioning effect observed in the case of uncoated specimens at dc and 60 Hz voltages. An explanation of the observed behavior is given in terms of a surface charging model. It is postulated that the improvement in the insulation by the coatings is due to a reduction in the secondary electron emission yield.

Sudarshan and Cross (1976) [7A-8] experimentally investigated the effect of chromium oxide coatings upon the surface flashover of high-density alumina in vacuum.

Abstract [7A-8]: It is shown that such coatings improve the withstand voltage for dc and 60 Hz, and impulse voltages by significant amounts. The coatings are simply applied, effective, and durable. It is postulated that the improvement in insulation is due to the elimination of positive surface charging by the use of a coating with a secondary-electron emission coefficient less than unity for all incident electron energies.

Gas Molecules on Insulator Surfaces

Watanabe et al. (1987) [7A-9] experimentally examined microdischarges on an electron gun (**Fig. 7A.6**) operated at 100 kV, by concentrating the greatest interest on gas molecules on the insulator and electrode surfaces.

Experiment [7A-9]

Microdischarge depends on the concentration of electric field in the gun chamber. Equipotentials around and on the electron gun, simulated by the finite-element method, are presented in **Fig. 7A.6**. The electric field is relatively high near the top and side of the Wehnelt electrode.

The holes of the Wehnelt and anode electrodes through which an electron beam passes are about 0.4 mm and 6 mm in diameter, respectively. The surfaces of the electrodes and chamber walls (304 stainless steel) are mirror-polished. The insulator is a kind of porcelain whose surface has been treated to be glassy. The junction of metal and insulator is covered by a guard ring (304 stainless steel, mirror polished) in electrical and mechanical contact with the Wehnelt electrode.

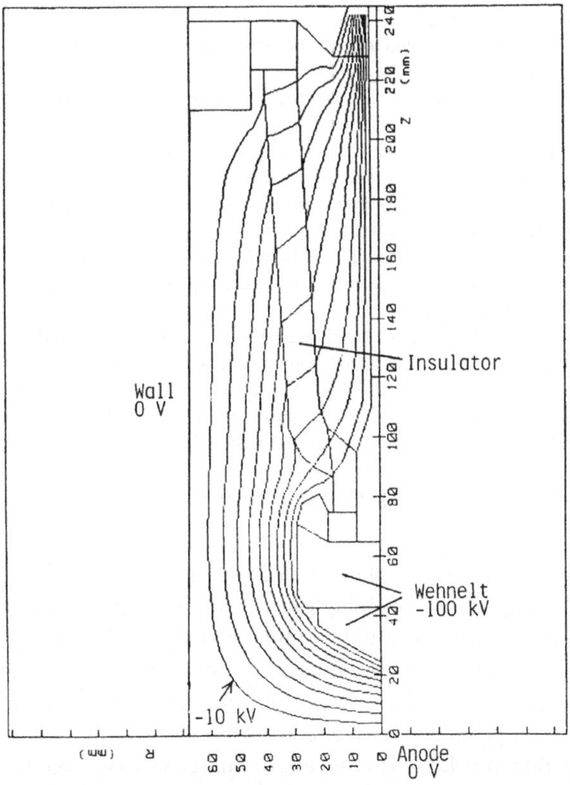

Fig. 7A.6 Equipotentials in a gun chamber with a 100 kV electron gun, which were simulated by the finite-element method (Watanabe et al., 1987) [7A-9]

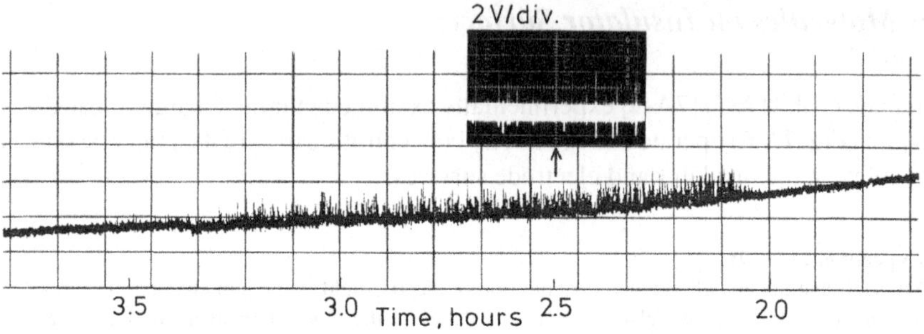

Fig. 7A.7 Microdischarges under about 10^{-4} Pa on "C-1" (1st day after cleaning) without AGC (Ar-glow conditioning) during an earlier period of the elapsed time. Microdischarges can be identified as voltage spikes on the chart line (Watanabe et al., 1987) [7A-9]

Experiments were conducted using a typical DP evacuation system. –100 kV is applied to the Wehnelt electrode with the anode electrode being kept at ground potential.

Microdischarges occurred most frequently on C-1 (1st day after CH_3CCl_3 cleaning) without AGC (Ar-glow conditioning), as presented in **Fig. 7A.7**, where microdischarges are identified as voltage spikes on an oscilloscope and on a recorder line. Microdischarges began to occur at about 2 h and occurred very frequently in the period from 2 h to 3.5 h.

We considered that microdischarges depending on the elapsed time must be related to the outgassing of the electron gun heated by the tungsten-filament emitter. During electron-beam extraction, the heat (about 10 W) from the filament emitter must be gradually raising the temperature of the insulator, resulting in increased outgassing from the insulator.

The outgassing from the insulator was examined by measuring pressure-increases caused by the lighted filament emitter and by analyzing evolved gases, in an SIP (sputter ion pump) evacuation system presented in **Fig. 7A.8**. Pressure-changes under various conditions in the system continuously evacuated by a 160 L/s SIP are presented in **Fig. 7A.9**.

Experimental results clearly show that the outgassing from the insulator on C-1 without AGC is the largest among those under all conditions.

The microdischarge characteristics depending on the elapsed time (**Fig. 7A.7**) are indeed analogous to the outgassing characteristics from the insulator depending on the elapsed time after starting to heat the filament emitter on C-1 (**Fig. 7A.9**). Microdischarges are enhanced by the outgassing from the insulator. The reason is that the insulator surface with a high density of gas molecules causes high-yield secondary-electron emission, leading to positive charging on the insulator surface.

Thermal degassing for the electrodes and insulators has a conditioning effect to reduce microdischarges [7A-9].

Saito et al. (1994) [7A-10] investigated breakdown phenomena in alumina windows for high-power rf use. "Multipactor due to high yields of secondary-electron emission causes F-center oxygen vacancies in the alumina, which induces surface melting. High-purity alumina without micro-porosities or pre-existing F-centers is durable under high-power rf operation. TiN coatings can effectively suppress the multipactor, when the thickness is optimized so as to reduce secondary electrons and to avoid any excessive heating due to rf loss. The surface discharge observed on the TiN-coated surface during higher-power operation indicates the electron-trapping in surface defects and the electron avalanche accompanied by luminescence."

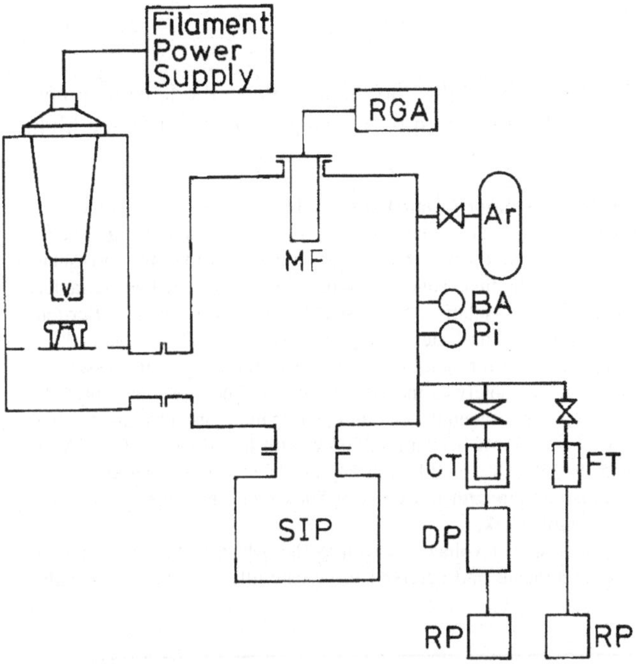

Fig. 7A.8 Ultrahigh-vacuum system with a sputter ion pump. Metal gaskets are used in the high-vacuum side, with the exception of a Viton O-ring for attaching the gun. RGA, residual gas analyzer; MF, quadrupole mass filter; SIP, sputter ion pump (160 L/s); BA, Bayard-Alpert ionization gauge; Pi, Pirani gauge; CT, cold trap; FT, foreline trap; DP, oil diffusion pump, and RP, mechanical rotary pump (Watanabe et al., 1987) [7A-9]

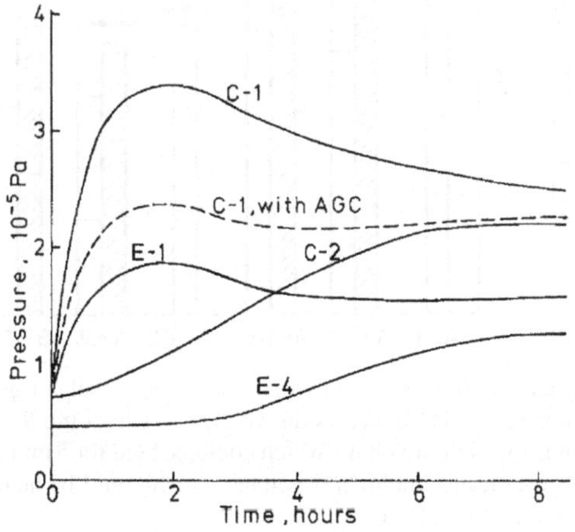

Fig. 7A.9 Pressure-changes in the SIP system after heating the filament emitter with about 10 W. The system was continuously evacuated by the SIP (160 L/s). C-1, 1st day after cleaning without AGC; E-1, 1st day after exposing the chamber to the atmosphere without AGC (Watanabe et al., 1987) [7A-9]

Triple Junction

Shannon et al. (1965) [7A-11] reported that the shape of the cathode triple junction and so the electric field around the junction is very important to obtain the highest insulating performance.

Flashover Strength of Various Insulator Geometries [7A-11]
The high-voltage performances of typical insulator designs are summarized in **Fig. 7A.10**. Both 7070 and 7740 Pyrex glasses were used. In each bar on the chart the thin horizontal line indicates the voltage at which the first spark occurred. Higher up on the bar, a heavier horizontal line indicates the level of insulating (5 min steady value) performance after ten conditioning sparks. The top of the bar indicates the highest insulating voltage obtained.

In this work on the comparative performance of various insulator designs, it is essential that the limitations recorded be due solely to insulator flashover. Therefore, as a preliminary step, it was demonstrated experimentally that an open 25 mm gap between the two stainless-steel electrodes was capable of insulating 250 kV with less than 5×10^{-10} A of inter-electrode current, and that 300 kV could be reliably insulated without sparking. The glass itself has a volume puncture strength in excess of 500 kV/cm. No signs of volume puncture ever were detected in this work.

It was found that the first flashover voltage, as well as the behavior during the initial phase of conditioning, was significant and deserved to be carefully observed, especially

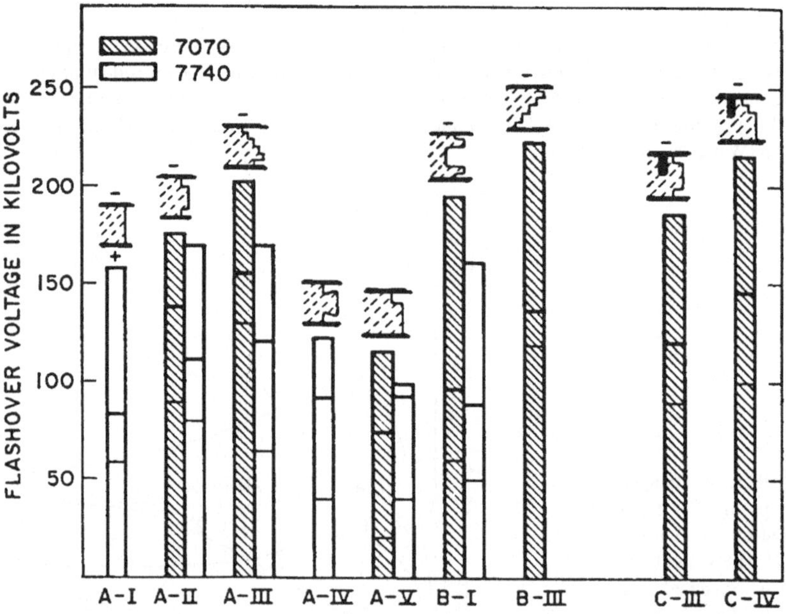

Fig. 7A.10 Flashover strength of various insulator geometries in vacuum. On each bar, the light horizontal line indicates the voltage at which the first flashover took place, the heavy horizontal line the highest voltage which could be held for 5 min after 10 conditioning sparks. The top of each bar indicates the highest voltage which could be held for 5 min after full conditioning (Shannon et al., 1965) [7A-11]

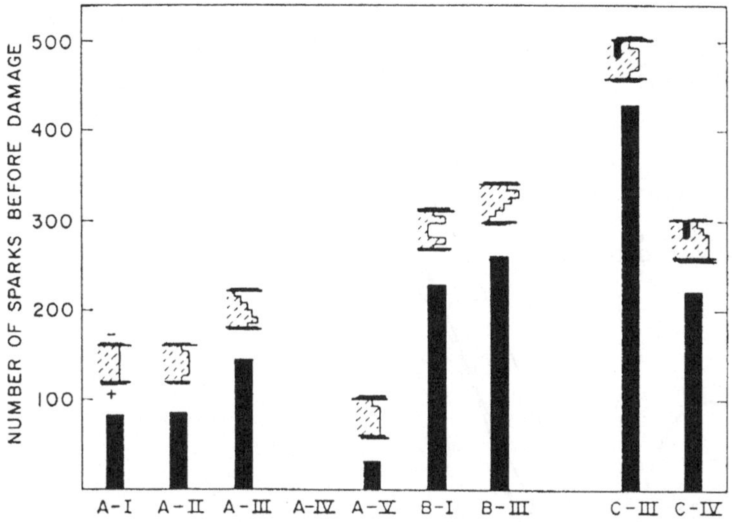

Fig. 7A.11 Number of sparks required to damage various different insulator geometries of 7070 Pyrex glass (Shannon et al., 1965) [7A-11]

since in practice many insulators cannot be conditioned by repeated flashover. The conditioned performance after 30 or more sparks was, in most cases, satisfactorily reproducible.

The ability of an insulator to undergo repeated flashover without damage is of obvious advantage in most applications. Moreover, this resistance to damage is related to its ability to condition to a high insulating strength.

Figure 7A.11 displays the total number of sparks required to damage irreversibly the various different insulator designs. A correlation appears to exist between good insulating performance after conditioning and the ability to experience repeated flashover without damage.

Watson (1967) [7A-12] presented an article, "Pulsed flashover in vacuum."

Abstract [7A-12]: Experiments on the vacuum flashover of dielectric frusta subjected to 35 and 75 ns pulses show the strong influence of cone angle on insulation strength. A theory is presented to explain the formative process leading to flashover for one polarity. It is proposed that positive surface charge grows by emission of hot electrons during the application of the electric field. When the surface field intensity is sufficient to draw other electrons into it, they will multiply by secondary emission until the critical breakdown condition is achieved. Flashover is thus initiated by secondary emission, but the formative time lag is governed by thermionic emission of electrons. The experimental data are analyzed and interpreted in the light of this theory.

Experimental results show the strong influence of cone angle on insulation strength (**Fig. 7A.12**).

Yamamoto et al. (1996) [7A-13] examined the charging of an insulator with double ribs or a recess by using a two dimensional Monte-Carlo simulation method based on the secondary emission electron avalanche (SEEA) mechanism.

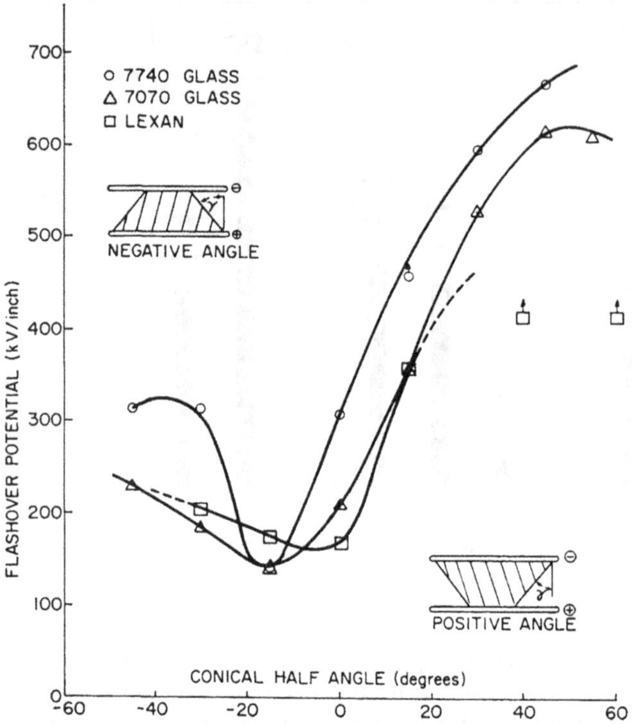

Fig. 7A.12 The flashover strength of glass and Lexan (G.E. polycarbonate resin) as a function of geometry. Tests were first made in the positive-angle configuration and the polarity was reversed to obtain negative-angle data (Watson, 1967) [7A-12]

Abstract [7A-13]: The electric field distribution on a cathode around a triple junction was analyzed taking the charge accumulation into account. Results were compared to those obtained for a simple cylindrical insulator and also insulators with a single rib. Flashover tests were also conducted by using dc voltage. The results of these investigations imply that the number of ribs is unimportant, and that the recess on an insulator is not so effective for the insulation of bridged vacuum gaps.

Surface Flashover in SF_6 Gas

Before the restriction on use of chlorofluorocarbons, high-voltage generator circuits were assembled in a gas tank containing pressurized Freon 12 (CCl_2F_2) gas. After that, they have been assembled in a gas tank containing pressurized SF_6 gas.

Nakanishi et al. (1982) [7A-14] presented an article, "Surface flashovers along insulators under non-uniform electric fields in SF_6 gas". "Many breakdown-voltage characteristics, such as ac-voltage breakdown characteristics and dc-voltage breakdown characteristics, are presented for the two cases, one with a negative-voltage protrudent electrode and the other with a positive-voltage protrudent electrode."

Review

Hawley (1968) [7A-15] reviewed 26 articles on solid insulators in vacuum.

Abstract [7A-15]: After reviewing the pre-breakdown phenomena (leakage currents, outgassing and luminosity) that arise across a solid insulator held between two electrodes in a vacuum gap, the factors that affect the magnitude of the breakdown voltage (conditioning, electrode and insulator material and surface finish, insulator length and shape, pressure, etc.), the phenomena occurring in the region of the cathode-insulator junction and the hypotheses put forward to explain the breakdown mechanism, are discussed.

Miller (1989) [7A-16] reviewed 99 articles on surface flashover of insulator.

Abstract [7A-16]: Surface flashover of insulators in vacuum generally is initiated by the emission of electrons from the cathode triple junction (the region where the electrode, insulator, and vacuum meet). These electrons then usually multiple as they traverse the insulator surface, either as a surface secondary electron emission avalanche, or as an electron cascade in a thin surface layer, causing desorption of gas which has been adsorbed on the insulator surface. This desorbed gas is then ionized, which lead to surface flashover of the insulator. Some suggestions are made regarding how to choose the material geometry, and processing when selecting an insulator for a particular application.

7B Microdischarges between High-Voltage Electrodes

Anode-Initiation Mechanism

High-voltage breakdown across electrode gaps is explained by the anode-initiation model for high-voltage, low electric-field systems with wide gaps.

Cranberg (1952) [7B-1] presented an article, "The initiation of electrical breakdown in vacuum."

Hypothesis of Cranberg [7B-1]

The hypothesis is that the initiation of breakdown is due to detachment by electrostatic repulsion of a clump of material loosely adhering to one electrode, but in electrical contact with it; traversal by the clump of most or all of the high-voltage gap, and impingement on an electrode at much lower, or at the lowest potential.

A quantitative formulation of this initiation hypothesis may be set down very simply as follows. Assume that breakdown will occur when the energy per unit area W delivered to the target electrode exceeds a value C', a constant, characteristic of a given pair of electrodes. This quantity W is just the product of the gap voltage V by the charge density on the clump. The latter is proportional to the field E at the electrode of origin so that the breakdown criterion becomes simply

$$VE \geq C, \tag{7B.1}$$

where C is a product of C', some numerical factors, and possibly a field-intensifying factor due to microscopic field inhomogeneities in the neighborhood of the clump during detachment from its parent electrode. For the case of plane-parallel electrodes where $E = V/d$,

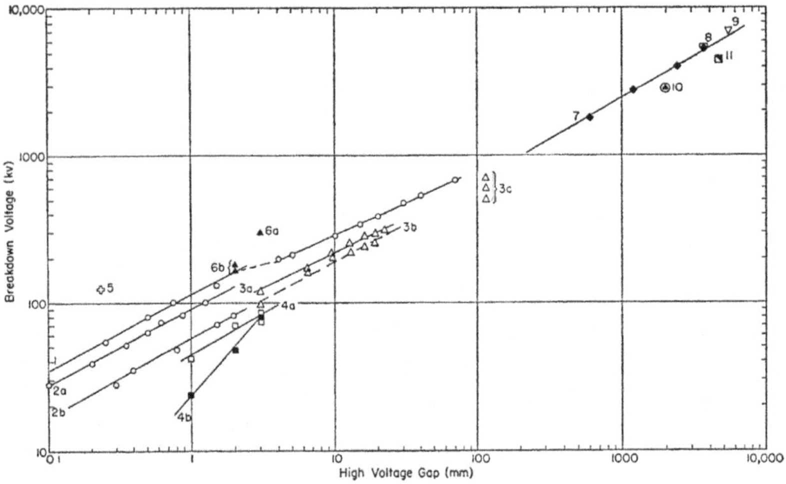

Fig. 7B.1 Plot of data from the literature of breakdown voltage vs. distance from highest to lowest potential electrode, for uniform-field and near-uniform-field geometry. Numbers on curves indicate sources as listed below (Cranberg, 1952) [7B-1]

this criterion predicts at once that the voltage which a gap can sustain is proportional to the square-root of the gap length for a given pair of electrodes.

$$V \propto \sqrt{Cd}. \tag{7B.2}$$

The prediction of this simple result has led to search of the literature for pertinent evidence, a summary of which is given in **Fig. 7B.1** in the form of a log-log plot of voltage vs. length of gap, for uniform and nearly uniform field conditions. The only contradictory evidence is that provided by a three-point curve given by Gleichauf [(P. Gleichauf, *J. Appl. Phys.* **22**, 766, (1951)] for copper electrodes in the range 1–3 mm, whereas all other data are consistent with this prediction.

References

1. J. G. Trump and Van de Graaff, [see J. G. Trump and R. J. Van de Graaff, J. Appl. Phys. **18**, 327 (1947)] 1-inch sphere steel anode, 2-inch steel disk, outgassed with glow discharge.
2. Wm. Parkins, "Vacuum Sparking Potentials under Surge Conditions", *MDDC* **858**, 18 February, 1946. Voltage applied in pulses of 3×10^{-7} sec rise time. (a) Tungsten hemispheres 2-inch diameter, outgassed by spark discharge. (b) Copper hemispheres 2-inch diameter, outgassed by spark discharge.
3. J. L. McKibben and R. K. Beauchamp, "Insulation-Flashover Tests in Vacuum and Pressure", *AECD* 2039. (a) Flat aluminum. (b) Flat cold-rolled steel. (c) Van de Graaff test-section $4\frac{1}{2}$ inches long, aluminum rings sandwiched between Mykroy rings, steel anode plate, negative end open to vacuum system in simulation of operation as beam tube, results on three test-sections.
4. P. Gleichauf [see *J. Appl. Phys.* **22**, 766 (1951)]. (a) Kovar cathode, 18–8 stainless-steel anode, flat with rounded ends. (b) Copper, flat, with rounded ends (hole in center of anode).
5. R. J. Piersol, British Assoc., *Advancement of Science*, Report 359 (1924). Molybdenum spheres after heating to 1400 °C.
6. J. L. Hayden, *Am. Inst. Elec. Engrs. J.* **41**, 852 (1922). (a) Molybdenum spheres 1 cm diameter outgassed to red heat, polished. (b) Molybdenum spheres 1 cm diameter.
7. Los Alamos big Van de Graaff, polished aluminum electrodes.
8. Robinson et al., *Phys. Rev.* (to be published).
9. J. G. Trump (private communication on performance of new 12 MeV Van de Graaff machine, polished aluminum electrodes.
10. Los Alamos small Van de Graaff machine (not limited by tube sparking), steel electrodes.
11. Wisconsin Van de Graaff machine (not limited by tube sparking), steel electrodes.

Chatterton et al. (1972) [7B-2] investigated the role of low-velocity microparticles (diameters$<$ 100 μm; velocities $\sim$ 10–50 m/sec) in inducing breakdown in a high-vacuum gap

Abstract [7B-2]: On the basis of a simple model, it is shown that as an anode initiated microparticle approaches the cathode, fields sufficient to cause appreciable field emission ($> 10^9$ V/m) are possible at the cathode surface. The emission current not only causes partial neutralization of the initial charge on the microparticle, but also raises the temperature (>2000 K) of the particle surface. As a consequence, a significant increase in gas pressure ($\sim$ 100 Torr) in the microvolume between the cathode and the particle is possible, resulting in a discharge between the two. Such a discharge could act as a trigger for the breakdown of the main gap. Other possible processes that could arise due to the onset of melting or boiling of the bombarded surface, and which may lead to breakdown of the entire gap, are also discussed.

Davies (1973) [7B-3] reviewed the recent investigations of prebreakdown conduction phenomena in vacuum both for steady and for step-function applied voltages.

Abstract [7B-3]: The results of these investigations have led to the formulation of models to explain the initiation of the vacuum discharge by ionization of a vapor medium, the vapor being composed of atoms of one or both of the electrodes. A detailed examination of the events leading to breakdown has allowed a more critical evaluation of the merits and deficiencies of the various breakdown models.

Menon and Srivastava (1974) [7B-4] presented an article, "Microparticle-initiated vacuum breakdown-Some possible mechanisms."

Abstract [7B-4]: It is known that micron- and submicron-sized metallic particles are released from the electrode surfaces when a vacuum gap is subjected to a high dc stress. It is also well known that larger particles ($> 10\,\mu m$) are generated within the inter-electrode gap when a vacuum gap is subjected to conditioning or severe prebreakdown current flow. This paper examines the role of such particles in inducing the breakdown of a vacuum gap. While the larger particles induce breakdown by way of a trigger discharge, it is shown that the smaller particles ($< 3\,\mu m$) can initiate breakdown because of effects associated with impact. The various effects associated with the high-speed impact of a metallic microparticle on a target electrode, viz., cratering, production of metal vapor, and production of thermally generated plasma and their relative significance on vacuum breakdown, are examined.

Davies and Biondi (1977) [7B-5] presented an article, "Emission of electrode vapor resonance radiation at the onset of dc breakdown in vacuum."

Abstract [7B-5]: Simultaneous time-resolved spectroscopic and current measurements have been made during the early stages of current growth leading to dc breakdown in vacuum between plane-parallel electrodes of dissimilar material. These measurements show that resonance-radiation characteristic of anode material is emitted prior to that of cathode material from all regions of the inter-electrode gap. Further, the radiation is emitted first from the cathode region of the gap. The results indicate that the initial current growth occurs in anode vapor followed at later times by the appearance of cathode vapor in the gap.

Allen et al. (1979) [7B-6] presented an article, "The source of high-β electron emission sites on broad-area high-voltage alloy electrodes."

Abstract [7B-6]: Two recently developed techniques have been used sequentially in an attempt to define the nature of high-β (β, field-increase factor) field-emission sites on a commercial-alloy, broad-area, electrode surface. The techniques involved were: (1) an in situ electron-optical method for locating and examining the emission site; and (2) a high-resolution spectrometer for determining the energy spectra of the electrons field emitted from this site. Furthermore, following these measurements, the elemental composition of the emission area was determined by electron microprobe x-ray analysis. Observations, which cast further doubt upon the traditional concept of field-enhancing micro-projections, indicate that the sites are non-metallic and probably consist of impurities located at cracks or grain boundaries in the surface. Tentative hypotheses are proposed for the emission mechanism and include provision for an unstable situation caused by the effects of adsorbed gas atoms.

Diamond (1998) [7B-7] presented an article, "New perspectives in vacuum high voltage insulation. I . The transition to field emission."

Abstract [7B-7]: Field emission is not present at all fields, but typically starts after some process occurs at the cathode surface. Three effects have been identified that produce the transition to field emission: work function changes; mechanical changes produced by the strong electrical forces on the electrode surfaces; and gas desorption from the anode with sufficient density to support an avalanche discharge. Localized regions of low work function can be produced on the cathode by the transfer of microparticles from the anode and by stripping small areas of the cathode. The regions of low work function then serve as the source of enhanced field emission, leading to secondary effects which produce breakdown. Gas desorption is produced at an unconditioned anode as the voltage is increased. None of these effects are significant for a point cathode opposite a broad-area anode, but account for much of the large difference between microscopic and macroscopic breakdown fields [7B-7].

Diamond (1998) [7B-8] also presented an article titled "New perspectives in vacuum high voltage insulation. II. Gas desorption."

Abstract [7B-8]: An examination has been made of gas desorption from unbaked electrodes of copper, niobium, aluminum, and titanium subjected to high voltage in vacuum. A very significant observation was that the gas desorption was more dependent on the total voltage between the electrodes than on the electric field. It was not triggered by field-emitted electrons but often led to field emission, especially at large gaps. The physical processes that lead to voltage-induced desorption are complex, but there is strong evidence that the microdischarges are the result of an avalanche discharge in a small volume of high-density vapor desorbed from the anode. The source of the vapor may be water or alcohol stored as a fluid in the many small imperfections of the polished metal surface. Microdischarges can then trigger field-emitted electrons which, in turn, heat a small area of the anode. As the temperature of this region of the anode reaches about 500 °C, some fraction of the desorption products are ionized positively and accelerated to the cathode, producing secondary electrons with a yield greater than unity per incident ion. The positive ions appear to originate from the bulk of the metal rather than from surface ionization and the yield increases exponentially with temperature, rapidly producing a runaway condition, i.e., electrical breakdown. These observations support a new perspective on vacuum-high-voltage insulation and produce new insight into vacuum outgassing of metals.

Ion-Exchange Process and Total-Voltage Effect

Cranberg and Henshall (1959) [7B-9] investigated to determine the conditions under which the voltage sustained by a 2 ft length of ion-accelerator tube may be made proportional to the length of the tube.

Abstract [7B-9]: It has been found that such linearity may be obtained if the tube is segmented at $4\frac{1}{2}$ in. intervals by diaphragms which are so arranged that no straight path is possible from one end of the tube to the other. No significant deterioration in performance of the tube was observed when axial holes were made in each diaphragm up to $3/4$ in. in diameter. The voltage gradient realized on these tests was 60 kV/cm.

Mansfield (1960) [7B-10] presented an article, "Pre-breakdown condition in continuously-pumped vacuum systems."

Abstract [7B-10]: Measurement have been made under impulse conditions of the coefficients A', the number of H^+ ions emitted per 250 keV H^- ion, and B', the number of H^- ions emitted per 250 keV H^+ ion, for metal surfaces covered with the contaminating layers likely to be formed in continuously pumped high-voltage apparatus. The values obtained for A' were 1.0, 1.1 and 0.54, and for B' 0.43, 0.24 and 0.44 for copper, aluminum and steel targets, respectively. The product of these coefficients is such as to make very probable the hypothesis that pulse discharge conduction in these system is due to the regenerative exchange of positive and negative ions of hydrogen. The transient nature of this form of conduction is thought to be due to the charging up of the insulating contaminant.

Powell and Chatterton (1970) [7B-11] examined the prebreakdown conduction between stainless steel, copper, aluminum and tungsten electrodes at small gaps (less than 1 cm) and in poor (10^{-6} Torr, unbaked) and clean (3×10^{-9} Torr, baked) vacuum systems.

Abstract [7B-11]: Emphasis has been placed on means of distinguishing between the various types of conduction found to occur. These were identified as (1) field emission and an associated current pulse structure, (2) field emission and activation/ignition effects, (3) microdischarges. This identification was possible directly as a result of using a fast response current measuring system. Equally, the determination of true field emission characteristics, especially in the poor vacuum system, in the presence of the pulse structure and activation/ignition effects was achieved only by means of an ageing process and a fast I-V plotter. The role of surface contamination in determining apparent emission law characteristics has been recognized and a simple, qualitative model proposed which can account for many of the observed effects.

Prichard (1973) [7B-12] studied high-voltage (up to 400 000 V) electrical discharge mechanisms in a stainless-steel and alumina-ceramic vacuum system evacuated by an orbitron getter-ion pump.

Abstract [7B-12]: Mass-spectrometer studies were made of ions emitted from the terminal during discharges with positive and negative terminal polarities. Ions ejected from the terminal by a beam of Na^+ ions were also examined. Results of these and other studies led to the conclusion that the regularly spaced discharge pulses were the result of an ion-exchange multiplication process. Measurements demonstrate that the gas hinders this mechanism by stripping electrons from negative ions. Electrons were found to be only by-products in the multiplication process. In contrast to the whisker mechanism, the ion-exchange multiplication was found to depend only on total voltage and not on electric-field strength. Experiments with gaps smaller than 2 in. supported the conclusions that the voltage is limited by field emitting whiskers and the presence of gas which damages these whiskers.

Takaoka et al. (1982) [7B-13] presented an article, "Electron energy analysis of vacuum discharge in high-voltage accelerator tube."

Abstract [7B-13]: Measurement of electron energy spectra is useful for diagnosis of vacuum discharge in accelerator tubes. This gives the starting positions of discharge electrons. One block of the accelerator tube of 3 MV electron microscope was tested. Experiments confirm that the main process of discharge is ion-exchange between the top and bottom plates. For three models of electrode shapes, the possibility of secondary-electron multiplication in the 3 MV tube was discussed from the standpoints of both the measured energy spectra and the calculated secondary-electron trajectories. The results were consistent with the actual features of discharge in the 3 MV tube.

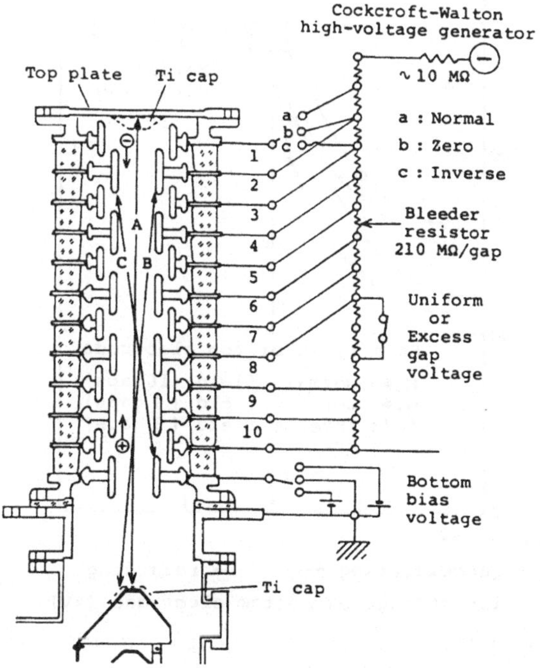

Fig. 7B.2 Application of gap and bias voltages and paths of ion-exchange (Takaoka et al., 1982) [7B-13]

Energy Spectra in the Case of Secondary-Ion Suppression [7B-13]
If the discharge process is ion-exchange, it may be effective, as mentioned by Eastham et al., [1] to suppress the secondary-ion emission due to ion bombardment at the top and bottom plates. Retarding potentials for secondary ions were applied as shown in the right-hand circuit of **Fig. 7B.2**. The potential of the top electrode can be switched to three different values which correspond to the different positions of the bleeder resistors. The bottom bias voltage is applied to the lowest electrode as shown in **Fig. 7B.2**. The switch in the 8th gap is usually closed.

Figure 7B.3 shows the relation between the threshold voltage and the bias voltages. In this figure, the legend "normal", "zero" or "inverse" corresponds to the switch position for the top plate as shown in **Fig. 7B.2**. The threshold voltage does not change for positive bias on the bottom electrode which corresponds to a retarding field for positive secondary ions, while it decreases greatly as the negative bias increases. The field inverse (field retarding secondary ions) to the top plate hardly improves the threshold voltage at all but relaxes the effect of the negative bottom bias as seen in **Fig. 7B.3**.

Figure 7B.4 shows the energy spectra corresponding to three different bias voltages applied to the top plate for the case of a uniform gap voltage. As seen from **Fig. 7B.2**, the odd-numbered electrodes are remote from the tube axis, compared to the even-numbered

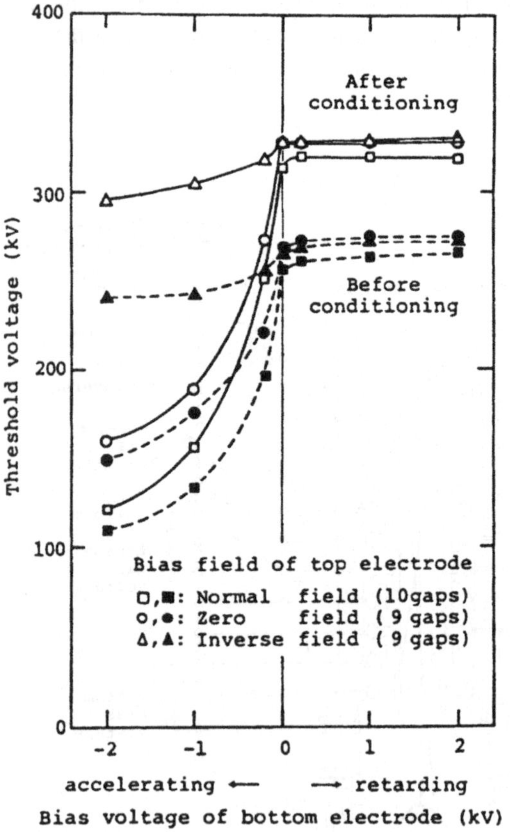

Fig. 7B.3 Threshold voltage as a function of the bias voltage at the top and bottom electrodes (Takaoka et al., 1982) [7B-13]

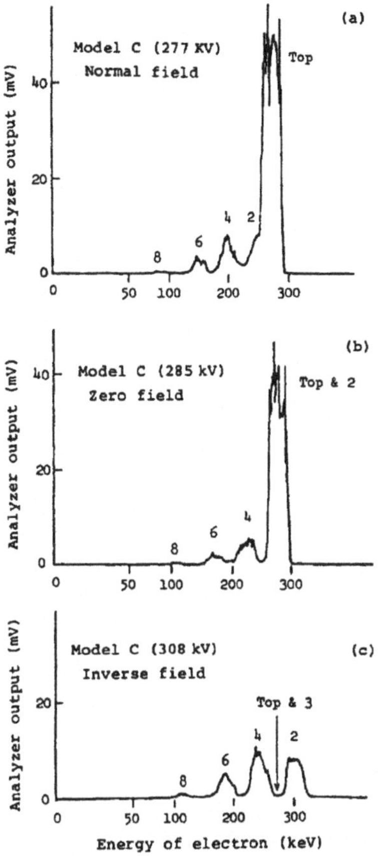

Fig. 7B.4 Electron-energy spectra in the case of uniform gap voltage; the top bias voltage is **(a)** normal field, **(b)** zero field and **(c)** inverse field (Takaoka et al., 1982) [7B-13]

electrodes. Hence, odd-numbered peaks except for the first are not observed in the energy spectra as seen in **Fig. 7B.4**; the first peak corresponds to the electrons from the top plate.

In the cases of the normal and zero fields, most of the electrons come from the top electrode. This implies that the path predominantly used for ion-exchange would be path "A" in **Fig. 7B.2**, that is, the longest path. On the other hand, in the case of the inverse field, the potential of the top plate is the same as that of the 3rd electrode. Therefore, if an electron comes from the top plate, it should be found on the same energy level as the 3rd electrode in the energy spectrum. It is seen in **Fig. 7B.4 (c)** that the energy of the electron from the top plate is drastically reduced. This implies that the secondary negative ion from the top plate would be similarly suppressed by the retarding field. That is, in this case, the most predominantly used path "A" of ion-exchange may disappear due to the retarding field.

Figure 7B.5 shows the energy spectra when a three-fold excess voltage is applied to the 8th gap by the opening of the switch in **Fig. 7B.2**. The excess voltage would act as a trigger for discharge. Here, the bias voltages are applied to both ends, whereas in the case of **Fig. 7B.4**, only the top plate is biased. **Figure 7B.5 (a)** shows the energy spectrum when the bias voltage for both ends is zero. The peak corresponding to the discharge at the 8th gap is clearly observed, although the highest peak is at the highest energy. **Figure 7B.5 (b)**

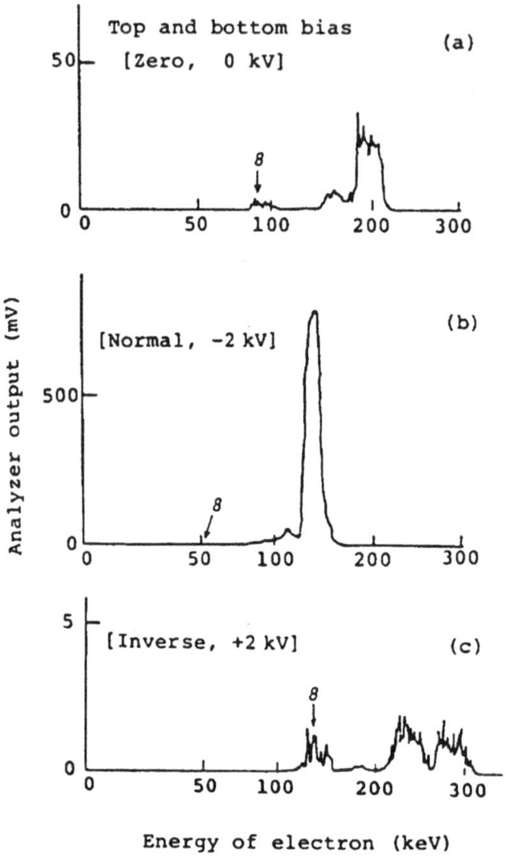

Fig. 7B.5 Electron-energy spectra in the case of excess voltage at the 8th gap; the top and bottom bias voltage are **(a)** zero, **(b)** normal and –2kV and **(c)** inverse and +2kV, respectively (Takaoka et al., 1982) [7B-13]

shows the energy spectrum in the case of the normal top bias and negative bottom bias; in this case the secondary-ion emission is enhanced. The peak corresponding to the 8th gap is scarcely observed, because the threshold voltage decreases to the same extent as the trigger discharge in the 8th gap disappears. It is notable that the peak values at the highest energy are one order higher than that of **Fig. 7B.5 (a)** and that the energy spectrum is more confined to the highest energy than the case in **Fig. 7B.4 (a)**. These features were commonly observed when the applied voltage was lower than the conditioning voltage. When the retarding fields for secondary ions are applied to both end plates, the peak value at the highest energy is depressed, as shown in **Fig. 7B.5 (c)**, by about one order as compared with **Fig. 7B.5 (a)**.

The above results are consistent with the supposition that the main discharge process occurring in the accelerator tube would be ion-exchange. If this is assumed, the reason why the threshold voltage did not increase in line with the retarding field as seen in **Fig. 7B.3** can be explained also as follows. At first, in **Fig. 7B.2** we (Takaoka et al.) consider some paths for ion-exchange, for example, paths "A", "B" and "C". We may assume that the ion-exchange multiplication factors of these paths are unity during the conditioning process. Indeed, if this factor for a certain path is more than unity, the electrode surfaces corresponding to this path would be selectively aged by the ion-exchange process until unity is achieved. On the contrary, if the factor is less than unity, the corresponding surfaces would not be aged and this factor would achieve unity as the conditioning voltage increased. Even if the ion-exchange for the longest path "A" disappeared due to the retarding field, another path "C", for example, would remain because the retarding field does not affect path "C". In this case, the threshold voltage is unaltered. Furthermore, it seems that the drastic reduction at the highest energy level in **Fig. 7B.5 (c)** corresponds to the reduction of surface area contributing to ion-exchange; the surface area contributing to path "C" is much smaller than that to path "A".

Reference

1. D. A. Eastham and R. Thorn, *J. Phys. D: Appl. Phys.*, **11**, 1149 (1978)

Projection (Whisker) on Cathode

Peter (1984) [7B-14] presented an article, "Vacuum breakdown and surface coating of rf cavities."

Abstract [7B-14]: Electrode surface coating may play an important part in overcoming power limitations in rf cavities for voltages far above the electron multipacting limit. In such cases, the principal use of the coating is not to reduce secondary emission but to isolate electrode whiskers from the cavity chamber and to serve as a trap for slow electrons. Restrictions on the layer thickness are derived theoretically, and calculated for conventional accelerator cavities.

Miller (1984) [7B-15] presented an article, "Influence of gap length on the field increase factor β of an electrode projection (whisker)."

Abstract [7B-15]: β, the increase factor of the macroscopic electric field at the tip of a projection, varies with the gap length. The sign and magnitude of this variation depends upon how the gap length is defined. If gap length is defined as x, the distance from the projection tip to the opposing electrode, then β is a strong function of x and may be approximated by $\beta(x) = \beta_\infty x/(x+h)$ [h = projection height] in the region where $x/h > 10/\beta_\infty$. If gap length is defined as d, the inter-electrode distance ignoring the projection, then β is a weak function of d and may be set equal to β_∞ in the region $d/h > 2$.

Gas Molecules on Electrode Surfaces

The breakdown voltage becomes higher after a voltage higher than the rated voltage has been applied for a while to a high-voltage electrode (this is called "high-voltage conditioning"). A similar effect is achieved after the surfaces of high-voltage electrodes have been bombarded with energetic ions and electrons. Baking of electrode and insulator surfaces also has a conditioning effect. So-called "conditioning" is considered a kind of degassing.

Yoshimura et al. (1987) [7B-16] examined the conditioning effect of Ar-glow conditioning (AGC; 350 Vac, 180 mA, 5 min in about 10 Pa of Ar) for microdischarges between high-voltage electrodes.

Ar-Glow Conditioning

Effect of Ar-Glow Conditioning [7B-16]

The Wehnelt electrode and insulator of the electron gun (**Fig. 7A.6**) were first cleaned using CH_3CCl_3 solvent, and microdischarges were examined under high vacuum before and after AGC. Experimental results, together with the experimental process, are presented in **Fig. 7B.6**.

When –100 kV was first applied to the Wehnelt electrode under a high vacuum in the 10^{-4} Pa range, microdischarges of about 10 V (peak to peak) occurred immediately. On the other hand, no microdischarge occurred after the gun had been treated with AGC. However,

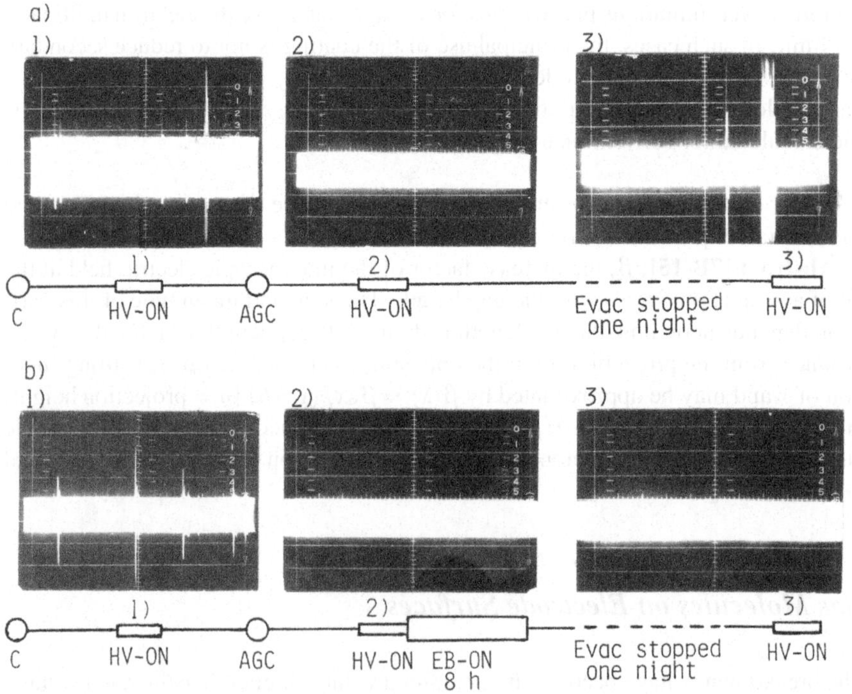

Fig. 7B.6 Microdischarges in the 10^{-4} Pa range before and after AGC. **(a)** Electron beam (EB) off, and **(b)** EB on (about 50μA, about 8 h). Sensitivity and scanning speed of CRT are 1 V/div. and 10 s/frame, respectively (Yoshimura et al., 1987) [7B-16]

even after AGC, microdischarges again occurred under a high vacuum on the next day after the gun had been held in the vacuum chamber isolated from the vacuum pump for one night (see **Fig. 7B.6 a**). Next, in another process, an electron beam (EB) of about 50 μA was extracted from the filament emitter for about 8 h under about 10^{-4} Pa following AGC. In this case, microdischarges did not occur on the next day (see **Fig. 7B.6 b**). The Wehnelt electrode and the insulator (a kind of porcelain whose surface is treated to be glassy) had to be thermally degassed by the heat (about 10 W) from the filament emitter. Thermal degassing for the electrodes and insulators has a conditioning effect to reduce microdischarges between the electrode and over the insulator surface.

High-Voltage Conditioning (HVC)

Effect of High-Voltage Conditioning [7B-16]

High-voltage conditioning (HVC) in high vacuum is a well-known treatment for high-voltage electrodes. Several kV ac superimposed on about 110 kV dc is generally used in HVC for an electron gun of 100 kV. However, high-voltage ion bombardment sometimes damages the electrode and the insulator. Also, the durability of the effect of HVC is rather short.

Example data on the durability of the effect of HVC, applied to an electron gun of an old electron microscope, are presented in **Fig. 7B.7**. The gun is similar in construction to the one shown in **Fig. 7A.4**. These data were obtained after evacuating the electron microscope for a short time.

As seen in **Fig. 7B.7**, the durability is considerably short, so repeated HVC is generally necessary. The short durability is caused by the fact that only a limited small area of the electrodes is bombarded with ions and electrons. Also, as is seen in **Fig. 7B.8**),

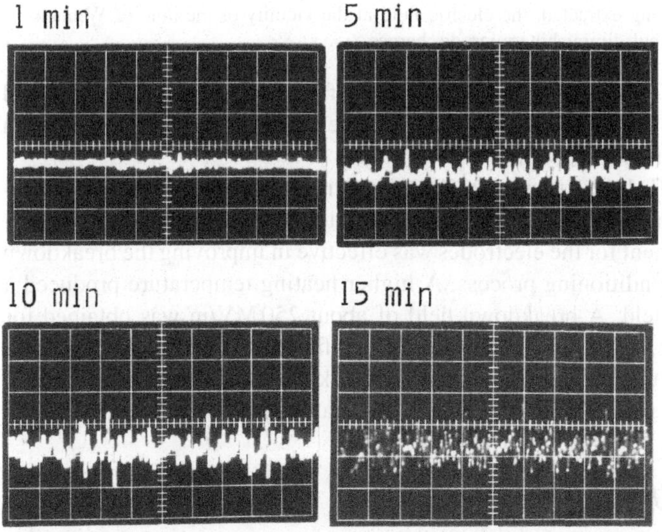

Fig. 7B.7 Durability of the effect of HVC applied to an electron-microscope gun. Sensitivity and scanning speed of CRT are 1 V/div. and 10 s/frame, respectively (Yoshimura et al., 1987) [4B-16]

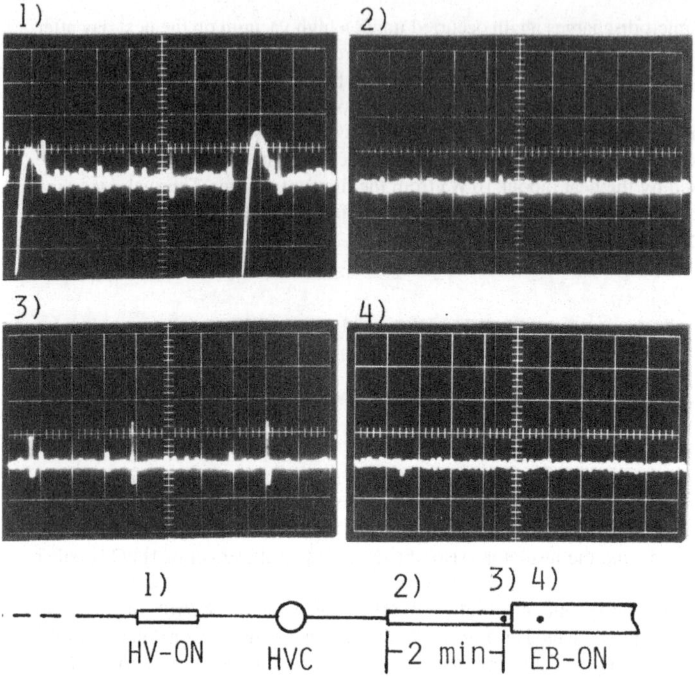

Fig. 7B.8 Microdischarges between the electrodes in 10^{-4} Pa range before and after HVC. Sensitivity and scanning speed of CRT are 1 V/div. and 10 s/frame, respectively (Yoshimura et al., 1987) [7B-16]

> microdischarges diminish during electron beam extraction. It is considered that when an electron beam is being extracted, the electric field at the vicinity of the hole of Wehnelt electrode becomes low, diminishing microdischarges.

Kobayashi et al. (1996) [7B-17] investigated the effects of heating electrodes, and diamond turning of electrode surfaces (oxygen-free copper) to a mirror finish upon electrical breakdown in a vacuum.

Abstract [7B-17]: It was found that the mirror finish obtained by diamond turning reduces the number of repetitive breakdowns to achieve higher hold-off voltages. Heating treatment for the electrodes was effective in improving the breakdown strength after the conditioning process. A higher heating temperature produced a higher breakdown field. A breakdown field of about 250 MV/m was obtained for electrodes heated at 700 °C in a vacuum and then finished by diamond turning. It has been concluded that an improvement in the breakdown field by heating is due to the reduction of the gas contents rather than recrystallization.

Conditioning Effect

There are the interesting views on the effect of conditioning, which were presented by Hawley (1968) [7A-15] and Peter (1984) [7B-14], respectively.

Hawley's View on Conditioning [7A-15]

As in a plain vacuum gap, the phenomenon of conditioning takes place whereby the electrical strength improves during successive breakdowns. Gleichauf [1] investigated the general behavior of the conditioning process, the part played by the electrodes and the part played by the insulators on the conditioning process and he (Gleichauf) also studied the effect on conditioning of the circumstances under which the breakdown arc extinguished. He (Gleichauf) found that with successive breakdowns the breakdown voltage followed a general trend towards higher voltages, the rate of increase diminishing with time. When the voltage was removed for a while, part of the conditioning was lost but the insulator subsequently reconditioned at a faster rate. The degree of conditioning lost was dependent on the previous history of the insulator. As the time interval between the successive series of tests was increased, the initial breakdown voltage on resumption of testing decreased but was usually of a higher value than the very first breakdown voltage for the insulator. The same loss of conditioning was found if the insulator was taken from the continuously pumped vacuum chamber and exposed to the atmosphere, the length of exposure to the atmosphere being immaterial.

Reference

1. P. H. Gleichauf, *J. Appl. Phys.* **22**, 535, 1951.

Peter's View on Conditioning [7B-14]

An important factor in breakdown-voltage levels appears to be the extent of conditioning of the electrode surfaces [1, 2, 3]. Conditioning is achieved by low-current glow discharges, repeated sparking, or gradual increases in the applied voltage. An interesting theory proposed by Halbritter [4] suggests that the reduced electron emission from conditioning is due to the polymerization of adsorbed hydrocarbons on the electrode surfaces. Such layers show strong inelastic scattering for slow electrons, effectively reducing the secondary emission and field emission out of excited states. Another theory is that conditioning is responsible for blunting the micron-length whiskers that grow on metal surfaces when high voltages are applied [3].

There is some evidence [4] that the polymerized hydrocarbon layers show strong inelastic scattering for slow electrons, thereby reducing emission out of this layer for both field-emitted and secondary electrons. According to this theory, voltage thresholds may be raised even higher by rinsing surfaces with hydrocarbon-based materials in a dust-free environment [4]. This decrease in emission by the application of a thin (<10 Å) absorbing medium for slow electrons is different from the use of a thick surface coating (about 1000 Å) to isolate electrode whiskers from vacuum or to decrease secondary emission. In the former case, for example, the actual yield of secondary emission is decreased; in the latter case, increased inelastic scattering suppresses secondary emission out of the material. It is possible, therefore, to apply the hydrocarbon layer over the initial (carbon-coated or dielectric) layer to further reduce the possibility of high-voltage discharge.

References

1. Yu. Nikolaev, *Zh. Tech. Fiz.* **33**, 479 (1963). [*Sov. Phys. Tech. Phys.* **8**, 354 (1963).]
2. D. Boehne, W. Karger, E. Miersch, W. Roske, and B. Stadler, *IEEE Trans. on Nucl. Sci.* **18**, 568 (1971).

3. D. W. Reid and R. A. Lohsen, *Los Alamos Scientific Laboratory report* LA-UR-82-2400, October, 1982.
4. J. Halbritter, *J. Appl. Phys.* **53**, 6475 (1982).

Comment: Peter's view, that the conditioning accompanied by discharge creates the surfaces of electrode with low emission efficiency of secondary electrons, is very interesting. As described in "5C Darkening in Secondary Electron Images" secondary-electron emission from the area pre-irradiated with a fine electron probe is much reduced. Also, surfaces covered with polymerized thin films show lower secondary-electron emission compared with the original surfaces.

Review

Oostrom and Augustus (1982) [7B-18] reviewed 46 articles on electrical breakdown between stainless-steel electrodes in vacuum.

Abstract [7B-18]: Electrical breakdown in vacuum is reviewed, particularly for the case of a small gap (<2 mm) between a pair of stainless-steel electrodes. Theoretical descriptions and experimental results are compared, and it is concluded that field-emitting sites on the cathode are responsible for the electron bombardment of the anode, the subsequent evaporation of anode material and the initiation of a breakdown.

References

Microdischarges over Insulator Surfaces

7A-1. P. H. Gleichauf, "Electrical breakdown over insulators in high vacuum", *J. Appl. Phys.* **22** (6), pp. 766–771 (1951).
7A-2. A. S. Pillai and R. Hackam, "Surface flashover of solid insulators in atmospheric air and in vacuum", *J. Appl. Phys.* **58** (1), pp. 146–153 (1985).
7A-3. C. H. De Tourreil and K. D. Srivastava, "Mechanism of surface charging of high-voltage insulators in vacuum", *IEEE Transactions on Electrical Insulation* **EI-8** (1), pp. 17–21 (1973).
7A-4. T. S. Sudarshan and J. D. Cross, "DC electric-field modifications produced by solid insulators bridging a uniform-field vacuum gap", *IEEE Transactions on Electrical Insulation* **EI-8** (4), pp. 122–128 (1973).
7A-5. R. A. Anderson and J. P. Brainard, "Mechanism of pulsed surface flashover involving electron-stimulated desorption", *J. Appl. Phys.* **51** (3), pp. 1414–1421 (1980).
7A-6. A. S. Pillai and R. Hackam, "Surface flashover of solid dielectric in vacuum", *J. Appl. Phys.* **53** (4), pp. 2983–2987 (1982).
7A-7. J. D. Cross and T. S. Sudarshan, "The effect of cuprous oxide coatings on surface flashover of dielectric spacers in vacuum", *IEEE Transactions on Electrical Insulation.* **EI-9** (4), pp. 146–150 (1974).

7A-8. T. S. Sudarshan and J. D. Cross, "The effect of chromium oxide coatings on surface flashover of alumina spacers in vacuum", *IEEE Transactions on Electrical Insulation* **EI-11** (1), pp. 32–35 (1976).
7A-9. H. Watanabe, N. Yoshimura, S. Katoh, and N. Kobayashi, "Microdischarges on an electron gun under high vacuum", *J. Vac. Sci. Technol. A* **5** (1), pp. 92–97 (1987).
7A-10. Y. Saito, S. Anami, S. Michizono, N. Matuda, A. Kinbara, S. Kobayashi, "Breakdown of alumina rf windows and its inhibition", *T. IEE Japan*, Vol. 114-A, No. 2, pp. 100–107 (1994) (in Japanese).
7A-11. J. P. Shannon, S. F. Philp, and J. G. Trump, "Insulation of high voltage across solid insulators in vacuum", *J. Vac. Sci. Technol.* **2** (5), pp. 234–239 (1965).
7A-12. A. Watson, "Pulsed flashover in vacuum", *J. Appl. Phys.* **38** (5), pp. 2019–2023 (1967).
7A-13. O. Yamamoto, T. Hara, H. Matsuura, Y. Tanabe, and T. Konishi, "Effects of corrugated insulator on electrical insulation in vacuum", *Vacuum* **47** (6–8), pp. 713–717 (1996).
7A-14. K. Nakanishi, Y. Shibuya, Y. Arahata, and T. Eura, "Surface flashovers along insulators under non-uniform electric fields in SF_6 gas", *Transactions of The Institute of Electrical Engineers of Japan A: Publication of Fundamentals and Materials Society* **102** (9), pp. 1–8 (1982) (in Japanese).
7A-15. R. Hawley, "Solid insulators in vacuum: A review", *Vacuum* **18** (7), pp. 383–390 (1968).

Microdischarges Between High-Voltage Electrodes

7A-16. H. C. Miller, "Surface flashover of insulators", *IEEE Transactions on Electrical Insulation* **24** (5), pp. 765–786 (1989).
7B-1. L. Cranberg, "The initiation of electrical breakdown in vacuum", *J. Appl. Phys.* **23** (5), pp. 518–522 (1952).
7B-2. P. A. Chatterton, M. M. Menon, and K. D. Srivastava, "Processes involved in the triggering of vacuum breakdown by low-velocity microparticles", *J. Appl. Phys.* **43** (11), pp. 4536–4542 (1972).
7B-3. D. K. Davies, "The initiation of electrical breakdown in vacuum–A review", *J. Vac. Sci. Technol.* **10** (1), pp. 115–121 (1973).
7B-4. M. M. Menon and K. D. Srivastava, "Microparticle-initiated vacuum breakdown–some possible mechanisms", *J. Appl. Phys.* **45** (9), pp. 3832–3835 (1974).
7B-5. D. K. Davies and M. A. Biondi, "Emission of electrode vapor resonance radiation at the onset of dc breakdown in vacuum", *J. Appl. Phys.* **48** (10), pp. 4229–4233 (1977).
7B-6. N. K. Allen, B. M. Cox, and R. V. Latham, "The source of high-β electron emission sites on broad-area high-voltage alloy electrodes", *J. Phys. D: Appl. Phys.* **12** (6), pp. 969–978 (1979).
7B-7. W. T. Diamond, "New perspectives in vacuum high voltage insulation. I. The transition to field emission", *J. Vac. Sci. Technol. A* **16** (2), pp. 707–719 (1998).
7B-8. W. T. Diamond, "New perspectives in vacuum high voltage insulation. II. Gas desorption", *J. Vac. Sci. Technol. A* **16** (2), pp. 720–735 (1998).
7B-9. L. Cranberg and J. B. Henshall, "Small-aperture diaphragms in ion-accelerator tubes", *J. Appl. Phys.* **30** (5), pp. 708–710 (1959).
7B-10. W. K. Mansfield, "Pre-breakdown conduction in continuously-pumped vacuum systems", *Brit. J. Appl. Phys.* **11** Oct., pp. 454–461 (1960).
7B-11. H. P. S. Powell and P. A. Chatterton, "Prebreakdown conduction between vacuum insulated electrodes", *Vacuum* **20** (10), pp. 419–429 (1970).
7B-12. B. A. Prichard Jr., "Mechanisms of electrical discharges in high vacuum at voltages up to 400 000 V", *J. Appl. Phys.* **44** (10), pp. 4548–4554 (1973).
7B-13. A. Takaoka, K. Ura, and K. Yoshida, "Electron energy analysis of vacuum discharge in high voltage accelerator tube", *J. Electron Microsc.* **31** (3), pp. 217–225 (1982).

7B-14. W. Peter, "Vacuum breakdown and surface coating of rf cavities", *J. Appl. Phys.* **56** (5), pp.1546–1547 (1984).
7B-15. H. C. Miller, "Influence of gap length on the field increase factor β of an electrode projection (whisker)", *J. Appl. Phys.* **55** (1), pp. 158–161 (1984).
7B-16. N. Yoshimura, H. Watanabe, S. Katoh, and N. Kobayashi, "Microdischarges on a high-voltage electron gun under high vacuum", *Shinku (J. Vac. Soc. Japan)* **30** (3), pp. 105–115 (1987) (in Japanese).
7B-17. S. Kobayashi, Y. Hashimoto, M. Maeyama, Y. Saito, and Y. Nagai, "Electrical breakdown strength of oxygen-free copper electrodes under surface and bulk treatment conditions", *Vacuum* **47** (6–8), pp. 745–747 (1996).
7B-18. A. van Oostrom and L. Augustus, "Electrical breakdown between stainless-steel electrodes in vacuum", *Vacuum* **32** (3), pp. 127–135 (1982).

Other Articles

Microdischarges over Insulator Surfaces

*7-1. J. P. Vigouroux, C. Le. Gressus, J. P. Duraud, "Electrical surface breakdown: Secondary electron emission and electron spectroscopy of insulators", *Scanning Electron Microscopy/1985/II*, pp. 513–520.
*7-2. C. Biscardi, H. Hseuh, and M. Mapes, "Application of porcelain enamel as an ultra-high-vacuum-compatible electrical insulator", *J. Vac. Sci. Technol. A* **18** (4), pp. 1751–1754 (2000).
*7-3. H. C. Miller and E. J. Furno, "The effect of Mn/Ti surface treatment on voltage-holdoff performance of alumina insulators in vacuum", *J. Appl. Phys.* **49** (11), pp. 5416–5420 (1978).
*7-4. C. Biscardi, H. Hseuh, and M. Mapes, "Application of porcelain enamel as an ultra-high-vacuum-compatible electrical insulator", *J. Vac. Sci. Technol. A* **18** (4), pp. 1751–1754 (2000).
*7-5. S. Michizono, Y. Saito, S. Anami, and A. Kinbara, "Multipactor phenomenon observed at high-power rf windows", *Shinku (J. Vac. Soc. Japan)* **37** (3), pp. 261–263 (1994) (in Japanese).
*7-6. T. Ishii, S. Kobayashi, Tumiran, M. Maeyama, and Y. Saito, "Studies on the changes of alumina surface state by vacuum surface flashover with cathode luminescence spectroscopy", *Shinku (J. Vac. Soc. Japan)* **38** (3), pp. 299–302 (1995) (in Japanese).
*7-7. Tumiran, S. Kobayashi, H. Imada, M. Maeyama, T. Ishii, and Y. Saito, "The measurement of charge distribution on alumina surface after voltage application in vacuum", *Shinku (J. Vac. Soc. Japan)* **38** (3), pp. 307–310 (1995) (in Japanese).
*7-8. H. Kawai, H. Matsuura, S. Michizono, Y. Saito, and A. Inagaki, "Influence of surface treatment on secondary electron emission of alumina ceramics", *Shinku (J. Vac. Soc. Japan)* **36** (3), pp. 256–259 (1993) (in Japanese).
*7-9. S. Michizono, Y. Saito, S. Anami, and A. Kinbara, "Dielectric property of high-power rf window and its breakdown phenomena", *Shinku (J. Vac. Soc. Japan)* **36** (3), pp. 260–262 (1993) (in Japanese).
*7-10. S. Michizono, Y. Saito, T. Sato, and S. Kobayashi, "Surface charging of rf windows", *Shinku (J. Vac. Soc. Japan)* **41** (3), pp. 231–234 (1998) (in Japanese).
*7-11. T. Sugimoto, S. Michizono, Y. Saito, and S. Kobayashi, "Residual stresses of aluminas and their surface flashover voltages in vacuum", *Shinku (J. Vac. Soc. Japan)* **41** (3), pp. 235–238 (1998) (in Japanese).

Microdischarges between High-Voltage Electrodes

*7-12. A. K. Chakrabarti and P. A. Chatterton, "Microparticle trigger discharges and impact damage in a high-voltage vacuum insulated gap", *J. Appl. Phys.* **47** (12), pp. 5320–5328 (1976).

*7-13. S. Kobayashi, Y. Hashimoto, Y. Saito, Y. Yamamoto, and Y. Nagai, "Vacuum breakdown properties of annealed oxygen free copper electrodes", *Shinku (J. Vac. Soc. Japan)* **37** (3), pp. 258–260 (1994).

*7-14. S. Kobayashi, Y. Hashimoto, Y. Saito, Y. Yamamoto, Y. Nagai, K. Takeuchi, and T. Sugano, "Electrical breakdown strength of vacuum gaps between vacuum degassed oxygen-free copper electrodes machined by diamond turning for mirror finish", *Shinku (J. Vac. Soc. Japan)* **37** (3), pp. 289–291 (1994).

*7-15. Y. Hashimoto, S. Kobayashi, Y. Saito, K. Takeuchi, T. Sugano, and Y. Nagai, "Vacuum breakdown properties of vacuum gaps consist of oxygen-free copper electrodes machined by diamond turning for mirror finish", *Shinku (J. Vac. Soc. Japan)* **38** (3), pp. 303–306 (1995).

*7-16. A. Iwai, S. Kobayashi, and Y. Saito, "Charges in copper electrode surface conditions caused by breakdown in ultra-high vacuum", *Shinku (J. Vac. Soc. Japan)* **41** (3), pp. 227–230 (1998).

Chapter 8
Emitters for Fine Electron Probes

Modern electron microscopes are usually provided with a field emission (FE) electron gun (FEG) using a W <310> cold FE emitter or ZrO/W thermal FE emitter. Also, an LaB_6 thermionic emitter is widely used for high-brightness electron guns for electron beam instruments such as electron microscopes (EM) and Auger electron spectrometers (AES). The LaB_6 emitter can work well in a high vacuum of the 10^{-5} Pa range.

The ZrO/W emitter is classified as a Schottky emission type, although the emitter is generally recognized as a thermal FE emitter. The ZrO/W emitter can work in the 10^{-7} Pa range, and the W <310> emitter works in the 10^{-8} Pa range. The W <310> emitter produces the smallest spot size among these emitters. The ZrO/W Schottky emission (SE) emitter can emit a higher current than the W <310> emitter.

Keywords

Remolding: The W <310> plane builds up when a negative, intense electric field is applied to a W <310> tip during heating.
Oxygen build-up treatment: The W <100> plane builds up when a positive, intense electric field is applied to a W <100> tip in ultrahigh vacuum containing oxygen gas of the 10^{-7} Pa range during heating.
Flashing: The filament with an emitter tip is flashed by running a current for about 1 second in order to heat the tip to about 1900 K.

Kim et al. (1994) [8B-7] compared a field emission (FE) cathode and Schottky emission (SE) cathode as follows.

Comparisons [8B-7]
Schottky emission (SE) tips have been widely used as an electron source for microscopy and electron beam lithography with high brightness, good emission stability, and long lifetime [1]. As shown in **Fig. 8.1**, the field emission (FE) of the cold emission tip occurs by electron tunneling through the potential barrier near the Fermi level (E_f. In the Schottky

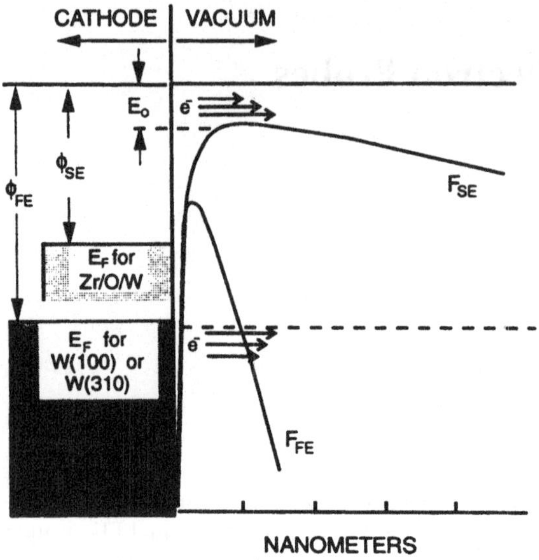

Fig. 8.1 Diagram of energy levels for the cold field emission (F_{fe}) and Schottky emission (F_{se}). E_0 is the field reduction of the work function for Schottky emission (Kim et al., 1994) [8B-7]

emission (SE) case, the work function is lower and electron transmission occurs primarily over the top of the potential barrier with a smaller electric field F_{se}. The current densities for the cold FE (J_{fe}) and SE (J_{se}) are given by

$$J_{fe} = a\left(F^2/\phi\right)\exp(10.4/\phi^{1/2})\exp(-b\phi^{3/2}/F)\ (\mathrm{A/cm^2}). \tag{8.1}$$

$$J_{se} = 120T^2\exp[-(\phi - 3.8F^{1/2})/kT]\ (\mathrm{A/cm^2}). \tag{8.2}$$

where $a = 1.54 \times 10^{-6}$, $b = 0.644$, $k = 8.617 \times 10^{-5}$eV/K, T is the tip temperature in K, F is the electric-field strength in V/Å, and ϕ the emitter work function in eV. From these equations, the SE (Schottky emission) is highly sensitive to the tip temperature, while both equations depend strongly on the electric field and the emitter work function.

Table 8.1 shows the comparison between cold FE and SE sources. The cold FE cathodes typically used are single crystal <111> and <310> tungsten. Most cold FE cathodes have a radius of about 100 nm after annealing. The SE cathode most commonly used is a ZrO coated <100> oriented tungsten emitter. The ZrO reduces the work function of W <100> from 4.5 to 2.7 eV. A disadvantage of an SE emitter is that the tip operates at 1800 K and requires a few watts of heating power. It also has a larger energy spread compared to a cold field emitter at a given angular emission-current density [2, 3]. However, SE emitters have a much more stable emission current than the cold FE source due to the lower electric fields and the larger and more stable emitting areas. Typically, it has less than 1% of noise fluctuation over several hours. The SE emitter can be operated under poorer vacuum conditions than the cold FE source. The operating pressure range for the SE emitter is 10^{-8}–10^{-9} Torr.

Table 8.1 Comparison of cold FE source and SE source (Kim et al., 1994) [8B-7]

Property	Cold FE source	SE source
Cathode	W single crystal <111>, <310>	Zr/O/W
Radius	≦100 nm	~ 1 μm
Temperature	300 K	1800 K
Work function	4.4 eV	2.7 eV
Virtual source size	About 5 nm	About 15 nm
Emission current	About 10 μA	About 100 μA
Angular emission	$< 100 \mu A/sr$	$> 100 \mu A/sr$
Energy spread	0.3-0.8 eV	0.6-1.2 eV
Heating power	none	About 2 W
Stability (noise)	A few %	$< 1\%$
Vacuum pressure	$10^{-10} - 0^{-11}$ Torr	$10^{-8} - 10^{-9}$ Torr

References

1. L. W. Swanson and N. A. Martin, *J. Appl. Phys.* **46**, 2029 (1975).
2. M. G. R. Thomson, *J. Vac. Sci. Technol. B* **12**, 3498 (1994).
3. A. E. Bell and L. W. Swanson, *Phys. Rev. B* **19**, 3353 (1979).

8A W FE Emitter

Crewe and coworkers [8A-1, 8A-2, 8A-3] must be pioneers in studying and developing an FE electron gun (FEG) for the electron microscope. Crewe et al. (1968) [8A-2] built a simple scanning transmission electron microscope using a field emission electron source, a new electron gun, and one lens to produce a high-contrast picture with 30 Å resolution. Crewe et al. (1969) [8A-3] built a simple SEM which uses an FEG alone, without the aid of auxiliary lenses.

Characteristics

Crewe et al. (1968) [8A-1] presented an article, "Electron gun using a field emission source." They studied almost all of W-FE emitter production techniques systematically.

Characteristics of the FE Tip [8A-1]
Dyke and Trolan [1] have used an average technique to define the current density in terms of the emission current (I). Using this approach, they have measured continuous current densities up to $10^6\,\mathrm{A/cm^2}$ compared to approximately $10\,\mathrm{A/cm^2}$ for a hot filament tip [2]. This feature alone is interesting, but there is also the fact that the apparent source size is much smaller than the actual tip size. If we consider a tip with a hemispherical end and also assume that electrons will be emitted within a finite voltage range 0 to V_T, then one can show that the apparent source radius (r) is approximately [3]

$$r = R\sqrt{V_T/V_1}, \tag{8A.1}$$

where R is the radius of the tip and V_1 is the potential applied to the tip. This approximation is good for a spherical source only. Reasonable values to insert in this equation are $R = 500\,\text{Å}$, $V_T = 0.5\,\mathrm{V}$, and $V_1 = 1\,\mathrm{kV}$, which lead to $r = 11\,\text{Å}$.

The Fowler-Nordheim theory [4, 5, 6] for field emission gives the following equation for the current density:

$$J = aE^2 \exp\left(-B\Phi^{1/2}/E\right), \tag{8A.2}$$

where J is the current density in $\mathrm{A/cm^2}$, E is the electric field applied at the tip in V/cm, Φ is the work function in eV, B is a constant equal to 6.12×10^8 in the above units, and a is a function of $\Phi (a = 3.5 \times 10^{-5}\,\mathrm{A/V^2}$ for tungsten with $\Phi \approx 4.5\,\mathrm{eV})$ [7]. Gomer [8] describes how the tip radius or the metal work function can be found experimentally using the slope of the Fowler-Nordheim equation. Charbonnier and Martin [9] present a unique method for obtaining information on tip geometry from the electrical data $I(V)$, the relationship between field-emitted current and applied voltage.

An FE source has at least two advantages when it is considered for use in a scanning electron microscope (SEM), namely, increased current density and very small spot size.

Making the Tip [8A-1]
We (Crewe et al.) fabricated our tips using the techniques outlined by Dyke et al. [10]. A piece of 0.125 mm diameter tungsten wire 1–3 mm long is spot-welded onto a preformed 0.2 mm diameter tungsten filament which is hairpin-shaped. The assembly is electropolished, and the tip is etched by immersing it in a one-normal sodium hydroxide (NaOH) solution and by applying 12 V dc between the tip and a remote electrode in the solution

Fig. 8A.1 Tip etching arrangement (Crewe et al., 1968) [8A-1]

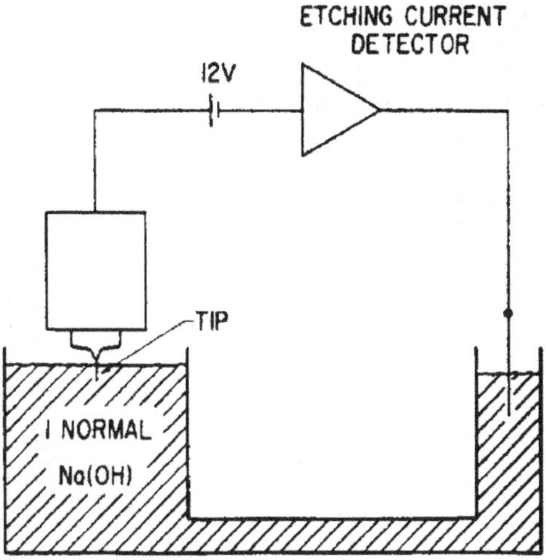

(see **Fig. 8A.1**). The second electrode is removed some distance so that gas given off at this electrode will not disturb the liquid surrounding the tip. It is important that the NaOH solution be clean and that the etching assembly be mounted on a vibration-free platform. A tip can be etched in about 3 min and a significant change in etching current occurs just as the unwanted part of the tip is etched off. An electronic sensing circuit is used to detect this change and to turn off the applied voltage within about 1 μsec. If the voltage is left on much longer than this, blunting of the tip will occur.

Tip operation [8A-1]

The etched tip is mounted in an enclosure which is evacuated to about 10^{-9} Torr. The tip is "formed" (rounded off) by sending a brief pulse of current through the filament. As the magnitude of the current pulse is increased, the filament begins to glow red. By this time the heat has usually driven off the contaminants so that the pressure no longer rises during the flash. The tip is then tested to see whether cold field emission is occurring. After emission is detected, a check is made to determine if the tip is properly formed by comparing an experimental voltage-current curve with that predicted by the Fowler-Nordheim equation. A typical Fowler-Nordheim plot for a tungsten tip having the plane with Miller indices <310> perpendicular to the axis is shown in the next section. This crystal orientation was selected because it produces intense emission along the axis [11].

The subsequent performance of the tip appears to be more dependent on the local gas pressure than on any other parameter. In general, the emission current at constant voltage appears to be a curve similar to that of **Fig. 8A.2** [12]. At first, there is a small decline in emission current as the surface of the tip becomes coated with contaminants which increase the work function. Thereafter the current rises until it becomes erratic, and the tip eventually destroys itself by a vacuum arc [13, 14, 15]. The time scale for this process can vary from seconds to thousands of hours depending on the pressure [12].

The stability of the emission current does not appear to have a severe dependence on the magnitude of this current. The principal effect is the rise in local pressure caused by the production of gas molecules as the electron beam strikes the surface of the first anode. This effect can be controlled by thoroughly preheating the anode. We have found that once the anode has been outgassed, cycling up to atmospheric pressure has almost no adverse effects as long as the system is not left at this pressure for many hours.

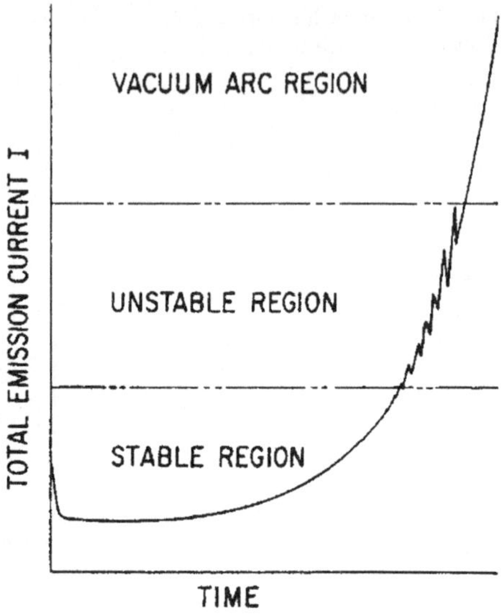

Fig. 8A.2 Typical dependence of emission current with time. As the tip becomes coated with contaminants, the emission first of all drops, and then begins a steady rise until the emission becomes erratic and the tip will eventually destroy itself with a vacuum arc (Crewe et al., 1968) [8A-1]

When the emission current becomes erratic, the performance of the tip can easily be restored to its original condition by providing a pulse of current through the filament to evaporate the contaminants. This "flashing" technique is used periodically and the frequency of flashing depends on the time scale in **Fig. 8A.2**. We have operated a tip at a continuous emission current of 1 mA $\pm$ 1.5% for over 70 h without flashing when the system was kept below a pressure of 10^{-9} Torr using a 400 L/s ion pump. This contamination problem is not severely limiting however, because the tip can be flashed with operating voltages left on, and if the usable electron beam (beam contained within the limiting aperture) is confined to relatively small angles (less than a few degrees) about the axis, the beam current within this solid angle is reasonably independent of the stability of the total emission current shown in **Fig. 8A.2**.

Remolding

Remolding the Tip [8A-1]

Field emission can be obtained by applying potentials of 2–5 kV to a tip with a radius of 1000 Å. In our application it was desirable to obtain an emission current of approximately 0.1 μA at a potential of 1–2 kV. If the tips are formed in the normal manner, the necessary potential for the required emission current frequently reaches 3–5 kV after they have been formed by heating. In order to obtain the same emission current but at a lower applied voltage (V_1), we flash the tip with a pulse of filament current equal in magnitude to that used to form it and at the same time apply a positive dc voltage (V_R) to the tip with respect to the anode [16]. Using this reverse bias technique, we have consistently reduced the potential necessary to achieve a given emission current. **Figure 8A.3** illustrates this phenomenon as a function of the reverse voltage (V_R) for several different tips. The curves are reproducible to within 5%. We have also observed a threshold temperature (flashing current) below which the remolding process does not occur even though the reverse voltage is applied. If the remolded tip is flashed at the forming current with no reverse bias, the operating conditions immediately return to those at the starting point ($V_R = 0$ for any given tip in **Fig. 8A.3**).

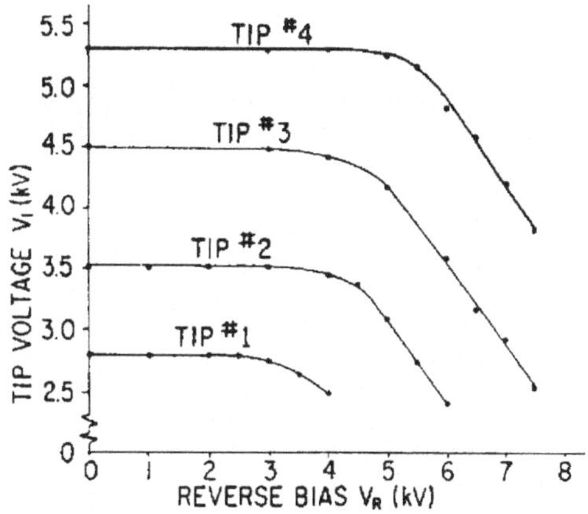

Fig. 8A.3 Graphs illustrating the remolding process as a function of the applied reverse bias. The four curves are for different tungsten tips. The tip voltage shown in the ordinate is the voltage required to achieve emission of 1 μA after applying the indicated reverse bias (Crewe et al., 1968) [8A-1]

Typical Fowler-Nordheim plots for the normal and remolded tip are shown in **Fig. 8A.4** with the tip radii found using the slope method [8].

The remolding process, then, is a method of adjusting the required voltage (V_1) to achieve a given emission current (I). This variability of tip potential is important since V_1 also affects the optical parameters of the electron gun. We have not yet examined the effect of tip remolding on the apparent source size for this electron gun, but no change in microscope resolution is observed when the tip is remolded.

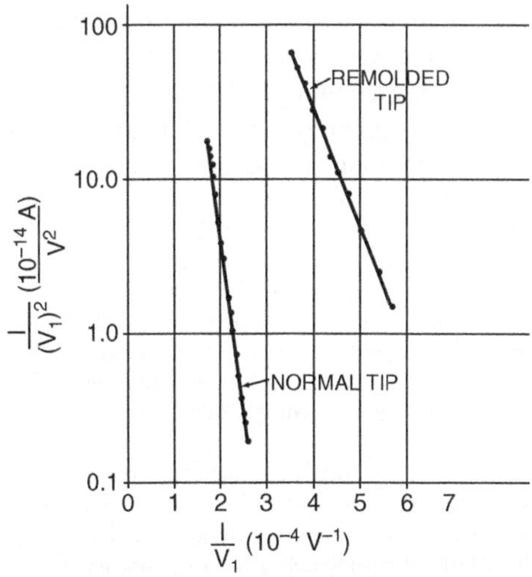

	NORMAL TIP	REMOLDED TIP
TIP RADIUS	1900 ± 400 Å	600 ± 120 Å
VOLTAGE FOR I = 0.1 μA	4300 V	1880 V

Fig. 8A.4 Typical Fowler- Nordheim plots for a normal tip and a remolded tip. The tip radii are calculated from the slopes of the curves (Crewe et al., 1968) [8A-1]

References

1. W. P. Dyke and J. K. Trolan, *Phys. Rev.* **89**, 799 (1953)
2. V. E. Cosslett and M. E. Haine, *Intern. Conf. Electron Microscopy* (London), p. 639 (July 1954).
3. M. Drechsler, V. E. Cosslett, and W. C. Nixon, *4th Intern. Conf. Electron Microscopy* (Berlin), p. 13 (Sept.1958).
4. R. H. Fowler and L. W. Nordheim, *Proc. Roy. Soc.* (London), **A119**, 173 (1928).
5. L. W. Nordheim, *Proc. Roy. Soc.* (London), **A121**, 626 (1928).
6. A. G. J. van Oostrom, *Thesis*, University of Amsterdam (June 1965).
7. W. P. Dyke and F. M. Charbonnier, *Advan. Electron Tube Tech.* **2**, 199 (1962).
8. R. Gomer, *Field Emission and Field Ionization*, (Harvard University Press, Cambridge, Massachusetts, 1961), Chap. 2, p. 47.
9. F. M. Charbonnier and E. E. Martin, *J. Appl. Phys.* **33**, 1897 (1962).
10. W. P. Dyke, J. K. Trolan, W. W. Dolan, and G. Barnes, *J. Appl. Phys.* **24**, 570 (1953).
11. <310> oriented tungsten wire may be obtained from Field Emission Corp. McMinnville, Ore.
12. E. E. Martin, J. K. Trolan, and W. W. Dolan, *J. Appl. Phys.* **31**, 782 (1960).
13. W. P. Dyke, J. K. Trolan, E. E. Martin, and J. P. Barbour, *Phys. Rev.* **91**, 1043 (1953).
14. W. W. Dolan, W. P. Dyke, and J. K. Trolan, *Phys. Rev.* **91**, 1054 (1953).
15. W. P. Dyke, and W. W. Dolan, *Advan. Electron Electron Phys.* **8**, 89 (1956).
16. I. L. Sokolovskaia, *Soviet Phys.-Tech. Phys.* (English Transl.) **26**, 1147 (1956).

Tamaki et al. (1976) [8A-4] remolded the <310>-oriented tungsten FE emitter in situ in the field-ion microscope (FIM).

Abstract [8A-4]: The remolding was most successful when the electric field of 100 MV/cm was applied to the emitter kept at about 1620 K. The micro-protrusion was formed on the <100> plane if the remolding was carried out in high vacuum (10^{-8} Torr) and on the <310> plane in ultrahigh vacuum (10^{-10}Torr). The dimensions of the protrusion were 10 to 20 Å both in height and diameter. The micro-miniature electron beam with the brightness higher than $10^{11}\,A \cdot cm^{-2} \cdot rad^{-2}$ could be extracted from the remolded emitter.

Tip Preparation [8A-4]

The tungsten wire several millimeters long is needed to obtain a tip by the conventional method of etching [1], so that it is not desirable for the expensive <310>-oriented tungsten. Another disadvantage is that the shape of the tip apex is affected, to a great extent, by the critical time at which the voltage is cut off in the final stage of etching. Therefore we made a little modification to the conventional method.

As shown in **Fig. 8A.5 (a)** and **(b)**, the thin layer of 5% NaOH aq. solution was formed within the ring of 0.3 mm diam. platinum wire. With AC 5 to 8 volts the <310>-tungsten wire was etched electrochemically in the layer, on both ends of which were spot-welded two hair pin loops of 0.15 mm diam. tungsten. The <310>- tungsten only 1 to 2 millimeters long was sufficient to obtain two tips. In the final stage the voltage was lowered to 2 to 3 volts gradually and was cut off at the moment when the loop B in **Fig. 8A.5 (a)** dropped into the holder. The etching solution should be changed as often as possible.

Figure 8A.5 (c) shows the optical micrograph of the tips thus obtained, in which the magnification is 350. The tip of ideal shape was made at a high rate of success by the method of etching stated here.

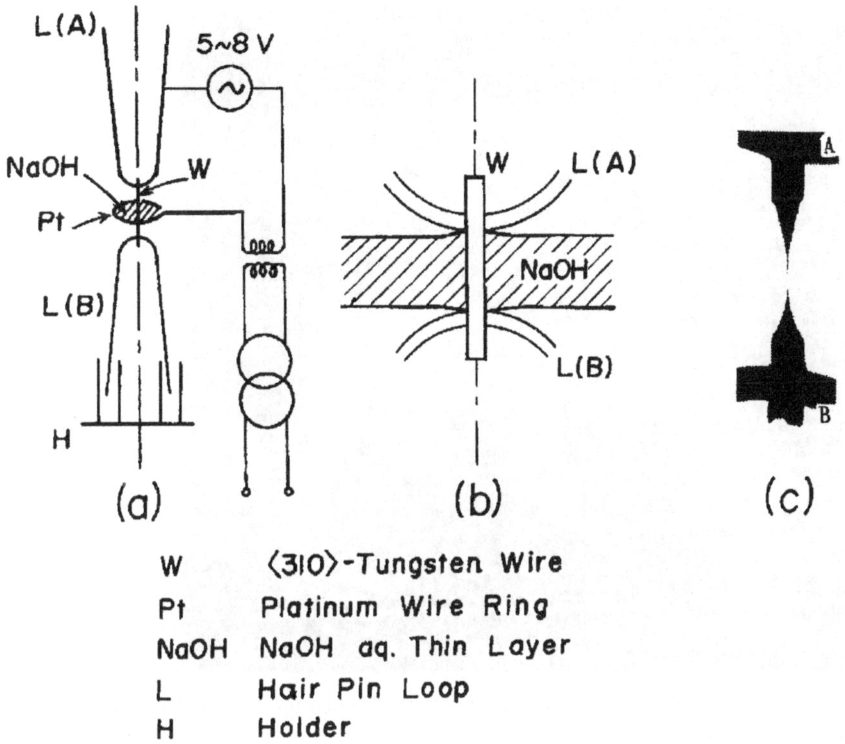

Fig. 8A.5 Method of tip preparation. <310>-tungsten wire was etched in thin layer of 5% NaOH aq. solution ((**a**) and (**b**)). (**c**) is tips thus obtained (Tamaki et al., 1976) [8A-4]

Experimental Results[8A-4]

Figure 8A.6 is the typical helium ion micrograph of the <310>-oriented tungsten tip field-evaporated at 78 K. All the following experiments of remolding were carried out using the tip of this figure.

Remolding in Ultrahigh Vacuum (10^{-10} Torr) [8A-4]

Figure 8A.7 shows the changes in the field emission micrograph and the emission current with the remolding time for the <310>-oriented tungsten tip. The tip was cleaned by flashing beforehand and the remolding was carried out in ultrahigh vacuum at about 10^{-10} Torr, keeping the tip at about 1620 K and applying an electric field of 100 MV/cm. The vertical axis is the emission current at the same tip voltage 3.0 kV, and the value of V_p given with each micrograph is the tip voltage at which the emission current reaches 0.15 μA. After a certain period from the beginning of the remolding (in this case after 50 seconds) protrusions near the {310} family in the micrograph (**d**) gather together to make one spot in several seconds. The result is shown in (**e**). In view of the lens aberration, the focused paraxial electron beam is most desirable for the electron microscope, so that the state (**e**) is the best one for the electron source. In addition it takes only 10 seconds from (**d**) to (**f**) and the best state (**e**) is realized in a short period. In other words, the best tip shape is kept only for a short period at the above stated condition.

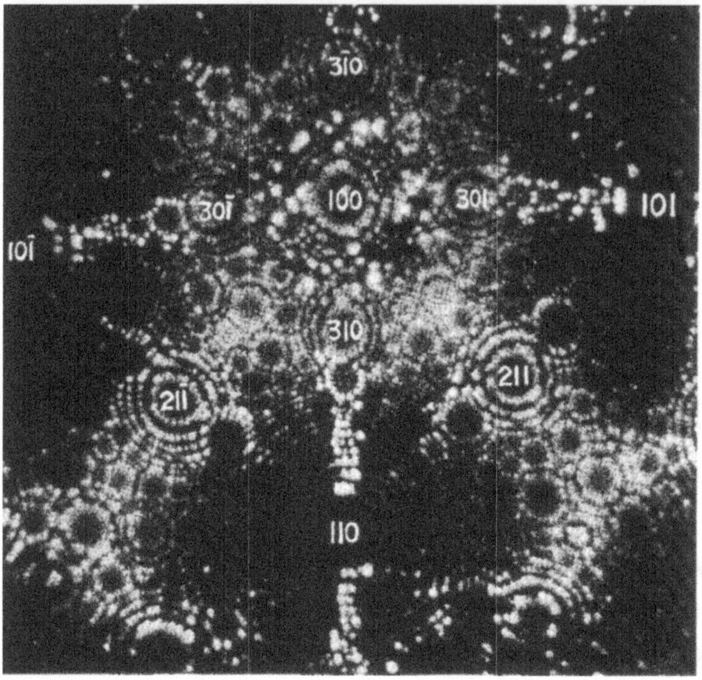

Fig. 8A.6 Typical helium ion micrograph of <310>-tungsten tip. Mean radius of curvature is about 520 Å (Tamaki et al., 1976) [8A-4]

In order to observe atomically the protrusion that gives the <310> spot of **Fig. 8A.7 (e)**, we utilized the FIM. The emission and the ion micrographs of the <310>-tungsten tip remolded in situ in FIM are shown in **Fig. 8A.8 (b)** and **(c)** respectively, in which the protrusion appears on the position of the <310> plane. **Figure 8A.8 (d)** is the enlarged view of **(c)**. It is interesting to note several atoms on the surface of the protrusion (**Fig. 8A.8 (d)**). By successive field-evaporation at 78 K the ion image changed to **(e)** and **(f)**. Removing 5 atomic layers from **(f)** by field-evaporation **(g)** was obtained, in which the low index planes <001>, <011> and <112> form flat faces. The field emission micrograph of the tip in **(g)** resembles that of the tip of field-evaporation endform at 78 K (see **(h)**).

From these ion micrographs, it can be seen clearly that the greater part of the <310> protrusion is removed by field-evaporation of about ten atomic layers. Since the interplaner distance of the <310> plane is nearly 1 Å, the <310> protrusion proved to be about 10 Å in height.

In addition, when the remolded tip in **Fig. 8A.8 (b)** to **(d)** was kept below the room temperature in ultrahigh vacuum (10^{-10} Torr) for several days, any change did not appear in the emission micrograph and the emission characteristics. This was also the case when the electron beam was extracted from such tip continuously for several hours. These facts would suggest that the remolded tip can be used as a stable electron source if it is kept below the room temperature in ultrahigh vacuum.

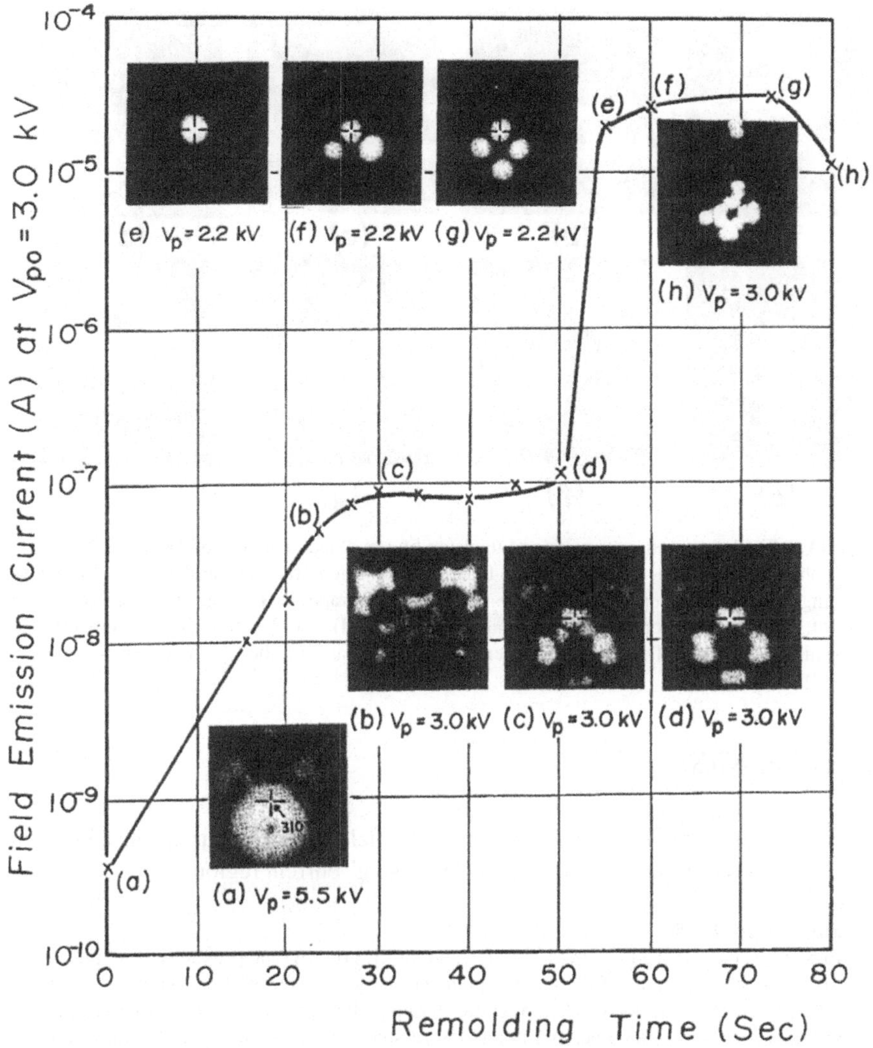

Fig. 8A.7 Field emission current and field emission micrographs of <310>-tungsten tip during remolding at about 1620 K and 100 MV/cm in ultrahigh vacuum (10^{-10} Torr) (Tamaki et al., 1976) [8A-4]

Reference

1. E. W. Müller and T. T. Tsong: *Field Ion Microscopy* (Elsevier, New York, 1969).

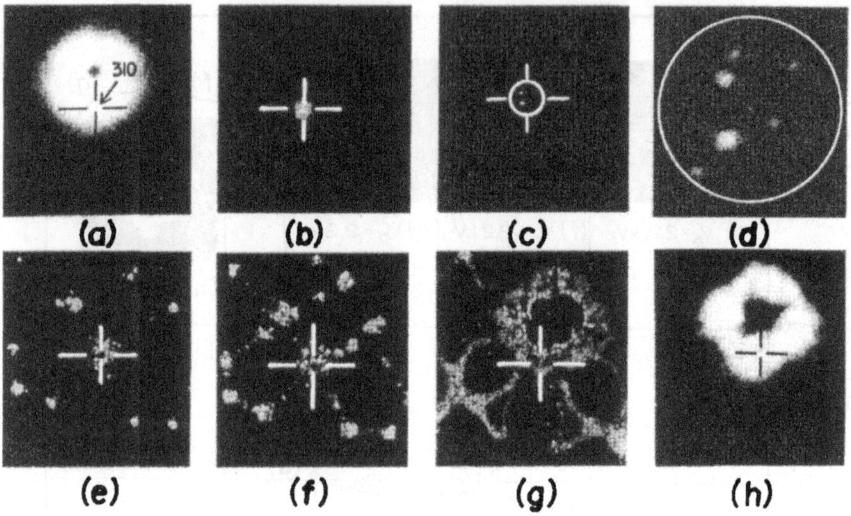

Fig. 8A.8 Field-evaporation sequence of tungsten tip remolded at about 1620 K and 100 MV/cm in ultrahigh vacuum (10^{-10} Torr). **(a)**; the field emission micrograph of the flashed tip before remolding and **(b)**; after remolding. **(c)**; the field-ion micrograph of **(b)**. **(d)**; the enlarged view of **(c)**. **(e)**; after field-evaporation of 2 atomic layers from **(d)**. **(f)**; 3 atomic layers from **(e)**. **(g)**; 5 atomic layers from **(f)**. **(h)**; the field emission micrograph of the tip in **(g)** (Tamaki et al., 1976) [8A-4]

Emission Noise

Yamamoto et al. (1976) [8A-5] measured the field emission current instability of various planes of tungsten in the "ever-decreasing" current region.

Experiment [8A-5]
Figure 8A.9 shows the local field emission current LFE (in nA) and peak to peak amplitude ΔI_{P-P} (in arbitrary unit) of the <310> plane taken by chart recorder and photo-recorder, respectively. All data were taken after the flashing. Except for right after the flashing, the LFE decreases with time, and this is the reason why this region is called the "ever-decreasing" current region. The ΔI_{P-P}, however, shows a maximum. The <100> plane shown in **Fig. 8A.10** also shows similar behavior, except that the maximum in ΔI_{P-P} appears much earlier.

Comment: As seen in these figures, the emission noise of W <310> and W <100> planes becomes small soon after their flashing treatments.
Swann and Smith (1973) [8A-6] measured the short-term fluctuation, or noise, in the emission current using a chart recorder.

Stability of Emission Current [8A-6]
The noise was measured as the fraction, $(I_{max} - I_{min})/I_{mean}$ where $I_{\max}$ was obtained by graphically averaging the peak levels along a typical length of the recorded trace, and $I_{\min}$ was obtained by similarly averaging the troughs.

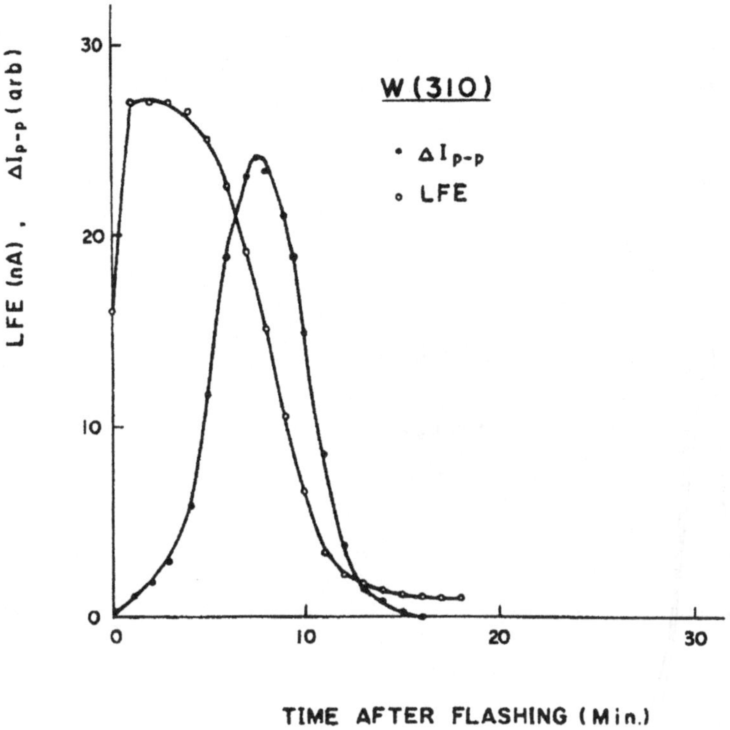

Fig. 8A.9 Change in time after flashing of local field emission (LFE) current and peak-to-peak amplitude of instability obtained from W <310> (Yamamoto et al., 1976) [8A-5]

Using the photomultiplier technique, the results shown in **Fig. 8A.11** were obtained. **Figure 8A.11 (a)** and **(b),** which show the fluctuations in the emission from <111> and <310> crystal faces respectively, were obtained by selecting the appropriate area by means of a mask containing a 6 mm diameter aperture. This defined a geometrical semi-angular aperture at the emitter of about 3×10^{-2} radian although the corresponding semi-angle of emission was somewhat larger because of curvature of the electron trajectories near the emitter. **Figure 8A.11 (c)**, showing the noise in the total emission current, was obtained by observing the whole emission pattern. Considerable variation was experienced in the experimental values measured under apparently similar conditions. The curves may therefore be regarded as indicating general trends only. It appears from the results that the <310> orientation generates lower noise levels than the <111> orientation at pressures in the region of 10^{-9} Torr. Generally as expected the noise increases with operating pressure and emission current, although at relatively high pressures the curves start to fall. This effect is almost certainly anomalous and may be connected with the nonlinear response of the recorder to the high frequencies associated with higher pressures.

Todokoro et al. (1982) [8A-7] clarified the role of ion bombardment in field emission current instability experimentally by varying the emission current level and pressure over a wide range.

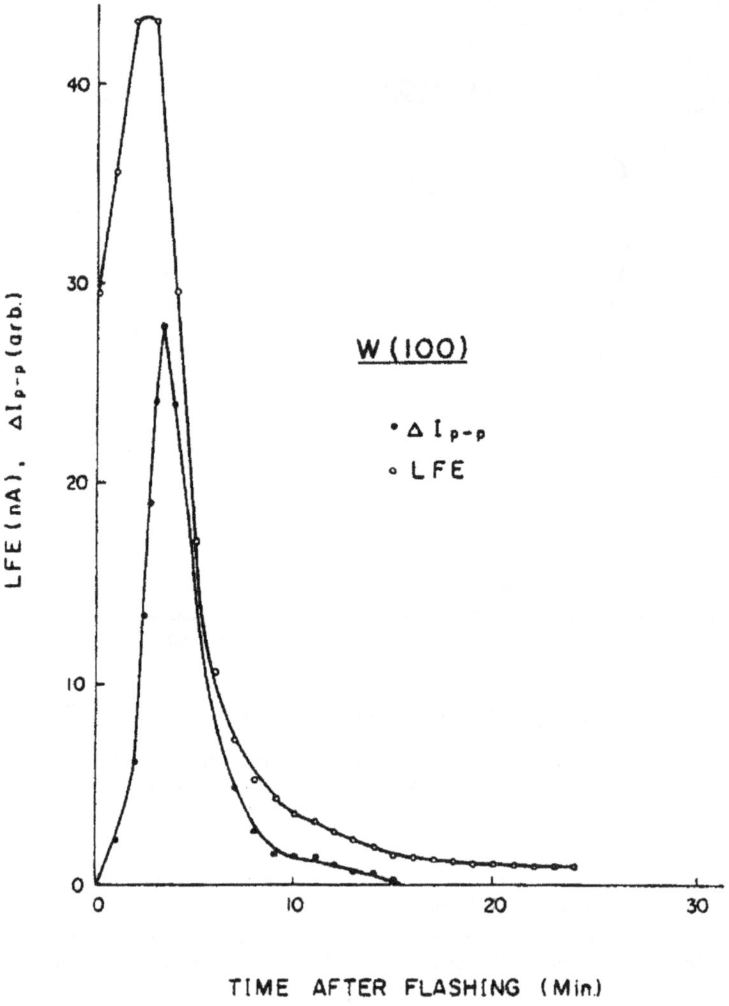

Fig. 8A.10 Change in time after flashing of local field emission (LFE) current and peak-to-peak amplitude of instability obtained from W <100> (Yamamoto et al., 1976) [8A-5]

Experiment [8A-7]

Measurements were carried out using two types of guns in a bakable stainless steel vacuum chamber evacuated by a 160 L/s ion pump (**Fig. 8A.12**). The lowest pressure in the chamber was about 6×10^{-8} Pa. To change the pressure, air could be introduced by opening a slow leak valve. The initial current decrease and relative fluctuation were measured using an A-type gun. A B-type gun was used to confirm the origin of the ions. The B-type gun had two emitters arranged to face the same anode at the same distance of 1 cm. The distance between the two emitters was about 5 mm.

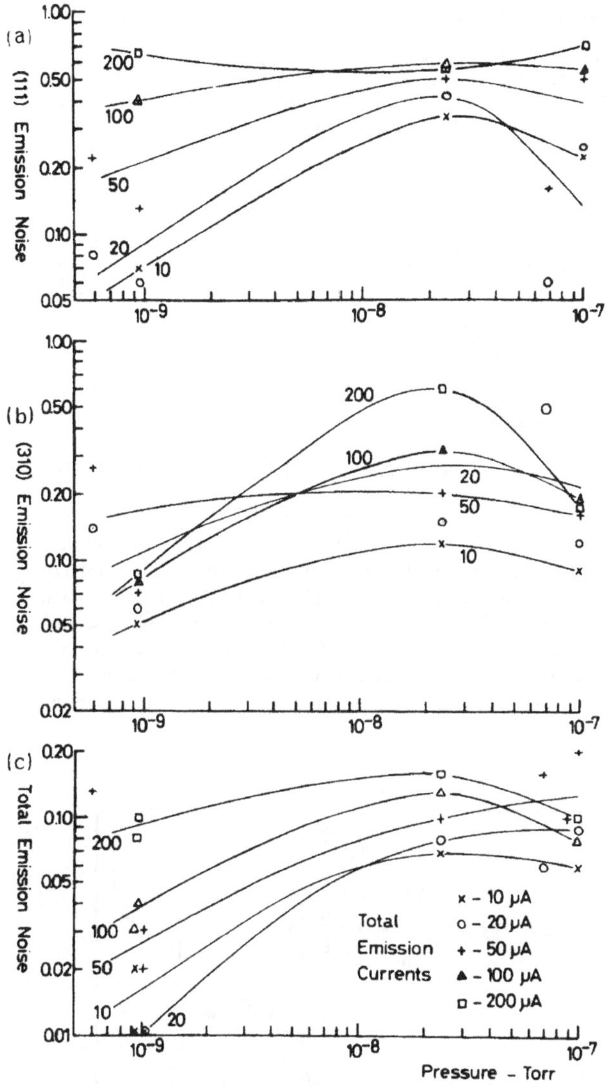

Fig. 8A.11 Variations of emission noise, $(I_{\max} - I_{\min})/I_{mean}$. **(a)** Emission from <111> face. **(b)** Emission from <310> face. **(c)** Total emission (Swann and Smith, 1973) [8A-6]

Relative Fluctuation [8A-7]

Relative fluctuation as a function of $P \times I$ was measured. The pressure was controlled by the air leak. The relative fluctuation was proportional to the logarithm of $P \times I$ above $7 \times 10^{-12}\,\text{Pa} \cdot \text{A}$, as shown in **Fig. 8A.13**. The result shows that a relative fluctuation increase is caused by bombardment of ions generated by electron collisions with residual gas molecules, because the relative fluctuation is a function of the product of $P \times I$, not of I. Below $7 \times 10^{-12}\,\text{Pa} \cdot \text{A}$, relative fluctuation is constant at about 1%. This small fluctuation could be caused by adsorbed gas molecules migrating on the emitter surface.

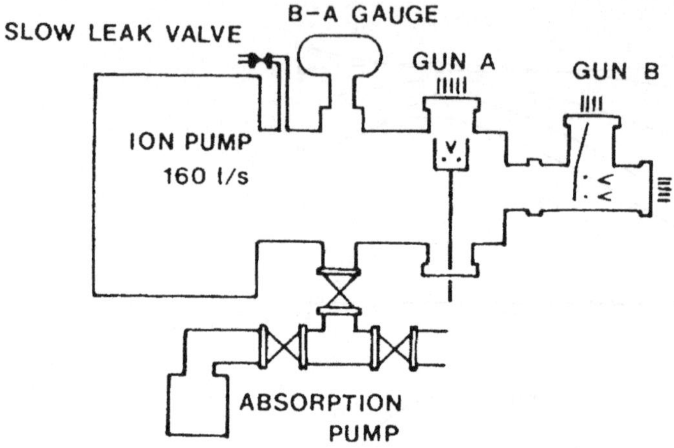

Fig. 8A.12 Schematic diagram of the experimental apparatus (Todokoro et al., 1982) [8A-7]

To confirm the origin of the ions, a sophisticated experiment has been done. Two emitters were operated at the same time in a B-type gun. Emitter I emitted a current in the range of 5 μA–250 μA, while emitter II was kept at a constant 5 μA current. The result is shown in **Fig. 8A.14**. Emitter II fluctuation did not change. The two emitters were so close that they have a common ambient gas pressure. The number of ions coming from the anode is the same for both emitters. This result shows that bombarding ions are produced closed to the emitters, not at the anode. Otherwise, fluctuation in emitter II would increase with the increase in emitter I current.

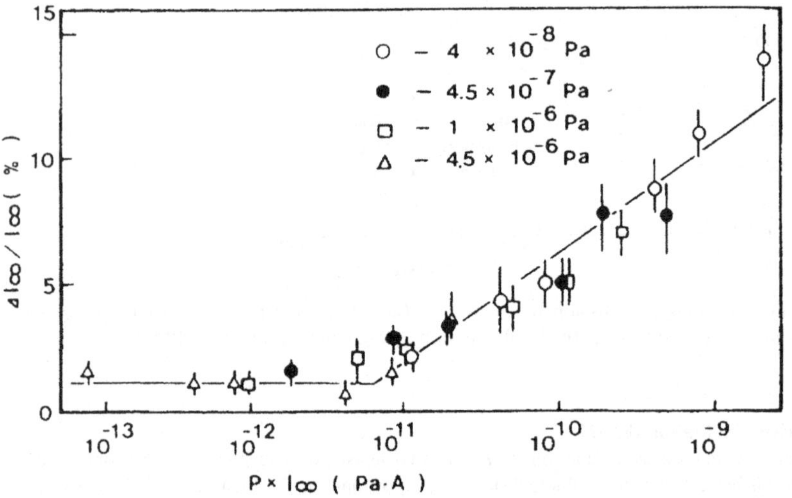

Fig. 8A.13 Relative fluctuation as a function of the product of P and I_∞ (Todokoro et al., 1982) [8A-7]

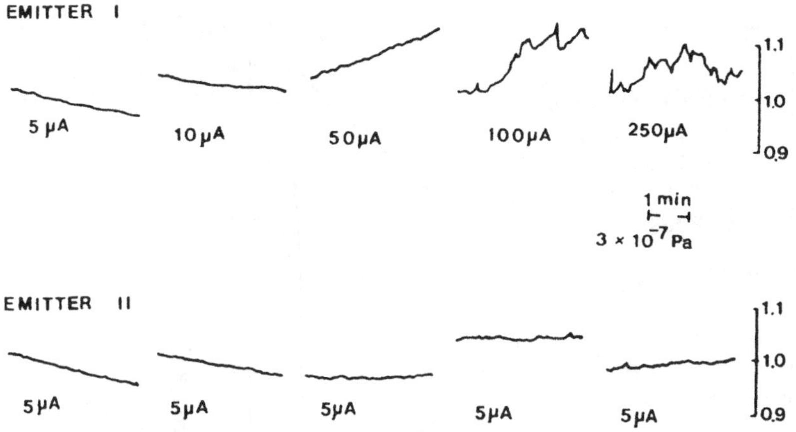

Fig. 8A.14 Experimental results confirming the origin of ions. Two emitters operated at the same time. Emitter II emitted a current in the range 5 μA–250 μA, while emitter II was controlled to emit a constant 5 μA current (Todokoro et al., 1982) [8A-7]

Build-up Treatment

Veneklasen and Siegel (1972) [8A-8] developed an oxygen-enhanced thermal-field shaping method for obtaining high angular confinement of the electron emission from a <100>-oriented tungsten tip.

Abstract [8A-8]: A source is obtained from which the beam is confined to a half-angle of ~ 0.15 rad. This strongly-forward-direction beam gives a source that can be used to obtain beams with currents as high as 5×10^{-8} A within an acceptance angle of 5×10^{-3} rad with a total emission curve of only 10 μA. The processed tip is operated at 900 °C to give a source with long-term stability and reliability in a moderate ultrahigh vacuum of $\sim 10^{-9}$ Torr. An interpretation of the shaping of the tip based on field-induced thermal migration is given.

Tamura (1979) [8A-9] obtained a brightness level of $\sim 10^8\,\mathrm{A \cdot cm^{-2} \cdot sr^{-1}}$ at 100 kV by using a W <100> build-up field emitter.

Abstract [8A-9]: In the W <100> build-up, oxygen plays an important role. This is provided by installing a silver tube oxygen leak device which permits the penetration rate of oxygen into the vacuum chamber to be controlled by temperature variation. With this device, the rate of W <100> build-up can be controlled automatically.

Oxygen Introduction [8A-9]
To facilitate oxygen introduction, diffusion of oxygen through a silver tube [1] was investigated. The tube is surrounded by a spiral nichrome heater, and the penetration rate of oxygen was controlled by varying the temperature of the silver tube. When the silver tube was heated to about 500 K, oxygen penetrated to the vacuum chamber. The total emission current I_t and the pressure were changed as shown in **Fig. 8A.15**, resulting in the completion of the <100> build-up. Upon completion of the build-up, oxygen was no longer necessary

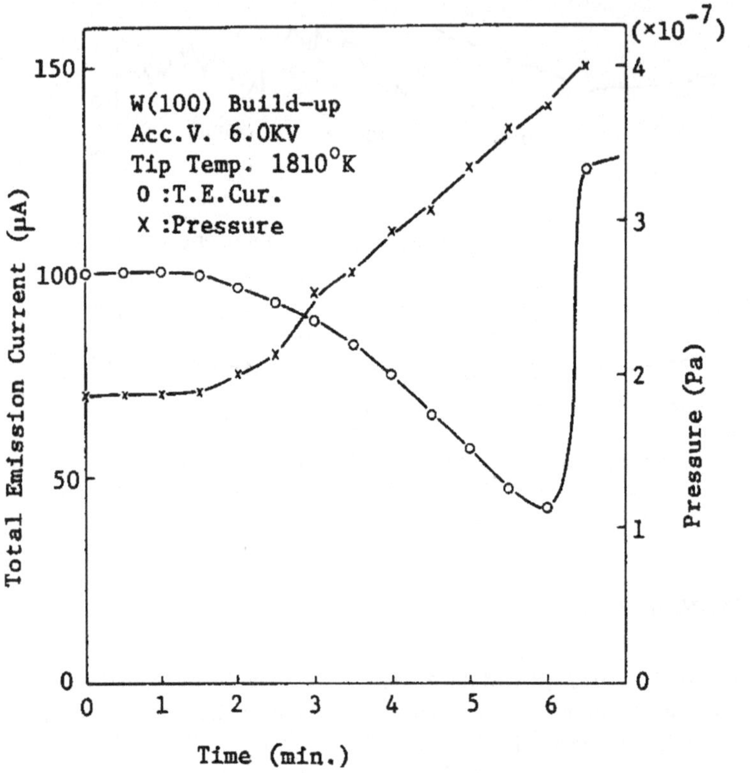

Fig. 8A.15 W <100> build-up process using the silver tube device (Tamura, 1979) [8A-9]

and the heating of the silver tube was stopped. By installing this silver tube onto the vacuum chamber, the <100> build-up processing became automatic.

Reference

1. N. R. Whetten and J. R. Young, "Use of a silver tube to admit oxygen to a vacuum system", *Rev. Sci. Instrum.* **30**, 1959, 472–473.

FE-Initiated Vacuum Arc

Dyke et al. (1953) [8A-10] presented an article, "The field emission initiated vacuum arc I: Experiments on arc initiation."

Abstract [8A-10]: The magnitude of the electric field, current density, and work function at the cathode, obtained by use of pulse electronic techniques, were simultaneously determined prior to breakdown. From this investigation it has been established that:

(1) the vacuum arc was initiated at a critical value of the field current density of the order of $10^8\,\mathrm{A/cm^2}$;
(2) breakdown was predictable and not random; in fact easily recognizable conditions preceding arc formation have been established; at current densities just below the critical value, an electron emission process was observed, which apparently involved both high electric fields and high temperatures;
(3) arc formation did not require cathode bombardment by material from the anode or from residual gases;
(4) breakdown was independent of the applied microsecond voltage in the range $5 < V < 60\,\mathrm{kV}$, provided the critical current density was not exceeded;
(5) the current during arc exceeded the initial field current by a factor of at least 100.

Dolan et al. (1953) [8A-11] presented an article, "The field emission initiated vacuum arc II: The resistively heated emitter."

Abstract [8A-11]: Electrical breakdown between clean metal electrodes in high vacuum was observed when the field current density at the single crystal tungsten cathode exceeded a critical value of the order of $10^8\,\mathrm{A/cm^2}$. At current densities just below the critical value, an electron emission process was observed which apparently involved both high temperature and high electric field. The calculations show that the resistive heating was sufficient to melt the emitter at the critical current density, assuming the accepted value of the physical constants for the polycrystalline metal.

Dyke et al. (1953) [8A-12] presented an article, "The field emitter: fabrication, electron microscopy, and electric-field calculations."

Abstract [8A-12]: An investigation of several methods of fabrication revealed that a smooth, simple, and relatively stable tungsten emitter geometry resulted from a refinement of the methods of Benjamin and Jenkins [M. Benjamin and R. O. Jenkins, *Proc. Roy. Soc.*, (London) (A) **176**, 262, 1940] which combines electrolytic etch and smoothing action of surface migration. The electric field at any point on an emitter surface was calculated when the emitter geometry was precisely fitted with one equipotential surface from the family surrounding a charged, isolated sphere-on-orthogonal-cone. A theoretical surface distribution of current density for a typical emitter was derived from the calculated field distribution and the wave-mechanical field emission theory. From this result was calculated a value of the emitting area which was in good agreement with experiment.

Dyke et al. (1960) [8A-13] presented an article, "Electrical stability and life of the heated FE cathode."

Abstract [8A-13]: Occurrence of a vacuum ark is proposed as the cause of cathode failure. Considerations based on this assumption lead to predictions concerning the dependence of average cathode life on temperature, and to quantitative expressions for the statistical distribution and average value of life for unbiased field cathodes operated at high temperatures.

Experimental results yielded by the operation of 85 cathodes show good agreement with the theory. The heated cathode shows constant performance over a long period after which cathode failure occurs abruptly, and the statistical nature of cathode life is confirmed. The statistical distribution of cathode life and the measured effect of various parameters on average life show good agreement with theoretical predictions.

Morphological Changes of Tip

Nakamura and Kuroda (1973) [8A-14] investigated the effect of polarization on the surface migration of metal atoms which is induced by a thermal-field during remolding.

Abstract [8A-14]: Experimental results show that the end-form of the thermal-field build-up of a crystal plane on tungsten and rhenium emitters is determined by the isotropic feature of the free energy of the surface, which in turn is influenced by the adsorption of gases and the polarization of surface metal atoms caused by the remolding field.

Nichols and Mullins (1965) [8A-15] derived the partial differential equation describing morphological changes of a surface of revolution due to capillarity-induced surface diffusion under the assumption of isotropy of surface tension and surface self-diffusion coefficient.

Abstract [8A-15]: Spheroidization of cylindrical rods, as well as field-emission tips with taper below a certain critical value, is predicted; for tapers above the critical value, steady-state shapes are predicted and equations describing the blunting and recession of the tips are presented. If the sintering results for spheres are represented by a plot of $\log x/a$ vs. $\log t$, it is found that the inverse slop varies from approximately 5.5 to approximately 6.5 for the range $0.05 \leq x/a \leq 0.3$, in contrast with the constant value of 7 found by Kuczynski [G. C. Kuczynski, *Trans. AIME* **185**, 169 (1949)] from an order-of-magnitude analysis. At higher values of x/a, n increases steadily and without bound."

Sugata and Mun (1969) [8A-16] observed the morphological changes of a tungsten cathode by heating over a wide range of temperatures (1600–2800 K) either continuously or intermittently.

Abstract [8A-16]: It was found from the experimental results that the receding rate of the tip depended on the cone angle. Furthermore, the process of typical changes in the shape of the tip was established, and why a neck formed in the range near the tip could be explained."

Pichaud et al. (1971) [8A-17] described a method to calculate the temperature distribution along the tip.

Abstract [8A-17]: Typical numerical results are given for tips of tungsten and some other metals at temperatures between 1000 and 2900 K. Considered are conical tips of different length (half cone angle between 0.5° and 6°) and some non-conical tips. The temperature gradient near the apex (which influences the shape change by surface diffusion) has a maximum for a particular angle of the conical tip. Very high gradients occur at tip constrictions.

Morikawa et al. (2001) [8A-18] studied the shape change of the tip-end of a tungsten field emitter due to the treatment of flashing and build-up process using a TEM and FE-SEM.

Abstract [8A-18]: A protrusion at every <111> pole has been obtained by annealing at $\sim 1500\,\mathrm{K}$ of the W-tip which in advance had been flashed at $\sim 2500\,\mathrm{K}$ and then kept in a vacuum for more than 4 hours. The <111> protrusion could be shaped up to a very sharp emitter, hopefully a single-atom emitter, by so-called T-F treatment using a positive high voltage. Some kind of surface contamination seems to play a very important role on the build-up process of a sharp <111> emitter.

FE-Related Technology

Morikawa and Goto (1988) [8A-19] improved the sharp-pointed tip preparation method by ac polishing, by controlling the applied voltage, wave shape, phase angle, frequency, and number of waves. This was performed by designing a controller in cooperation with a function generator and an audio amplifier.

Emons and Hagemans (1972) [8A-20] presented an article, "Use of a field-electron emitter as a pressure indicator in ultrahigh vacuum."

Abstract [8A-20]: Research has proved that for hydrogen which is the dominant component of the residual gas in such systems, there is a linear relationship between the measured rate of adsorption and the pressure reading on a Bayard-Alpert gauge. The highest pressure that can be determined by this method is limited by the maximum measuring speed; the lowest by the necessity to have a certain number of particles adsorbing on the small emitter surface within a reasonable course of time so that adsorption can be measured with a certain accuracy. This method can be used in the pressure range from 10^{-9} to 10^{-13} Torr.

Chernatony and Yarwood (1979) [8A-21] presented an article, "Problems in the production and measurement of very high vacuum, especially in applications, and a new approach to measurement based on the use of field emission."

Abstract [8A-21]: Opinions are expressed about the needs in the development of vacuum systems in which a very high vacuum is to be created. Topics considered are sealing materials, new pumps, especially for an oil-free vacuum environment, a new type of gauge based on field electron emission able to indicate surface gas coverage as well as extend downwards the partial pressure range of a mass spectrometer and the further automation of industrial vacuum plant in which repetitive processes are undertaken.

Chernatony (1979) [8A-22] presented an article, "Problems in the production and measurement of atomically clean surface environments and their confirmation based on the use of field emission."

Abstract [8A-22]: The superior performance obtained with highly pre-cleaned titanium sublimation pumps generates conditions below the working region of most mass spectrometers. For measurement in this region and to monitor surface conditions a simple field electron emission microscope based technique is developed. Changes in surface coverage are analyzed by color superposition techniques and in later development by computerized techniques.

Yamamoto et al. (1974) [8A-23] proposed a new semi-empirical model for predicting the work function of binary compounds.

Abstract [8A-23]: The model is a modification of the Gordy and Thomas relation (W. Gordy and W. Thomas, *J. Chem. Phys.* **24**, 439, 1956) and is applicable to binary compounds. The work functions of binary compounds, ϕ_{AB} are found to be primarily determined by the properties of the element of low work function, A, perturbed by the other element of high work function, B, according to $\phi_{AB} = \phi_A + (1.41/d_A)\left[(2\phi_A - \phi_B - 0.34)/(\phi_A + \phi_B - 0.68)\right]$. Experimental values tabulated by Fomenko (V. Fomenko, *Handbook of Thermionic Properties*, Plenum, New York, 1966) are compared with the predicted work functions, and the quantitative agreement is good in spite of the simplicity of the model.

Yamamoto and Miyokawa (1998) [8A-24] measured and discussed the emission characteristics of a conical field emission gun.

Abstract [8A-24]: The measured time dependent mean emission current was interpreted as a weighted superposition of local field emission from the W<310>, <100>, <110>, and <112> planes. This interpretation resulted in a mean work function of 4.6 eV in good agreement with the value determined from the FN plot. For an accelerating voltage of 15 keV and an emission current of 8 μA after flashing, the dependence of the probe current on the extraction voltage was measured. For $V_{acc}/V_{ex} = 1$, an angular current density of (41 ± 4) μA/sr was found. Similar measurements of the dependence of the probe current on the ratio V_{acc}/V_{ex} (accelerating voltage/extraction voltage) made 3 h after flashing yielded (20 ± 6) μA/sr for $V_{acc} = V_{ex}$. The transfer efficiency defined as the ratio between probe current and emission current was found to be 0.4%–0.3%.

Experiment [8A-24]

A W<310> single crystal rod 0.125 mm in diameter was welded to a polycrystalline tungsten wire and sharpened by chemical etching to form an emitter. This emitter was installed in an FEG chamber, and the chamber was baked at about 300 °C for about 20 h. After baking, the chamber pressure was kept at less than 2×10^{-8} Pa, and the emitter was cleaned by flashing. The emitter cleaning ("flashing") was done by heat conduction from the W wire heated by direct current for about 1 s. **Figure 8A.16** shows a schematic diagram of a conical FEG. Here the extraction voltage V_{ex} and the accelerating voltage V_{acc} are the voltages between the emitting cathode and the first and second anodes, respectively. The distance d_s between the apex surface of the cathode and the virtual source position varies with the voltage ratio V_{acc}/V_{ex}. If $V_{acc}/V_{ex} = 1$, then the virtual source position is close to the emitter apex but it moves upwards if $V_{acc}/V_{ex} \neq 1$. For $V_{acc}/V_{ex} = 0.1$, the displacement amounts to about 35 mm.

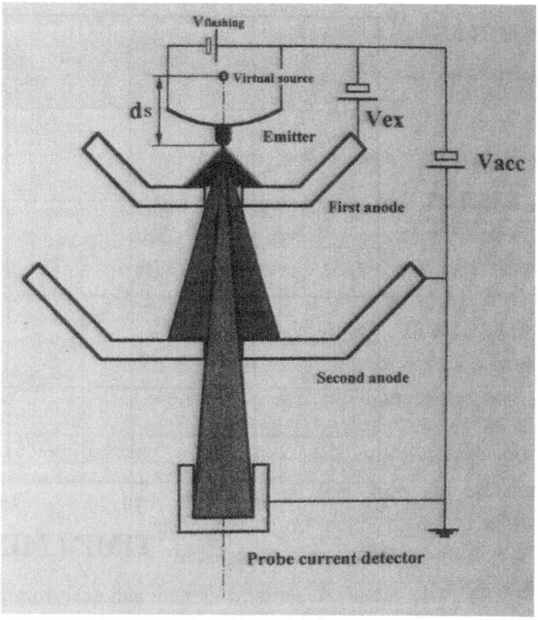

Fig. 8A.16 Schematic diagram of the conical field emission gun. The extraction voltage V_{ex} and accelerating voltage V_{acc} are the voltage applied between the cathode (emitter) and the first and second anodes, respectively (Yamamoto and Miyokawa, 1998) [8A-24]

Time Variation of Emission Current [8A-24]
For a constant extraction voltage $V_{ex} = 3.5\,\text{kV}$ and an initial emission current of 8 μA, a decrease of the emission current with time was observed as shown in **Fig. 8A.17**. The curve may be divided into three parts: After a slow decrease during the first six minutes (solid line) a faster decay (broken line) follows for the next eight minutes, and then a "tail" represented by a curved solid line. The decrease of the emission current is due to an increase of the mean work function caused by the adsorption of gases such as H_2, CO, N_2, and O_2 on the W emitter surface.

We assume that the vertex of the emitter is composed of <310>, <100>, <110>, <111>, and <112> planes and that the total emission current I_{total} is a weighted superposition of the local field emission (LFEs), i.e., the local current densities from these planes:

$$I_{total} = w(310) \times J_{310} + w(110) \times J_{110} + w(100) J_{100} + w(111) \times J_{111} + w(112) \times J_{112}. \quad (8.3)$$

Here, J_{310}, J_{100}, J_{110}, J_{111} and J_{112} are the current densities from the respective planes, and the $w(hkl)$ are the respective area having $\{hkl\}$ directions. Applying the FN current densities $J_{hkl} = k_0 \times \exp\left(-k_1 \phi_{hkl}^{3/2}\right)$ from the FN equation,

$J = BF^2 \exp\left(-\frac{6.83 \times 10^7 \phi^{3/2}}{F}\right)$, the following equation is obtained. B and F are assumed to have the same value for each plane:

$$\begin{aligned} I_{total} &= k_0 \times [w(310)\exp(-k_1 \phi_{310}^{3/2}) + w(110)\exp(-k_1 \phi_{110}^{3/2}) \\ &\quad + w(100)\exp(-k_1 \phi_{100}^{3/2}) \quad + w(111) \times \exp(-k_1 \phi_{111}^{3/2}) \\ &\quad + w(112) \times \exp(-k_1 \phi_{112}^{3/2})]. \end{aligned} \quad (8.4)$$

Here, $k_0 = BF^2$ and $k_1 = 6.83 \times 10^7/F$ are constants, and ϕ_{310}, ϕ_{110}, ϕ_{100}, ϕ_{111}, and ϕ_{112} are the work functions for each plane. Their values for clean surfaces are given in **Table 8A.1**.

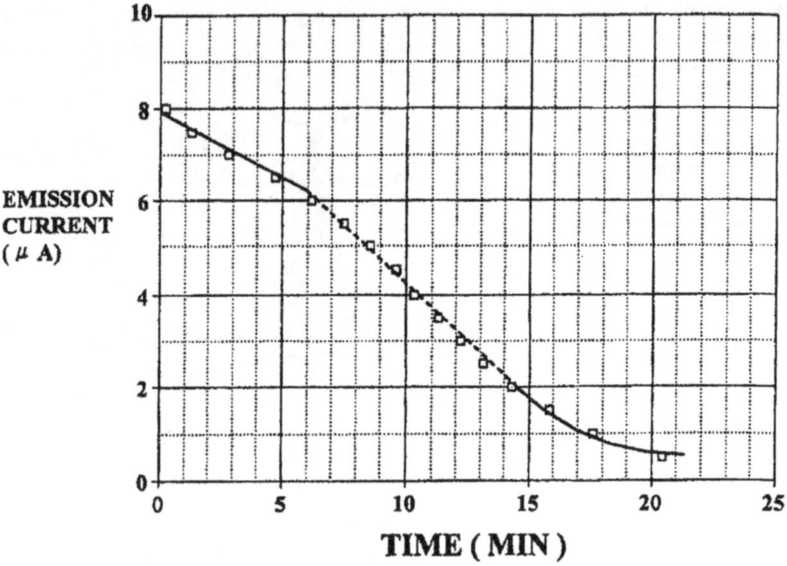

Fig. 8A.17 Decay of the emission current with time from an initial value of 8 μA. The constant extraction voltage equals 3.5 kV Yamamoto and Miyokawa, 1998) [8A-24]

Review

Hainfeld (1977) [8A-25] presented a review, "Understanding and using field emission sources."

Abstract [8A-25]: Field emission tips are growing in popularity due to their high brightness, small source size, minimal energy spread and potentially long lifetime. Disadvantages include the necessity of a moderate ultrahigh vacuum, beam current that fluctuates 2–5% and additional maintenance of the source such as processing or flashing. Several practical modes of operation have evolved including cold (room temperature) operation, oxygen processing and thermal (heated) use. Stability and lifetime depend on vacuum, tip current, first-anode outgassing and temperature. Anyone using an instrument with a field emission source should be aware of how field emission tips operate and how to optimize their performance.

Table 8A.1 Work function for crystallographic plane (Yamamoto and Miyokawa, 1998) [8A-24]

Crystal graphic plane	< 310 >	< 111 >	< 100 >	< 112 >	< 110 >
Work function (eV)	4.25	4.47	4.65	4.76	5.25

8B ZrO/W Emitter

Swanson and Martin (1975) [8B-1] presented an article, "Field electron cathode stability studies: Zirconium/tungsten thermal-field cathode," in which the characteristics of the ZrO/W emitter are systematically described in detail. Swanson and Martin must be pioneers for developing a ZrO/W thermal-field emitter.

Abstract [8B-1]: The emission angular distribution, total energy distribution, noise spectrum, and emitter life have been measured for the Zr/W thermal-field cathode. The results give an emitter life in excess of 1000 h at T = 1350 to 1450 K, $P \geq 2 \times 10^{-8}$Torr, and a total current of $\sim$ 100 μA. A low-noise, highly confined beam along the <100> direction can be obtained. A source brightness of 10^{10} $A/(cm^2sr)$ at 3 kV and an energy spread of less than 0.9 eV were measured under operational conditions.

Characteristics

Factors Influencing FE Cathode Life and Stability [8B-1]

Two basic parameters control the time fluctuations of the FE current at constant applied voltage: (1) the work function ϕ and (2) the surface geometry. This can be seen more clearly by considering the relationship between current density J and applied field F as given by the well-known Fowler-Nordheim equation:

$$J = \frac{1.54 \times 10^{-6} F^2}{\phi t(y)} \exp\left(\frac{-6.83 \times 10^7 \phi^{3/2} v(y)}{F}\right) \ (\mathrm{A/cm^2}), \tag{8B.1}$$

where F and ϕ are in V/cm and eV, respectively. Noting that the image-potential correction terms $t(y) \approx 1$ and $v(y) \approx 0.943 - 1.525 \times 10^{-7} F/\phi^2$ in the F and ϕ range of interest, Eq. (8B.1) becomes

$$J = \frac{1.54 \times 10^{-6} F^2}{\phi} \exp\left(\frac{10.4}{\phi^{1/2}}\right) \times \left[\exp\left(\frac{-6.83 \times 10^7 \phi^{3/2}}{F}\right)\right] \ (\mathrm{A/cm^2}). \tag{8B.2}$$

In order to more clearly show the role of emitter geometry in Eqs. (8B.1) and (8B.2), one can further note that

$$F = \beta V, \tag{8B.3}$$

where V is the diode voltage and

$$\beta = f(r, \alpha, R). \tag{8B.4}$$

The relationship between the emitter geometric factor β and the emitter radius r, emitter cone half-angle α, and emitter-to-anode spacing R is complex but can be given for typical emitter shapes by the following approximation [1]:

$$\beta \approx 1.7/r^{o.87} R^{0.13} \alpha^{0.33} \ (\mathrm{cm^{-1}}). \tag{8B.5}$$

This expression is valid in the range $r = 10^{-5}$ to 10^{-4}cm, $R = 0.1$ to 1 cm, and $\alpha = 5$ to $10°$. A less accurate approximation is

$$\beta = 1/5r. \tag{8B.6}$$

Thus, one can observe from the foregoing relationships that J is extremely sensitive to variations in ϕ and/or β. Ignoring the pre-exponential term variations, this can be put more quantitatively by noting that

$$\frac{dI}{I} \approx \frac{6.8 \times 10^7 \phi^{3/2}}{F} \frac{d\beta}{\beta} \tag{8B.7}$$

$$\approx \frac{10.2 \times 10^7 \phi^{3/2}}{F} \frac{d\phi}{\phi}. \tag{8B.8}$$

According to the approximation given in Eq. (8B.6) one can further deduce that

$$\frac{dI}{I} \approx \frac{6.8 \times 10^7 \phi^{3/2}}{F} \frac{dr}{r}. \tag{8B.9}$$

In the case of the Zr/W cathode, where typically ϕ =2.5 eV and $F = 2 \times 10^7$ V/cm, Eq. (8B.9) becomes

$$\frac{dI}{I} \approx 15.3 \frac{dr}{r}. \tag{8B.10}$$

We shall now examine factors which cause time variations in ϕ and β.

Adsorbed Surface Layers [8B-1]

The fluctuation in I due to variation in ϕ is primarily attributable to time variation of the average adsorbate coverage $\bar{\sigma}$ and/or adsorbate dipole moment μ. This can be expressed by the well-known Helmholtz equation:

$$\Delta\phi = g\pi\mu\bar{\sigma}, \tag{8B.11}$$

where g is a constant varying from 2 to 4 and $\Delta\phi$ is the change $\phi - \phi_0$ in work function from the zero coverage value ϕ_0. There are two basic types of time variations of $\overline{\sigma}$ and μ that are of interest: One is a long-term change in the value of σ due to adsorption or desorption; the other (which applies to both $\overline{\sigma}$ and μ) pertains to higher-frequency (flicker) fluctuations about an average value or between two or more values. Short-term fluctuation of $\overline{\sigma}$ can be diffusion-controlled or, in the case of an equilibrium gas pressure, it can be controlled by the adsorption-desorption rates. On the other hand, long-term variation in σ is primarily controlled by adsorption or desorption due to ambient pressure change.

If $P \leq 10^{-10}$ Torr after initial emitter-surface cleaning, one has the desirable condition that $\bar{\sigma} \approx 0$ and $d\sigma/dt \approx 0$, at least for reasonably long time periods. For the case $P > 10^{-9}$ Torr then $\bar{\sigma} \neq 0$ and $d\sigma/dt \neq 0$. After an equilibrium coverage $\bar{\sigma}_{eq}$ is established (the time to reach equilibrium will increase with decreasing P) the value of $\bar{\sigma}_{eq}$ will be determined by T, P, and the binding energy of the adsorbate. At $T \leq 1500$ K and $P \leq 10^{-8}$ Torr the value of $\bar{\sigma}$ for most residual gases will be near zero.

Fortunately, the adsorbed zirconium, though becoming mobile at high temperature, requires an excessively high temperature ($T > 1800$ K) to become irreversibly desorbed. Because of the large work function change due to zirconium adsorption it would be indeed a serious problem for the long-term current stability if it slowly desorbed.

Surface Geometry

Ion Bombardment [8B-1]

According to Eqs. (8B.7) and (8B.10) the time variation of the cathode surface geometry can seriously affect the current stability. There are two basic mechanisms that can alter the macro and micro surface features:

First, we consider microscopic surface roughness caused by ion bombardment of the cathode surface. The formation of ions that "sputter" or roughen the cathode surface arises from gas-phase electron ionization and from electron-stimulated desorption from the anode surface. The latter ions can be nearly eliminated from reaching the cathode by proper design of the anode lens structure. The Zr/W <100> cathode is particularly helpful in this respect since, as will be shown in a succeeding section, the angular divergence of the electron beam is 14° as opposed to 40° for an uncoated cathode.

The effect of thermal annealing of surface roughness caused by ion bombardment of a tungsten FE cathode has shown that rapid surface annealing occurs at temperatures above 800 K. Clearly, TF operation of a W field cathode above 1000 K should rapidly anneal surface roughness caused by ion bombardment and allow relaxation of vacuum and anode cleanliness requirements.

Thermal-Field Buildup [8B-1]

Heating an FE cathode in the absence of an applied field causes the emitter radius to increase according to [2]

$$\left(\frac{dr}{dt}\right)_0 = \left[1.25\gamma\Omega_0 D_0 \alpha \exp\left(\frac{-E_d}{kT}\right)\right]\left(A_0 kTr^3\right)^{-1}, \tag{8B.12}$$

where γ is the surface tension ($\gamma = 2900$ dyn/cm for W), Ω_0 is the volume per atom, A_0 is the surface area per atom, D_0 is the diffusivity constant ($D_0 = 4\ \mathrm{cm}^2/\mathrm{sec}$ for W), and E_d is the activation energy for surface diffusion($E_d = 3.14$ eV for W). For clean W we obtain

$$\left(\frac{dr}{dt}\right)_0 = \left[2.6\times 10^{-11}\alpha \exp\left(\frac{-36300}{T}\right)\right]\left(Tr^3\right)^{-1}\ (\mathrm{cm/s}). \tag{8B.13}$$

According to Barbour et al. [2], in the presence of an applied field

$$\left(\frac{dr}{dt}\right)_F \approx \left(1 - \frac{rF^2}{8\pi\gamma}\right)\left(\frac{dr}{dt}\right)_0. \tag{8B.14}$$

Hence, the theory predicts that the emitter dulling rate will decrease with dc applied fields of either polarity. It can be seen that dr/dt varies according to the following conditions:

(a) $\frac{dr}{dt} > 0$, if $F < \left(\frac{8\pi\gamma}{r}\right)^{1/2}$

(b) $\frac{dr}{dt} = 0$, if $F = \left(\frac{8\pi\gamma}{r}\right)^{1/2}$

(c) $\frac{dr}{dt} < 0$, if $F > \left(\frac{8\pi\gamma}{r}\right)^{1/2}$.

Case (a) leads to emitter dulling and hence to a decrease in emitted current with time. Case (b) ideally causes the emitter dulling due to surface tension forces to be balanced by the electrostatic force opposing the dulling force. This case is difficult to realize over the total emitting area due to the nonuniform field over the emitter surface. Case (c) does not lead to emitter sharpening because of the difficulty of nucleating new atomic layers in certain crystallographic directions. Consequently a more complex process called "field buildup" occurs in which the current increases erratically and gradually assumes a polyhedral shape [3].

For clean W the condition for $dr/dt \approx 0$ is given by

$$F_0 = 8.1\times 10^4 r_m^{-1/2}\ (\mathrm{V/cm}), \tag{8B.15}$$

where r_m and F_0 are the values giving condition (b) above. The useful range of the current densities ($J = 10^4$ to $10^8\ \mathrm{A/cm^2}$) corresponds to fields in the 4×10^7–8×10^7 V/cm range for a clean W cathode. In view of Eq. (8B.15) the necessity to avoid buildup precludes dc

TF operations of cathodes with $r > 8 \times 10^{-6}$ cm, since such radii correspond to F_0 values lower than 4×10^7 V/cm. Thus, for emitter radii in the 10^{-5}–10^{-4} cm range F_0 is considerably below that required for normal emission levels thereby eliminating the possibility of stabilizing the smooth emitter shape during TF dc operation of smooth and clean W emitters.

Fortunately, for the Zr/W cathode with $\phi \approx 2.6$eV a value of $F = 2 \times 10^7$ V/cm gives a useful current density of about 5.3×10^5 A/cm^2 and a value of $r_m = 1.64 \times 10^{-5}$ cm. If one assumes the emitting area is πr^2 one obtains a total current of $I = 443\,\mu$A; since the emitting area of the Zr/W <100> cathode is reduced by a factor about 0.1, a more realistic value of $I \approx 44\,\mu$A should be realized.

It should be pointed out that the preceding analysis is based on thermodynamic arguments and that the emitter temperature determines the rate at which geometric changes will

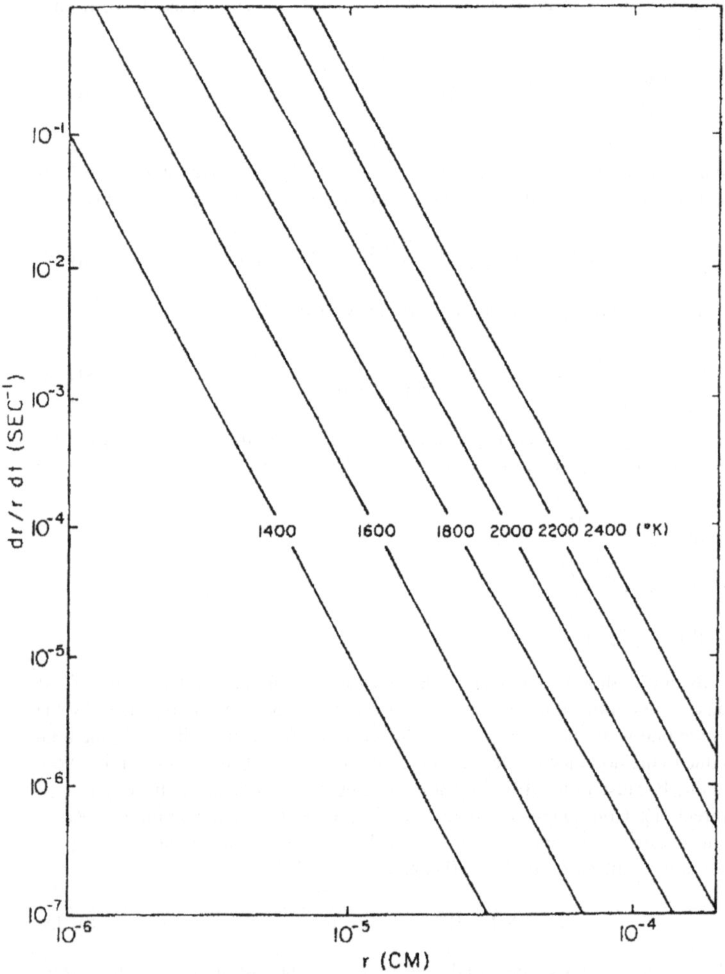

Fig. 8B.1 Plot of emitter dulling rate dr/rdt vs. emitter radius for various emitter temperatures for an emitter half-angle $\alpha = 0.10$rad (Swanson and Martin, 1975) [8B-1]

occur. In other words, according to Eqs. (8B.3), (8B.6), and (8B.15) a Zr/W cathode of $r = 1.64 \times 10^{-5}$ cm operated above 2000 V will be unstable with respect to field-induced geometric change. On the other hand, the rate of this change for clean W is controlled by (8B.13). Strictly speaking Eq. (8B.13) describes the dulling rate of a clean W surface; however, for the purpose of this discussion we assume that buildup is the reverse of dulling. **Figure 8B.1** shows a plot of Eq. (8B.13). Since $dr/dt \propto 1/r^3$, the rate of geometric change is very sensitive to both temperature and radius. Clearly, if r or E_d is sufficiently large, dr/dt can become negligible at a specific temperature. The rate of change of r is related to the change in I via Eq. (8B.10) and if V exceeds 2000 V at $r = 1.64 \times 10^{-5}$ cm and $T = 1400$ K, we can determine from **Fig. 8B.1** and Eq. (8B.10) that

$$\frac{dI}{Idt} = 1.53 \times 10^{-5}(\text{sec}^{-1}) = 5.51\%\,(\text{h}^{-1}).$$

In practice dI/Idt is even smaller because of an increase in E_d due to the presence of Zr.

Making use of the fact that dr/rdt decreases rapidly with increasing r, it should also be possible to operate a 5000 Å emitter under buildup conditions without incurring a significant rate of buildup. Not only would the rate of change of current be small, but the total emission current would be much larger for these operating conditions.

References

1. W. P. Dyke, J. K. Trolan, W. W. Dolan, and G. Barnes, *J. Appl. Phys.* **24**, 570 (1953).
2. J. P. Barbour, F. M. Charbonnier, W. W. Dolan, W. P. Dyke, E. E. Martin, and J. K. Trolan, *Phy. Rev.* **117**, 1452 (1960).
3. P. C. Bettler and F. M. Charbonnier, *Phys. Rev.* **119**, 85 (1960).

Properties

Swanson (1975) [8B-2] measured angular distribution, total energy distribution, and noise on both a zirconium coated and built-up W <100> field cathode operating at high temperature (1400–1800 K).

Summary [8B-2]: This study showed that the Zr/W and built-up TF cathodes are capable of long life ($\sim$ 1000 h) operation at 10 nTorr residual gas levels. Total energy distribution half-widths were less than 1.5 eV and noise levels for $P < 10$ nTorr did not exceed several percent for small solid angles and over practical bandwidths. The small angular divergence of the beam reduces the total current requirements for a specified transmitted current compared with conventional field cathodes. In addition, the smaller emitting area of the built-up cathode leads to a slightly larger noise level and current density for a specified current transmitted through a small aperture. Based on the calculated emitting areas, brightness values of $\sim 10^{10}\,\text{A} \cdot \text{cm}^{-2} \cdot \text{sr}^{-1}$ can be achieved by these cathodes at 2–3 kV. Thus, it can be concluded that these two TF cathodes are compatible with microprobe applications where high-brightness cathodes are required.

Wolfe (1979) [8B-3] presented an article "Operational experience with zirconiated T-F emitters."

Abstract [8B-3]: The beam current stability is better than $\pm 0.5\%$ at pressures in the region of 10^{-8} Torr. The source brightness is $10^9\,\mathrm{A \cdot cm^{-2} sr^{-1}}$. The operating point is typically $10^7\,\mathrm{V/cm}$ electric field, 1800 K and $10^5\,\mathrm{A/cm^2}$. Life tests have indicated lifetimes of 5000 h or greater, which may be expected at the above operating point and 10^{-8} Torr vacuum. The virtual source size for our geometry and operating point and has been calculated and measured at about 100 Å for an acceptance half angle of 0.015 rad. The zirconiated cathode is also highly recommended for thermionic and Schottky aided thermionic applications. It is more stable and longer lived than LaB_6.

Structure of the Electron Gun [8B-3]
The cathode structure is mounted in the electron gun as shown in **Fig. 8B.2**. A grid disk is provided 0.25 mm behind the tip of the cathode. Its sole function is to reduce the amount of thermionic emission from the shank of the needle in order to reduce anode heating. The anode is placed 0.5 mm from the tip of the cathode. With this geometry and a 1 μm tip radius one obtains $10^7\,\mathrm{V/cm}$ at the tip with 7500 V applied [8B-3].

Danielson (1981) [8B-4] studied the properties of the ZrO-W <100> emitter and the coadsorption of zirconium and carbon monoxide on W <100> by Auger electron spectroscopy and work-function measurements.

Abstract [8B-4]: The thermionic work function of the Zr-O-W <100> emitter from 1450 to 1800 K is 2.54 eV with a pre-exponential value of 4. The field-emission retarding potential work function is 2.75–2.80 eV. A Zr-O-W <100> emitter free of bulk carbon possesses a longer lifetime than an emitter containing bulk carbon. Heating the emitter in carbon monoxide causes diffusion of zirconium into the bulk, and vacuum annealing restores the zirconium to the surface. The coadsorption of zirconium and carbon monoxide on W <100> followed by vacuum annealing produces a minimum work function of 2.3 eV. The adsorption sequence is important and lower work functions correlate with higher carbon concentrations. The stability of the 2.3 eV surface under background gas adsorption and high-temperature heating is discussed.

Kang et al. (1983) [8B-5] presented an article, "A numerical analysis of the electric field and trajectories with and without the effect of space charge for a field electron source."

Abstract [8B-5]: Calculations show that the emission properties of a field emission gun are very sensitive to a small change in emitter shape. Space charge effects are examined numerically for various values of emitter radius and compared with experimental results for a Zr/W field emitter. It is shown that the decrease in cathode field due to space charge is the probable reason for the deviation of emission current from theoretical values at high current density. The commonly used planar diode approximation is shown to be inadequate for describing space charge effects of field emitters. Trajectory analysis shows that contributions to the field emitter virtual source size due to spherical aberration are similar for both a rounded and faceted emitter shape.

Speidel et al. (1985) [8B-6] presented an article, "The Zr-O-W-field emitter in the low temperature mode."

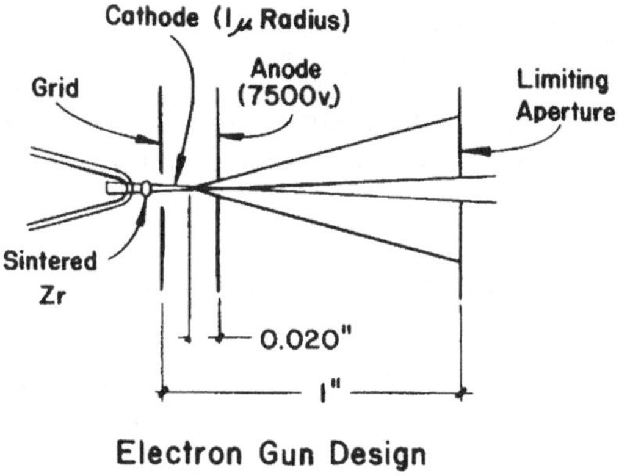

Fig. 8B.2 The geometry used in the electron gun used for this discussion (Wolfe, 1979) [8B-3]

Abstract [8B-6]: This paper describes accurately the production of the cathode. The Zr-O-W-cathode emits 100 nA of probe current in a 10 mrad aperture at $T = 1300\,\mathrm{K}$ and $p_{O_2} = 10^{-6}\,\mathrm{Pa}$. The DC-stability is less than 10% over 6 hours. The FWHM of the total electron energy distribution is 1.5 eV for 0.3 mA/sr of angular beam intensity.

Kim et al. (1994) [8B-7] evaluated Zr/O/W Schottky emission tips with microlenses for applications in scanning tunneling microscope aligned field emission microcolumns.

Abstract [8B-7]: Operation of Zr/O/W <100> Schottky emission tips at 1800 K with an axial separation of 50–100 μm between the tip and a microlens has been successfully tested. The microlens consists of an extraction electrode with a 5 μm diameter hole in a 1 μm-thick silicon membrane. The preliminary results of this study show that thermal field emission tips can be operated continuously in close proximity to a microlens over a long period of time, and that good emission stability of less than 1% noise fluctuation over 10 h is achieved with emission currents up to at least 100 μm.

Kim et al. (1997) [8B-8] measured the emission stability and energy distributions from a Zr/O/W Schottky emitter at a tip temperature range from 1330 to 1800 K."

Abstract [8B-8]: The changes of emission properties have been observed at reduced tip temperature. Noise fluctuations of the probe current increase with decrease of the tip temperature at a constant extraction voltage. The work function of the Schottky emitter increases with decrease of tip temperature. The energy width at a given temperature increases with increasing angular emission current density. The energy width also increases with decreasing tip temperature at a given angular emission current density. The results indicate that the energy broadening is mainly contributed by electron tunneling. A comparison of the measured energy width with the theoretical predictions is discussed.

8C LaB_6 Emitter

The LaB_6 emitter is still widely used as a bright, thermionic electron emitter in electron microscopes. It works well in a high vacuum in the 10^{-5} Pa range.

Characteristics

Lafferty (1951) [8C-1] investigated the thermionic emission properties of the borides of the alkaline-earth and rare-earth metals and thorium. Lafferty must be a pioneer for developing a LaB_6 emitter.

Abstract [8C-1]: Thermionic emission measurements made on these materials show the rare-earth metal borides to be superior to the others. The highest emission was obtained from lanthanum boride. Its emission constants for the Dushman equation were $\phi = 2.66$ volts and $A = 29\,\text{amps}\cdot\text{cm}^{-2}\cdot\text{degK}^{-2}$. This is higher than the emission normally obtained from thoria. Lanthanum boride has a relatively low evaporation rate corresponding to a latent heat of evaporation of 169 kilocalories per mole. If the hexaborides are operated at high temperature in contact with the refractory metals, boron diffuses into their metal lattices forming interstitial boron alloys with them. When this occurs, the boron framework which holds the alkaline-earth or rare-earth metal atoms collapses, permitting the latter to evaporate. However, the hexaboride cathodes may be operated at high temperatures in contact with tantalum carbide or graphite.

Electron Emission Constants [8C-1]

The electron emission from the various borides was measured as a function of their temperature with a constant potential of 500 V between cathode and collector. Richardson plots were made from these data, using the thermionic emission equation

$$\log_{10} I/T^2 = \log_{10} A - 11600\phi/2.303T, \qquad (8C.1)$$

where I is the current density in amperes per cm^2, T the absolute temperature in degrees Kelvin, A is an empirical constant expressed in amperes per cm^2 per K^2, and ϕ is the conventional work function expressed in volts.

Figure 8C.1 shows a series of Richardson plots for the alkaline-earth, rare-earth, and thorium borides. Straight lines were obtained in all cases, showing ϕ to be substantially independent of temperature. It will be noted that the rare-earth borides are better emitters than the alkaline-earth or thorium borides. Values of A and ϕ obtained are given in **Table 8C.1**.

Figure 8C.2 shows the emission obtained for some mixed borides of the rare-earth metals. X-ray analysis of the crystal structure of these mixed borides shows that they form solid solutions over the entire composition range. Misch metal, which is an alloy of all the rare-earth metals, forms a solid solution of mixed borides with an electron emission only slightly less than that of LaB_6. This is of commercial interest since misch metal is considerably cheaper than lanthanum metal.

It is of interest to compare the emission from LaB_6, the best boride emitter, with some of the conventional cathode materials. This is shown in **Fig. 8C.3**. The emission from LaB_6 is surpassed only by barium oxide at low temperatures. In the high current-density range, LaB_6 is superior to any of the conventional materials now available.

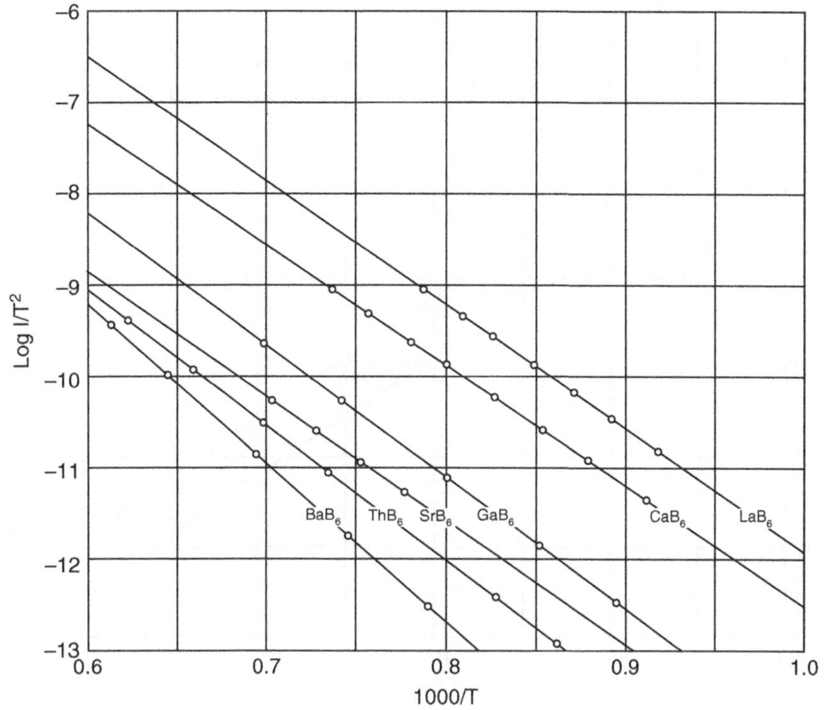

Fig. 8C.1 Richardson plots for the alkaline-earth, rare-earth, and thorium borides. The emission constants are given in **Table 8C.1** (Lafferty, 1951) [8C-1]

Mounting Methods

Nakagawa and Yanaka (1975) [8C-2] developed a LaB_6-cathode electron gun to obtain a highly stable electron probe.

Table 8C.1 Hexaboride emission constants as determined from the data in **Figs. 8C.1** and **8C.2**. (Lafferty, 1951) [8C-1]

Boride	A $(A \cdot cm^{-2} \cdot K^{-2})$	ϕ (V)
CaB_6	2.6	2.86
SrB_6	0.14	2.67
BaB_6	16	3.45
LaB_6	29	2.66
CeB_6	3.6	2.59
Misch metal boride	14	2.64
ThB_6	0.5	2.92

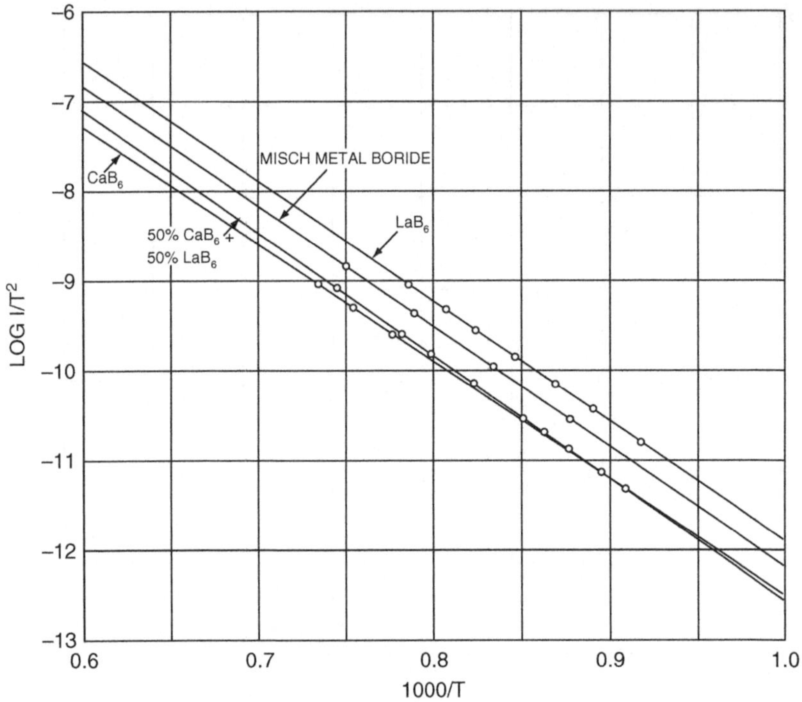

Fig. 8C.2 Richardson plots for some mixed borides of the rare-earth metals. Misch metal boride is a hexaboride made from misch metal, an alloy of the rare-earth metals (Lafferty, 1951) [8C-1]

Electron Gun [8C-2]
Figure 8C.4 shows a cutaway drawing of the LaB_6 cathode electron gun assembly with the cathode (sintered LaB_6 rod; tip curvature: 10 μm) sandwiched between two heating blocks of pyrolytic graphite. These blocks are supported by stainless steel props, employing screws built into their outer portions. Thus, the tip supports are symmetrical with regard to the tip center, thereby preventing any lateral drift of the tip due to thermal expansion [8C-2].

The diameter of the hole on the top of the Wehnelt is 1.5 mm (**Fig. 8C.4**), and the cathode tip is set 0.2–0.6 mm deep in the Wehnelt. Moreover, this gun assembly can be easily exchanged with a tungsten filament assembly. The anode can be adjusted for height and centered from outside the gun chamber. Thus, optimum brightness is attainable even when the accelerating voltage has been changed. Variable apertures and gun alignment coils are provided beneath the gun anode.

Emission is obtained by heating the cathode with graphite blocks heated by direct current (stability: $\leq 1 \times 10^{-4}/\text{min}$). The voltage can be varied from 2 kV to 35 kV (stability: $\leq 3 \times 10^{-6}/\text{min}$) and bias is of the self-bias type (resistance: 1–64 MΩ).

Crawford (1979) [8C-3] evaluated several LaB_6 mounting techniques which are currently in widespread use for high brightness applications.

Mounting Methods [8C-3]
Four methods of mounting boride cathodes (shown in **Fig. 8C.5**) are currently in widespread use in scanning microscopes and similar high-brightness applications.

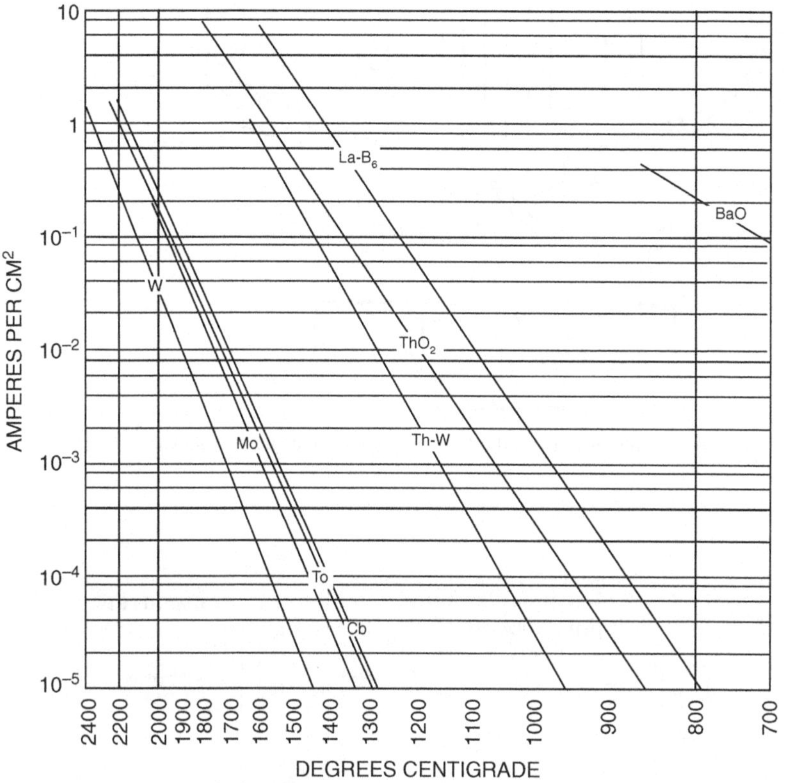

Fig. 8C.3 Electron emission of some common cathode materials and lanthanum boride plotted as a function of the temperature (Lafferty, 1951) [8C-1]

The earliest high-brightness LaB_6 cathode developed for SEM use is the electron-bombardment-heated rod type shown in **Fig. 8C.5 (a)** [1, 2, 3, 4, 5, 6, 7]. Here the material compatibility problem is avoided by having no material in contact with the hot LaB_6. One end of a LaB_6 rod is heated by electron bombardment, using an auxiliary tungsten coil heater which surrounds the hot end of the rod. The other end of the rod is held in a small collet at low temperature, with the full temperature gradient along the rod. The advantages of this method are first, the chemical reaction problem is solved in a simple elegant way. Second, the cathode is inherently rigid. Third, since a large quantity of LaB_6 is used, typically 0.1 g for each cathode, there is considerable LaB_6 available to evaporate. Unfortunately the disadvantages are also numerous. The design is unnecessarily complicated. An additional power supply (floating at high voltage) is required to provide the electron bombardment power, as is a completely different and more complex electron gun, with an additional wire in the high voltage cable.

Next, in development for high-brightness use is the carbon block mounted design as shown in **Fig. 8C.5 (b)** [8, 9, 10]. Here, a small, carefully lapped piece of LaB_6 is mechanically clamped between two small blocks of carbon, and the resulting assembly heated resistively by passing a current through all three pieces. Carbon is one of the few high temperature materials which does not significantly react with hot LaB_6 (a carbon crucible can be used to contain molten LaB_6). The advantages of this method are first a much simpler design. Porperly constructed, this design is also quite rigid. On the negative side, the car-

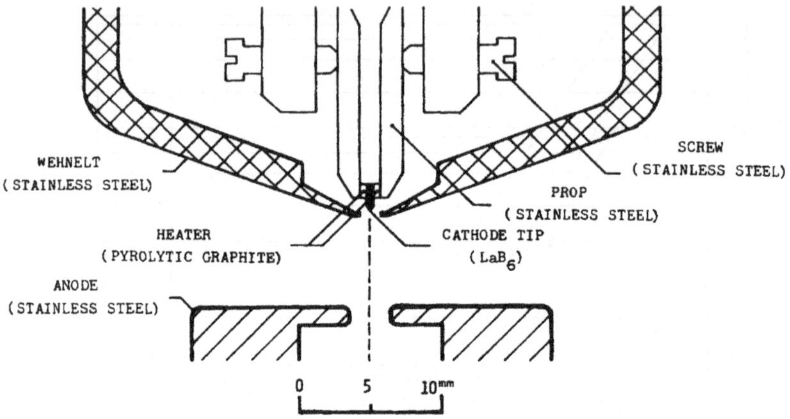

Fig. 8C.4 Cutaway drawing of LaB_6-cathode assembly (Nakagawa and Yanaka, 1975) [8C-2]

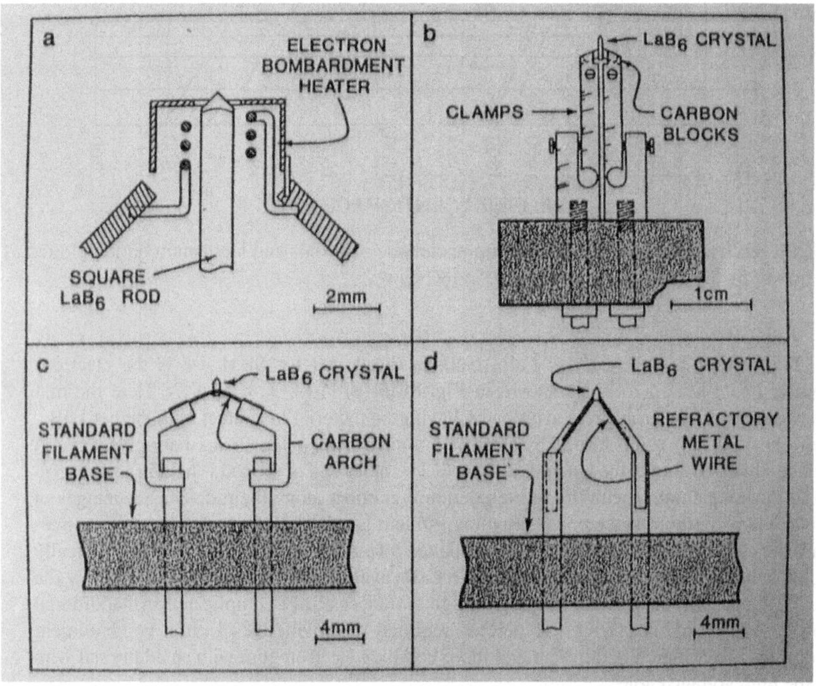

Fig. 8C.5 Major high-brightness mounting methods: **(a)** electron- bombardment-heated rod type; **(b)** carbon block mounted; **(c)** carbon arch mounted; **(d)** refractory metal mounted. All of these mounting methods are capable of high-brightness operation if properly implemented. Relative advantages differ according to the application. Note that the drawing scales are not equal (Crawford, 1979) [8C-3]

bon block method requires a number of machined parts which must be accurately made and carefully assembled. The clamps have a fairly high thermal mass; and standard filament bases cannot be used. Some units have a stability problem apparently due to the large heating current which passes through the carbon-lanthanum hexaboride interfaces.

The third mounting method is the carbon arch technique, which uses the compatibility of carbon and LaB_6 in a different way. A carefully machined tiny block of LaB_6 is held between two thin carbon strips bent into small arches, as shown in **Fig. 8C.5 (c)** [11, 12]. The resistive heating current passes through the arches only, and not through the LaB_6. The major advantage is that it can be made on standard filament bases which allow direct replacement of tungsten in many instruments. It also runs at a current and voltage, which are similar to those for tungsten, which means that the pre-existing filament supply need not be changed. Installation is simple. Because of the stability of LaB_6 mounted on carbon, and the fact the current does not pass directly through the LaB_6, this design has a good over-temperature tolerance. The worst drawback to this design is its fragility. The carbon arches will immediately break if touched with fingers, or if the unit is dropped. Another problem is that the angle of the arch is determined by the carbon properties, and is too shallow to fit many existing Wehnelt caps. Lastly, this type of cathode shares with the carbon block type, having the need for very accurately made parts, and also the same minor problem in that the carbon must be protected from oxidation.

The fourth mounting method is the refractory metal mounting method typified in **Fig. 8C.5 (d)** [13, 14]. Here, a small piece of LaB_6 is mounted on a refractory metal hairpin; the hairpin typically having dimensions similar to a standard tungsten filament. The LaB_6 is bonded to the refractory metal either by some type of brazing process, or by a melting-welding type process. The major advantages here are simplicity, adaptability, and a lack of fragility. The cathodes mount on standard bases and take up no more room than standard tungsten filaments; no special Wehnelt or other modifications need be made. Finally, their electrical impedance is easily adjusted and they have the potential of running at the lowest power of any of the presently used types. The major disadvantage of this type is the reintroduction of high-temperature reactions as a major life-limiting mechanism. It has been known for many years that the reaction rate of LaB_6 with two of the refractory metals, rhenium and tantalum, is relatively slow (rhenium is the slowest), and that the reaction rates are strongly temperature dependent. Hence if the temperature is carefully controlled, high brightness and reasonable life can both be obtained. However the over-temperature tolerance is not as good as with the other mounting methods. The situation is similar in those refractory-metal-mounted designs where an in-between material, such as a carbide or silicide is used to braze or sinter the cathode to the wire. Here the limiting reaction may occur between the wire and the braze rather than between the braze material and the LaB_6. At the present time, from tests made in our laboratories, none of the refractory metal mounts seem to have achieved the chemical stability of the carbon mounted cathodes. Nonetheless this method is very attractive because of its other advantages; it also probably has the best potential for improvement. Variations on this method were among the earliest used, back before scanning microscopes were invented. The other methods now in general use were all proposed (separately) as ways of overcoming the chemical reaction problem.

Conclusions [8C-3]

The redesign of cathodes emphasizing smaller size and lower power should both improve performance and ultimately lower costs. All of the presently used cathode types are capable of some miniaturization; and even without miniaturization all could benefit from application of heat flow calculations.

The best design to miniaturize is probably the refractory-metal mounted type. The major problem of this type is, as it has always been, destructive chemical reactions and material diffusion. However this problem is both less serious in the current designs, and more tractable to better solutions with the advent of modern surface physics techniques.

References

1. A. N. Broers, "Electron gun using long-life lanthanum hexaboride cathode", *J. Appl. Phys.* **38** (4), 1967, 1991–1992.
2. A. N. Broers, "Some experimental and estimated characteristics of the lanthanum hexaboride rod cathode electron gun", *J. Phys. E.* **2**, 1969, 273–276.
3. A. N. Broers, "Recent advances in scanning electron microscopy with lanthanum hexaboride cathodes", *SEM*/1974, IIT Research Institute, Chicago, IL, 60616, 9–18.
4. J. D. Verhoeven and E. D. Gibson, "Evaluation of a LaB_6 cathode electron gun", *J. Phys. E.* **9**, 1976, 65–69.
5. E. D. Gibson and J. D. Verhoeven, "The preparation of single crystal LaB_6 cathodes for electron microscopes", *J. Phys. E.* **8**, 1975, 1003–1004.
6. H. Ahmed, W. Blair, and R. Lane, "A low power heater for the LaB_6 rod cathode", *Rev. Sci. Instrum.* **43**, 1972, 1048–1049.
7. J. D. Verhoeven and E. D. Gibson, "On the design of the Broers type LaB_6 gun", *SEM*/1977/I, IIT Research Institute, Chicago, IL., 60616, 9–16.
8. S. F. Vogel, "Pyrolytic graphite in the design of a compact inert heater of a lanthanum hexaboride cathode", *Rev. Sci. Instrum.* **41**, 1970, 585–587.
9. S. Nakagawa and T. Yanaka, "A highly stable electron probe obtained with LaB_6 cathode electron gun", *SEM*/1975, IIT Research Institute, Chicago, IL., 60616, 19–26.
10. P. H. Schmidt, L. D. Longinotti, D. C. Joy, S. D. Ferris, H. J. Leamy, and Z. Fisk, "Design and optimization of directly heated LaB_6 cathode assemblies for electron-beam instruments", *J. Vac. Sci. Technol.* **15** (4), 1978, 1554–1560.
11. C. K. Crawford and H. I. Smith, "Directly heated LaB_6 cathodes interchangeable with standard SEM filaments". Presented at the August 1973 Joint EMSA/EPASA Meeting in New Orleans–Late Paper Session.
12. S. D. Ferris, D. C. Joy, H. J. Leamy, and C. K. Crawford, "A directly heated LaB_6 electron source", *SEM*/1975, IIT Research Institute, Chicago, IL., 60616, 11–18.
13. K. N. Ramachandran, "Rhenium Bonded LaB_6 Electron Source", *Rev. Sci. Instrum.* **46**, 1975, 1662–1663.
14. P. B. Sewell and K. N. Ramachandran, "Grid aperture contamination in electron guns using directly heated lanthanum hexaboride sources", *SEM*/1978/I, SEM Inc., AMF O'Hare, IL., 60666, 221–232.

Material Loss

Sewell (1980) [8C-4] presented an article, "High brightness thermionic electron guns for electron microscopes."

Abstract [8C-4]: A consideration of the small area around the tip which is responsible for electron emission leads to the conclusion that further progress towards achieving predictable and reproducible performance with single crystal LaB_6 emitters will depend on control of the microstructure around the cathode tip. Loss of material from the emitter, by processes such as oxidation, evaporation and ion bombardment, is a prime factor in limiting the useful life of the gun.

Evaporation and Oxidation [8C-4]

Figure 8C.6 shows the total loss of material from LaB_6 due to evaporation and oxidation. In the absence of oxidation, the loss of evaporation alone is shown by the line A-B. Curves above this line show the combined effect of evaporation and oxidation at various oxygen

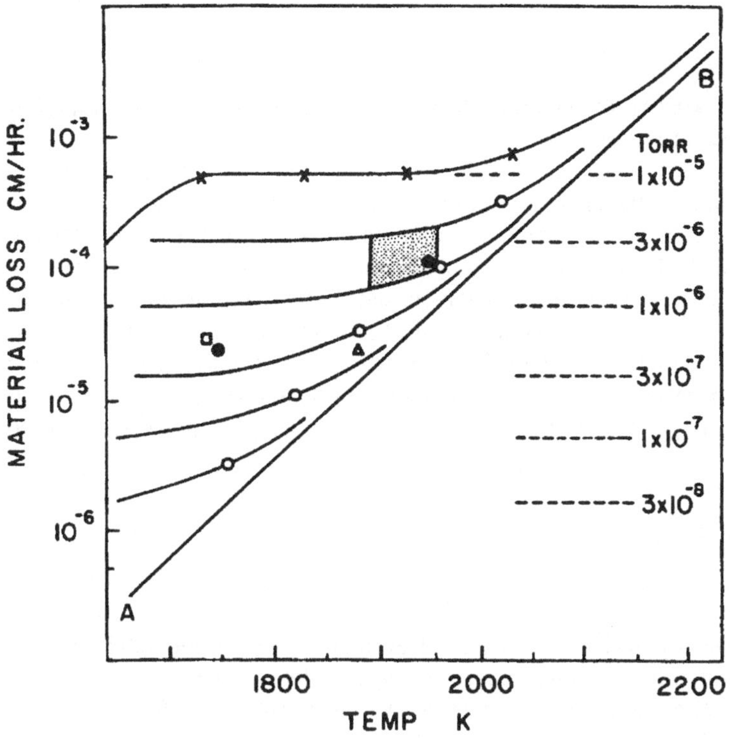

Fig. 8C.6 Data on material loss from LaB_6 due to evaporation and oxidation. Solid line A-B, evaporation data from Verhoeven [1]. × – – – ×, oxidation data at 1×10^{-5} Torr. Other curves are calculated losses due to evaporation and oxidation at the pressures indicated. For details see Sewell and Ramachandran [2] (Sewell, 1980) [8C-4]

pressures. The open circle on each curve represents the condition where losses by evaporation and oxidation are equal. It is seen at low temperatures (1700 K) which might be used to achieve long filament life, significant oxidation occurs even at pressures as low as 3×10^{-8} Torr. The gun performance will be particularly poor in pressures above 10^{-7} Torr. Even with filaments operating at 1900 K, the loss mechanism will be mainly oxidation at pressures above 3×10^{-7} Torr. At temperatures above 2000 K, gun pressures around 10^{-6} Torr might be tolerated as the evaporation of a conductive layer of LaB_6 predominates. For general use, it is strongly recommended that the gun pressure be in the low 10^{-7} Torr region. Much improved life and reliability will result when gun pressures are routinely in the low 10^{-8} Torr range.

References

1. J. D. Verhoeven, "On the problem of obtaining optimum brightness from your LaB_6 gun", *SEM*/1977/I, IIT Research Institute, Chicago, IL 60616, pp. 581–90.
2. P. B. Sewell and K. N. Ramachandran, "A source imaging detector for the SEM", *SEM*/1977/I, IIT Research Institute, Chicago, IL 60616, pp. 17–23.

Properties

Futamoto et al. (1980) [8C-5] presented an article, "Thermionic emission properties of a single-crystal LaB_6 cathode."

Abstract [8C-5]: A new type of LaB_6 cathode has been developed. It consists of a <001> oriented LaB_6 single-crystal tip and a glassy carbon filament. The influence of gas pressure on the thermionic emission properties is investigated for O_2, H_2O, CH_4, and H_2. Evaporation rates and surface composition of the LaB_6 cathode are also examined. Thermionic emission current decreased in all gases examined here, depending on cathode temperature and gas pressure. In O_2 and H_2O gas atmosphere, the cathode surface composition appears to change to La_2O_3 as gas pressure is increased. The evaporation rate of the LaB_6 single crystals is represented by equation $V = 4 \times 10^{14} \exp\left(-\frac{570\times10^3}{RT}\right)$ µm/h, for the temperature range between 1800 and 2200 K.

Review

Hohn (1985) [8C-6] presented an article, "Development and use of high brightness lanthanum hexaboride electron guns," reviewing 51 related papers.

Abstract [8C-6]: Important parameters such as crystal orientation, tip radius, cone angle, tip facetting, orientation dependent work function, electron gun geometry, vacuum conditions, tip positioning, tip temperature and others, have been investigated and reported in the literature. Reviewing several studies, it becomes apparent that a vacuum of better than 1×10^{-6} Torr is required for best performance and long lifetime. Preferable crystal orientations are the <100> and <346> rod axis which result in practical brightness values in excess of $10^6\,A \cdot cm^{-2} \cdot sr^{-1}$, at 20 kV acceleration voltage and 200 µA emission current.

The emission pattern and cross-over intensity distribution of single crystal LaB_6 cathodes are of high symmetry and depend strongly on the temperature and the crystal orientation. Both can be related to the crystal lattice structure as derived from the unit cell.

8D Other FE Emitters

The emission characteristics of TiC FE cathodes have been investigated [8D-1, 8D-2]. A micro FE electron gun has been applied to a very small electron-beam system [8D-3].

Adachi (1985) [8D-4] presented an article, "Approach to a stable field emission electron source," reviewing 96 related papers.

Abstract [8D-4]: New materials which have low work functions and high melting points are examined in the view of a stable field electron emitter, and it is shown that carbides of transition metals have potential as a stable field emitter. Very stable field emission has been reported for TiC single crystals. Operation in the thermal-field emission mode is examined and it is shown that a ZrO/W <100> emitter gives stable emission, whose fluctuation is less than 0.23% in the frequency interval 1–5000 Hz. The only disadvantage of the ZrO/W thermal field emitter is its rather high level of instability at very low frequencies.

Ichinokawa (1984) [8D-5] described his experience on various kinds of electron sources, cold FE sources, thermal FE sources, and thermionic electron sources (W and LaB_6).

Seiler (1983) [8D-6] presented a review, "Secondary electron emission in the scanning electron microscope."

Abstract [8D-6]: The energy distribution, the angular distribution, and the yield of secondary electrons from metals and insulators are reviewed as well as the escape depth of the secondary electrons and the contribution of the backscattered electrons to the secondary electron yield. The different detectors for secondary electrons in the scanning electron microscope are described. The contrast mechanisms in the scanning electron microscope, material, topography, voltage, magnetic, and crystallographic orientation contrast based on secondary electron emission, as well as the lateral resolution, depending among other things on the spatial distribution of the emitted secondary electrons, are discussed.

References

W FE Emitter

8A-1. A. V. Crewe, D. N. Eggenberger, J. Wall, and L. M. Welter, "Electron gun using a field-emission source", *Rev. Sci. Instrum.* **39** (4), pp. 576–583 (1968).

8A-2. A. V. Crewe, J. Wall, and L. M. Welter, "A high-resolution scanning transmission electron microscope", *J. Appl. Phys.* **39** (12), pp. 5861–5868 (1968).

8A-3. A. V. Crewe, M. Isaacson, and D. Johnson, "A simple scanning electron microscope", *Rev. Sci. Instrum.* **40** (2), pp. 241–246 (1969).

8A-4. S. Tamaki, T. Inoue, E. Sugata, and H. W. Kim, "Remolding of <310>-tungsten tip in field-ion microscope", *Japan. J. Appl. Phys.*.**15** (2), pp. 265–272 (1976).

8A-5. S. Yamamoto, S. Fukuhara, H. Okano, and N. Saito, "Field emission current instability in the "ever-decreasing" region, *Japan. J. Appl. Phys.*. **15** (9), pp. 1643–1646 (1976).

8A-6. D. J. Swann and K. C. A. Smith, "Lifetime and noise characteristics of tungsten field emitters", *Scanning Electron Microscopy*/1973/I, pp. 42–48.
8A-7. H. Todokoro, N. Saitou, and S. Yamamoto, "Role of ion bombardment in field emission current instability", *Japan. J. Appl. Phys.*. **21** (10), pp. 1513–1516 (1982).
8A-8. L. H. Veneklasen and B. M. Siegel, "Oxygen-processed field emission source", *J. Appl. Phys.* **43** (4), pp. 1600–1604 (1972).
8A-9. N. Tamura, "Basic study of TF emission", *Scanning Electron Microscopy*/1979/I, pp. 31–38.
8A-10. W. P. Dyke, J. K. Trolan, E. E. Martin, and J. P. Barbour, "The field emission initiated vacuum arc. I. Experiments on arc initiation", *Phys. Rev.* **91** (5), pp. 1043–1054 (1953).
8A-11. W. W. Dolan, W. P. Dyke, and J. K. Trolan, "The field emission initiated vacuum arc. II. The resistively heated emitter", *Phys. Rev.* **91** (5), pp. 1054–1057 (1953).
8A-12. W. P. Dyke, J. K. Trolan, W. W. Dolan, and G. Barnes, "The field emitter: fabrication, electron microscopy, and electric field calculations", *J. Appl. Phys.* **24** (5), pp. 570–576 (1953).
8A-13. W. P. Dyke, F. M. Charbonnier, R. W. Strayer, R. L. Floyd, J. P. Barbour, and J. K. Trolan, "Electrical stability and life of the heated field emission cathode", *J. Appl. Phys.* **31** (5), pp. 790–805 (1960).
8A-14. S. Nakamura and T. Kuroda, "Remolding of field emission cathode", *Oyo Buturi* (*Journal of the Japan Society of Applied Physics*) **42** (10), pp. 975–982 (1973) (in Japanese).
8A-15. F. A. Nichols and W. W. Mullins, "Morphological changes of a surface of revolution due to capillarity-induced surface diffusion", *J. Appl. Phys.* **36** (6), pp. 1826–1835 (1965).
8A-16. E. Sugata and D. Mun, "Morphological changes by heating of tungsten cathode for field emission microscope", *Oyo Buturi* (*Journal of the Japan Society of Applied Physics*) **38** (11), pp. 1024–1031 (1969) (in Japanese).
8A-17. M. Pichaud, A. Müller, and M. Drechsler, "Temperature distribution along metal tips (for field emission microscopy and the study of surface phenomena)", *Surface Science* **26**, pp. 14–20 (1971).
8A-18. H. Morikawa, K. Matsusaka, H. Kurata, Y. Yoshino, "The build-up process of (111) pole of a tungsten tip and the shape change of a tip cap", *Shinku (J. Vac. Soc. Japan)* **44** (1), pp. 36–41 (2001) (in Japanese).
8A-19. H. Morikawa and K. Goto, "Reproducible sharp-pointed tip preparation for field ion microscopy by controlled ac polishing", *Rev. Sci. Instrum.* **59** (10), pp. 2195–2197 (1988).
8A-20. A. J. Emons and K. L. Hagemans, "Use of a field-electron emitter as a pressure indicator in ultrahigh vacuum", *J. Vac. Sci. Technol.* **9** (1), pp. 112–116 (1972).
8A-21. L. de Chernatony and J. Yarwood, "Problems in the production and measurement of very high vacuum, especially in applications, and a new approach to measurement based on the use of field emission", *Vacuum* **29** (3), pp. 125–128 (1979).
8A-22. L. de Chernatony, "Problems in the production and measurement of atomically clean surface environments and their confirmation based on the use of field emission", *Vacuum* **29** (11/12), pp. 389–403 (1979).
8A-23. S. Yamamoto, K. Susa, and U. Kawabe, "Work functions of binary compounds", *J. Chem. Phys.* **60** (10), pp. 4076–4080 (1974).
8A-24. Y. Yamamoto and T. Miyokawa, "Emission characteristics of a conical field emission gun", *J. Vac. Sci. Technol. B* **16** (5), pp. 2871–2875 (1998).
8A-25. J. F. Hainfeld, "Understanding and using field emission sources", *Scanning Electron Microscopy*/1977/I, pp. 591–604.

ZrO/W Emitter

8B-1. L. W. Swanson and N. A. Martin, "Field electron cathode stability studies: Zirconium/tungsten thermal-field cathode", *J. Appl. Phys.* **46** (5), pp. 2029–2050 (1975).
8B-2. L. W. Swanson, "Comparative study of the zirconiated and built-up W thermal-field cathode", *J. Vac. Sci. Technol.* **12** (6), pp. 1228–1233 (1975).

8B-3. J. E. Wolfe, "Operational experience with zirconiated T-F emitters", *J. Vac. Sci. Technol.* **16** (6), pp.1704–1708 (1979).
8B-4. L. R. Danielson, "The Zr-O-W <100> emitter and coadsorption of zirconium and carbon monoxide on W <100>", *J. Appl. Phys.* **52** (11), pp. 6769–6776 (1981).
8B-5. N. K. Kang, D. Tuggle, and L. W. Swanson, "A numerical analysis of the electric field and trajectories with and without the effect of space charge for a field electron source", *Optik* **63** (4), pp. 313–331 (1983).
8B-6. R. Speidel, P. Brauchle, B. Kramer und U. Schwab, "Die Zr-O-W-Feldemissionskathode im Niedertemperaturbetrieb", *Optik* **71** (4), pp. 167–172 (1985).
8B-7. H. S. Kim, E. Kratschmer, M. L. Yu, M. G. R. Thomson, and T. H. P. Chang, "Evaluation of Zr/O/W Schottky emitters for microcolumn applications", *J. Vac. Sci. Technol. B* **12** (6), pp. 3413–3417 (1994).
8B-8. H. S. Kim, M. L. Yu, M. G. R. Thomson, E. Kratschmer, and T. H. P. Chang, "Performance of Zr/O/W Schottky emitters at reduced temperatures", *J. Vac. Sci. Technol.B* **15** (6), pp. 2284–2288 (1997).

LaB_6 Emitter

8C-1. J. M. Lafferty, "Boride cathodes", *J. Appl. Phys.* **22** (3), pp. 299–309 (1951).
8C-2. S. Nakagawa and T. Yanaka, "A highly stable electron probe obtained with LaB_6 cathode electron gun", *Scanning Electron Microscopy*/1975/I, pp. 20–26.
8C-3. C. K. Crawford, "Mounting methods and operating characteristics for LaB_6 cathodes", *Scanning Electron Microscopy*/1979/I, pp. 19–30.
8C-4. P. B. Sewell, "High brightness thermionic electron guns for electron microscopes", *Scanning Electron Microscopy*/1980/I, pp. 11–24.
8C-5. M. Futamoto, M. Nakazawa, K. Usami, S. Hosoki, and U. Kawabe, "Thermionic emission properties of a single-crystal LaB_6 cathode", *J. Appl. Phys.* **51** (7), pp. 3869–3876 (1980).
8C-6. F. J. Hohn, "Development and use of high brightness lanthanum hexaboride electron guns", *Scanning Electron Microscopy*/1985/IV, pp. 1327–1338.

Other FE Emitters

8D-1. I. Ishizawa, S. Aoki, C. Oshima, and S. Otani, "Field emission properties of surface-processed TiC <110> field emitter", *Shinku (J. Vac. Soc. Japan)* **29** (12), pp. 578–584 (1986) (in Japanese).
8D-2. Y. Ishizawa, S. Aoki, C. Oshima, and S. Otani, "Design and operation of extremely high vacuum field emission gun", *Shinku (J. Vac. Soc. Japan)* **29** (11), pp. 544–548 (1986) (in Japanese).
8D-3. Y. Endo, I. Honjo, and A. Ito, "Application of micro field emitters to a miniature electron beam column", *Shinku (J. Vac. Soc. Japan)* **43** (2), pp. 112–118 (2000) (in Japanese).
8D-4. H. Adachi, "Approach to a stable field emission electron source", *Scanning Electron Microscopy/1985/II*, pp. 473–487.
8D-5. T. Ichinokawa and Y. Kamiya, "Electron source", *Electron Microscopy* **19** (2), pp. 71–76 (1984) (in Japanese).
8D-6. H. Seiler, "Secondary electron emission in the scanning electron microscope", *J. Appl. Phys.* **54** (11), pp. R2–R18 (1983).

Other Articles

ZrO/W Emitter

*8-1. J. B. McGinn, L. W. Swanson, N. A. Martin, M. A. Gesley, M. A. McCord, R. Viswanathan, F. J. Hohn, A. D. Wilson, R. Naumann, and M. Utlaut, "100 kV Schottky electron gun", *J. Vac. Sci. Technol.* B **9** 6), pp. 2925–2928 (1991).

LaB_6 Emitter

*8-2. R. Shimizu, T. Shinike, T. Tanaka, C. Oshima, S. Kawai, H. Hiraoka, and H. Hagiwara, "Brightness of single crystal LaB_6 cathodes of <100> and <110> orientations", *Scanning Electron Microscopy/1979/I*, pp. 11–18.

*8-3. M. Futamoto, M. Nakazawa, K. Usami, S. Hosoki, and U. Kawabe, "Thermionic emission properties of a single-crystal LaB_6 cathode", *J. Appl. Phys.* **51** (7), pp. 3869–3876 (1980).

*8-4. M. Yamabe, Y. Furukawa, and T. Inagaki, "Electron emission from <100> LaB_6 cathodes with large cone angles and flat tips", *J. Vac. Sci. Technol. A* **2** (3), pp.1361–1364 (1984).

*8-5. H. Adachi, "Approach to a stable field emission electron source", *Shinku (J. Vac. Soc. Japan)* **29** (1), pp. 13–25 (1986) (in Japanese).

Other

*8-6. W. I. Karain, L. V. Knight, D. D. Allred, and A. Reyes-Mena, "Emitted current instability from silicon field emission emitters due to sputtering by residual gas ions", *J. Vac. Sci. Technol. A* **12** (4), pp. 2581–2585 (1994).

*8-7. S. Albin, W. Fu, A. Varghese, A. C. Lavarias, and G. R. Myneni, "Diamond coated silicon field emitter array", *J. Vac. Sci. Technol. A* **17** (4), pp. 2104–2108 (1999).

*8-8. D. Chen, W. Y. Cheung, S. P. Wong, Y. M. Fung, J. B. Xu, I. H. Wilson, and R. W. M. Kwok, "Field emission characteristics of SiC capped Si tip array by ion beam synthesis", *J. Vac. Sci. Technol. A* **17** (4), pp. 2109–2112 (1999).

*8-9. D. Temple, W. D. Palmer, L. N. Yadon, J. E. Mancusi, D. Vellenga, and G. E. McGuire, "Silicon field emitter cathodes: Fabrication, performance, and applications", *J. Vac. Sci. Technol. A* **16** (3), pp. 1980–1990 (1998).

*8-10. C. Lea, "Field emission from carbon fibres", *J. Phys. D: Appl. Phys*, **6**, pp. 1105–1114 (1973).

*8-11. N. de Jonge, Y. Lamy, K. Schoots, and T. H. Oosterkamp, "High brightness electron beam from a multi-walled carbon nanotube", *Nature* **420**, pp. 393–395 (2002).

*8-12. C. A. Spindt, I. Brodie, L. Humphrey, and E. R. Westerberg, "Physical properties of thin-film field emission cathodes with molybdenum cones", *J. Appl. Phys.* **47** (12), pp. 5248–5263 (1976).

*8-13. J. Itoh and H. Hiroshima, "Fabrication and theoretical study of micro-size vertical vacuum triodes", 2nd Int. Conf. on Vac. Microelectron., Bath, 1989, pp. 231–234 (Inst. Phys. Conf. Ser. No 99: Section 7).

*8-14. K. K. Chin and R. B. Marcus, "Field emitter tips for vacuum microelectronic devices", *J. Vac. Sci. Technol. A* **8** (4), pp. 3586–3590 (1990).

*8-15. T. R. Albrecht, S. Akamine, M. J. Zdeblick, and C. F. Quate, "Microfabrication of integrated scanning tunneling microscope", *J. Vac. Sci. Technol. A* **8** (1), pp. 317–318 (1990).

*8-16. W. N. Carr, H. J. Wang, K. K. Chin, and R. B. Marcus, "Vacuum microtriode characteristics", *J. Vac. Sci. Technol. A* **8** (4), pp. 3581–3585 (1990).

*8-17. A. D. Feinerman, D. A. Crewe, D. C. Perng, S. E. Shoaf, and A. V. Crewe, "Sub-centimeter micromachined electron microscope", *J. Vac. Sci. Technol. A* **10** (4), pp. 611–616 (1992).

*8-18. T. Durakiewicz and S. Halas, "Thermal relaxation of hot filaments", *J. Vac. Sci. Technol. A* **17** (3), pp. 1071–1074 (1999).

*8-19. S. Yamamoto, "Recent development of cathodes used for cathode ray tubes", *Shinku (J. Vac. Soc. Japan)* **40** (5), pp. 423–430 (1997) (in English).

*8-20. K. Yokoo, J. Ikeda, K. Tahara, Y. Abe, and H. Mimura, "Field emission characteristics of semiconductor cathodes", *Shinku (J. Vac. Soc. Japan)* **41** (4), pp. 428–433 (1998) (in Japanese).

*8-21. S. Kanemaru and J. Itoh, "Fabrication of integrated field emitters", *Shinku (J. Vac. Soc. Japan)* **41** (4), pp. 434–439 (1998) (in Japanese).

*8-22. U. Staufer, L. P. Muray, D. P. Kern, and T. H. P. Chang, "Investigation of emitter tips for scanning tunneling microscope-based microprobe systems", *J. Vac. Sci. Technol. B* **9** (6), pp. 2962–2966 (1991).

Index

The manufacturer's authorised representative in the EU is Springer Nature Customer Service Centre GmbH, Europaplatz 3, 69115 Heidelberg, Germany. If you have any concerns regarding our products, please contact ProductSafety@springernature.com

Printed and bound by CPI Group (UK) Ltd, Croydon, CR0 4YY
30/09/2026
02225599-0015